Gastronomy, Hospitality, and the Future of the Restaurant Industry:

Post-COVID-19 Perspectives

Ana Pinto Borges
ISAG - European Business School, Portugal & Research Center in Business Sciences and Tourism (CICET-FCVC), Portugal & COMEGI, Portugal

António Lopes de Almeida
ISAG - European Business School, Portugal & Research Center in Business Sciences and Tourism (CICET-FCVC), Portugal & GOVCOPP, Portugal

Elvira Pacheco Vieira
ISAG - European Business School, Portugal & Research Center in Business Sciences and Tourism (CICET-FCVC), Portugal & IPVC, Portugal & UNIAG, Portugal

Rui Rosa Dias
ISAG - European Business School, Portugal & Research Center in Business Sciences and Tourism (CICET-FCVC), Portugal

Paula Rodrigues
Universidade Lusíada, Portugal & COMEGI, Portugal & BRU-IUL, ISCTE - Instituto Universitário de Lisboa, Portugal

IGI Global
PUBLISHER of TIMELY KNOWLEDGE

A volume in the Advances in Hospitality, Tourism, and the Services Industry (AHTSI) Book Series

Published in the United States of America by
IGI Global
Business Science Reference (an imprint of IGI Global)
701 E. Chocolate Avenue
Hershey PA, USA 17033
Tel: 717-533-8845
Fax: 717-533-8661
E-mail: cust@igi-global.com
Web site: http://www.igi-global.com

Library of Congress Cataloging-in-Publication Data

Names: Borges, Ana Pinto, 1980- editor. | Almeida, Antonio Lopes de, 1966- editor. | Vieira, Elvira, 1977- editor. | Dias, Rui, 1970- editor. | Rodrigues, Paula Cristina Lopes, 1970- editor.
Title: Gastronomy, hospitality, and the future of the restaurant industry : post-COVID-19 perspectives / Ana Pinto Borges, António Lopes de Almeida, Elvira Vieira, Rui Dias, and Paula Rodrigues, editors.
Description: Hershey PA : Business Science Reference, [2022] | Includes bibliographical references and index. | Summary: "The main objective of the book is to contribute to the dissemination of knowledge, both to academics and professionals, in the areas of Gastronomy and Hospitality, and consequently Tourism, defining strategies and planning options"-- Provided by publisher.
Identifiers: LCCN 2021052901 (print) | LCCN 2021052902 (ebook) | ISBN 9781799891482 (hardcover) | ISBN 9781799891499 (paperback) | ISBN 9781799891505 (ebook)
Subjects: LCSH: Food tourism. | Hospitality industry. | COVID-19 Pandemic, 2020---Economic aspects.
Classification: LCC TX631 .G375 2022 (print) | LCC TX631 (ebook) | DDC 641.01/3--dc23/eng/20211122
LC record available at https://lccn.loc.gov/2021052901
LC ebook record available at https://lccn.loc.gov/2021052902

This book is published in the IGI Global book series Advances in Hospitality, Tourism, and the Services Industry (AHTSI) (ISSN: 2475-6547; eISSN: 2475-6555)

British Cataloguing in Publication Data
A Cataloguing in Publication record for this book is available from the British Library.

For electronic access to this publication, please contact: eresources@igi-global.com.

Advances in Hospitality, Tourism, and the Services Industry (AHTSI) Book Series

Maximiliano Korstanje
University of Palermo, Argentina

ISSN:2475-6547
EISSN:2475-6555

Mission

Globally, the hospitality, travel, tourism, and services industries generate a significant percentage of revenue and represent a large portion of the business world. Even in tough economic times, these industries thrive as individuals continue to spend on leisure and recreation activities as well as services.

The Advances in Hospitality, Tourism, and the Services Industry (AHTSI) book series offers diverse publications relating to the management, promotion, and profitability of the leisure, recreation, and services industries. Highlighting current research pertaining to various topics within the realm of hospitality, travel, tourism, and services management, the titles found within the AHTSI book series are pertinent to the research and professional needs of managers, business practitioners, researchers, and upper-level students studying in the field.

Coverage

- Leisure & Business Travel
- Service Training
- Casino Management
- Hotel Management
- International Tourism
- Sustainable Tourism
- Cruise Marketing and Sales
- Service Design
- Customer Service Issues
- Health and Wellness Tourism

IGI Global is currently accepting manuscripts for publication within this series. To submit a proposal for a volume in this series, please contact our Acquisition Editors at Acquisitions@igi-global.com or visit: http://www.igi-global.com/publish/.

The Advances in Hospitality, Tourism, and the Services Industry (AHTSI) Book Series (ISSN 2475-6547) is published by IGI Global, 701 E. Chocolate Avenue, Hershey, PA 17033-1240, USA, www.igi-global.com. This series is composed of titles available for purchase individually; each title is edited to be contextually exclusive from any other title within the series. For pricing and ordering information please visit http://www.igi-global.com/book-series/advances-hospitality-tourism-services-industry/121014. Postmaster: Send all address changes to above address.

Titles in this Series

For a list of additional titles in this series, please visit: www.igi-global.com/book-series/advances-hospitality-tourism-services-industry/121014

Mobile Computing and Technology Applications in Tourism and Hospitality
Agustinus Borgy Waluyo (La Trobe University, Australia) and Ling Tan (Australian Council for Educational Research, Australia)
Business Science Reference • © 2022 • 289pp • H/C (ISBN: 9781799869047) • US $215.00

Optimizing Digital Solutions for Hyper-Personalization in Tourism and Hospitality
Nuno Gustavo (Estoril Higher Institute for Tourism and Hotel Studies, Portugal) João Pronto (Estoril Higher Institute for Tourism and Hotel Studies, Portugal) Luísa Carvalho (Polytechnic Institute of Setúbal, Portugal & and CEFAGE, University of Évora, Portugal) and Miguel Belo (Lisbon University, Portugal & Portuguese National Funding Agency for Science, Research, and Technology, Portugal)
Business Science Reference • © 2022 • 379pp • H/C (ISBN: 9781799883067) • US $195.00

Challenges and Opportunities for Transportation Services in the Post-COVID-19 Era
Giuseppe Catenazzo (ICN Business School, France)
Business Science Reference • © 2022 • 320pp • H/C (ISBN: 9781799888406) • US $225.00

Handbook of Research on Digital Communications, Internet of Things, and the Future of Cultural Tourism
Lídia Oliveira (University of Aveiro, Portugal)
Business Science Reference • © 2022 • 587pp • H/C (ISBN: 9781799885283) • US $325.00

New Governance and Management in Touristic Destinations
Marco Valeri (Niccolò Cusano University, Italy)
Business Science Reference • © 2022 • 357pp • H/C (ISBN: 9781668438893) • US $225.00

Global Perspectives on Literary Tourism and Film-Induced Tourism
Rita Baleiro (University of the Algarve, Portugal & CiTUR, Portugal) and Rosária Pereira (University of the Algarve, Portugal & CinTurs, Portugal)
Business Science Reference • © 2022 • 371pp • H/C (ISBN: 9781799882626) • US $215.00

Planning and Managing the Experience Economy in Tourism
Rui Augusto Costa (GOVCOPP, University of Aveiro, Portugal) Filipa Brandão (GOVCOPP, University of Aveiro, Portugal) Zelia Breda (GOVCOPP, University of Aveiro, Portugal) and Carlos Costa (GOVCOPP, University of Aveiro, Portugal)
Business Science Reference • © 2022 • 407pp • H/C (ISBN: 9781799887751) • US $195.00

IGI Global
PUBLISHER of TIMELY KNOWLEDGE

701 East Chocolate Avenue, Hershey, PA 17033, USA
Tel: 717-533-8845 x100 • Fax: 717-533-8661
E-Mail: cust@igi-global.com • www.igi-global.com

Table of Contents

Detailed Table of Contents

to promote the Mediterranean diet as a solution to the improvement of food habits in the post-COVID-19 era. It will explore the future trends of Mediterranean diet concept use, specially focused on food industry/distribution and tourism as an inspiration to impulse more healthy and sustainable consumption in Portuguese restaurants. Gastronomy and tourism also will be presented as a sustainability safeguard possibility for the Mediterranean diet.

COVID-19, which is thought to be transmitted from infected animals to humans as a zoonotic disease, plays an important role in food safety and public health. Inappropriate, defective, and unsuitable food safety systems ignoring it play a key role in the spread of the disease. Controlling food safety from farm to fork provides access to healthy and high quality foods in the protection of health and the prevention of diseases. Making new effective and applicable standards and principles in the food safety system worldwide may be preventing possible diseases, risks, especially at epidemic level. In this context, official authorities, food businesses, and supply chain officials should take an active role to reform the process. Consequently, both food safety and free trade principles may be more efficient under common global food safety systems for public health.

In today's world, tourism is one of the world's leading economic sectors in revenue generation with the number of jobs and clients in addition to its cultural services. To get a share from this economic cake, many countries highlight their gastronomic values to enhance their attractiveness, competition, and sustainability of the destination. Countries promote their global, regional, and local cuisines by employing a variety of methods. In the last 20 years, adding a gastronomic component to historical touristic trips has accelerated the development of gastronomy by attracting not only tourists but also residents of the region. Despite these encouraging developments, climate change, desertification, biodiversity loss, increasing population pressure, and pandemic processes, all of which are global issues threatening food security, pose threats to gastronomy culture. In this context, immediate steps should be taken to ensure the survival of the newly emerging gastronomy community.

In today's competitive market environment, restaurateurs realize that they should differentiate their products and services by offering unique tastes and experiences. This is also true for restaurants that promote and deliver local foods to customers. Local food which is considered as a cultural identity in every destination can create a positive image about a city and even help the local authorities to establish a city brand. Despite this realization, there is limited research about local food tourism and its services,

Preface

The important role of gastronomy in human life is undeniable since it is a vital necessity, which can influence individuals' mental, physical, and social wellbeing as well as being a driver in the path for worldwide sustainable development. Moreover, gastronomy can be used as a form of culture, as a touristic tool, and represents a fundamental component of travel and experience in tourist destinations, becoming a differentiating factor of a destination compared to others and, consequently, increasing its attractiveness and competitiveness. On the other hand, tourism also improves the gastronomic and restaurant industries, through its capacity of creating suitable job opportunities, of attracting more customers to the restaurants and, as a result, enhance the economic growth.

The bidirectional relationship between gastronomy and tourism can bring mutual benefits, namely, the development of social and cultural awareness, promotion of vulnerable groups' inclusion in local communities, and/or decrease seasonality effects, minimising those sectors' negative externalities and improving economic development. Thus, the Gastronomy Food Tourism is a rising concept, which has made remarkable progress over the past 10-15 years in both academia and in its own sector, thus being mandatory to extend and fill the gaps in the literature concerning this thematic.

Bearing this in mind, this book comprises the recent research topic of food and gastronomy in hospitality and tourism, highlighting the challenges faced, emerging research themes and different methods of approach. More specifically, it constitutes a timely and relevant compendium of chapters that enlightens the readers on relevant issues in gastronomy and F&B management strategies in the hospitality industry, where tourism is discussed as a central performing action contextualized by trends and new perspectives.

Furthermore, this book contributes to the dissemination and sharing of knowledge in the areas of Gastronomy and Hospitality, and, consequently, Tourism, therefore it can be very useful among academics, practitioners (e.g., managers, policy makers, consulting professionals, marketing research), and other interested groups and individuals.

The book comprises 13 chapters covering different topics within gastronomy, hospitality, and restaurant industries' relationships.

Chapter 1 analyses the Olive oil tourism in Mirandela (a small Municipality in the Northeast of Portugal). More precisely, it intends to understand strategic alliances, which could allow the creation of composite tourism products, which would be properly articulated to fight against seasonality.

Chapter 2 intends to evaluate the Chinese government policies to support gastronomy tourism and promote the regional food culture in the context of post-COVID-19 tourism for visitors and travellers.

Chapter 3 studies the traditional gastronomic offer from Guayas-Ecuador, and the relationship between gastronomy and the development of local tourism. It also presents the main results that enhance the importance of a well-structured offer to influence the motivation of a tourist to go to a specific destination.

Chapter 4 conceptualises the Mediterranean Type Nutrition and its important role in tourism businesses, analysing the current approaches in the developing world.

Chapter 5 highlights the opportunity to promote the Mediterranean Diet as a solution to the improvement of food habits in the post COVID-19 era. It explores the future trends of Mediterranean Diet in Portuguese Restaurants, specially focused on food and tourism.

Chapter 6 aims to evaluate the precautions and new generation technologies related to the risks and dangers in the food industry in food safety during the COVID-19 period.

Chapter 7 focus on the current and future effects of the pandemic on food safety and gastronomy, bearing in mind some of the current challenges, such as, climate change, biodiversity loss and the socio-economic pressures.

Chapter 8 assesses Bonab city as a popular food tourism destination in Iran and aims to ascertain the factors that motivate customers to buy kebabs, using a qualitative research approach.

Chapter 9 provides a critical reflection on how individuals, families, tourists are looking for alternative forms of touristic and gastronomic experiences, very much rooted on the spiritual quest for peace, security, hope, self-awareness, and emotional wellbeing.

Chapter 10 intends to evaluate the evolution of taxation in the accommodation and restaurant industry, analysing the taxation of this sector in Portugal and post-COVID-19 perspectives.

Chapter 11 studies, for the first time, one mass masstige index for wine brands. The authors apply an empirical analysis to evaluate the relevance of that index in the restaurants' communications.

Chapter 12 focuses on post-COVID-19 revitalization of the restaurant industry in Lviv (Eastern Europe), which is a sector that plays a significant role in recovering the negative economic and socio-psychological consequences of the COVID-19 crisis.

Chapter 13 intends to contribute to the food tourism literature, combining the concept of co-creation and image building in order to turn Malaysian tourism industry into a vital sector in the economy in the endemic period.

Ana Pinto Borges
ISAG - European Business School, Portugal & Research Center in Business Sciences and Tourism (CICET-FCVC), Portugal & COMEGI, Portugal

António Lopes de Almeida
ISAG - European Business School, Portugal & Research Center in Business Sciences and Tourism (CICET-FCVC), Portugal & GOVCOPP, Portugal

Elvira Pacheco Vieira
ISAG - European Business School, Portugal & Research Center in Business Sciences and Tourism (CICET-FCVC), Portugal & IPVC, Portugal & UNIAG, Portugal

Rui Rosa Dias
ISAG - European Business School, Portugal & Research Center in Business Sciences and Tourism (CICET-FCVC), Portugal

Paula Rodrigues
Universidade Lusíada, Portugal & COMEGI, Portugal & BRU-IUL, ISCTE – Instituto Universitário de Lisboa, Portugal

Acknowledgment

All the chapters included in this book were double-blind review. The editors would like to acknowledge the help and availability of the authors and reviewers that took part in the review process and to all people involved in this process. Without their support, the accomplishment of this book would not be possible.

First, the editors would like to express their sincere gratitude to each one of the chapter's authors, who dispensed their time and expertise to this book.

Second, the editors wish to acknowledge the valuable contributions of the reviewers regarding the improvement of quality, coherence, and content presentation of chapters. Most of the authors also served as referees; we highly appreciate their double task.

Finally, a special acknowledgement to Daniela Castilho, the technical support of Research Center in Business Sciences and Tourism of Consuelo Vieira da Costa Fundation (CICET-FCVC), who help and provided assistance throughout all process.

Chapter 1
Olives and Touristic Experiences:
The Case of the Olive Tree and Olive Oil Museum

Ricardo F. Correia
https://orcid.org/0000-0002-0132-4002
Instituto Politécnico de Bragança, Portugal & CiTUR, Portugal

Palmira Felgueiras
Instituto Politécnico de Bragança, Portugal

Aida Carvalho
https://orcid.org/0000-0001-8997-9195
Instituto Politécnico de Bragança, Portugal & CiTUR, Portugal

Dominyka Venciute
University of Management and Economics, Lithuania

ABSTRACT

Gastronomy is one of the main motivations that inspires and triggers the tourist journey. It is frequently interconnected with intangible dimensions attached with a specific place. Olive oil tourism is a specific product inside gastronomic tourism with the potential to generate a diversity of experiences for the tourist. As olive oil tourism depends on a multiplicity of different actors, whose actions need to be integrated, its current situation is still far way from its potential in regions that are not able to implement this integration practice. That is exactly the case of Mirandela (a small municipality in the northeast of Portugal) examined in this chapter. Even with an actor (the Olive Tree and Olive Oil Museum) with a strong potential to act as an articulator of the offer, that integration is not even in the beginning yet, losing the destination the opportunity of generating valuable olive oil experiences to the tourists.

DOI: 10.4018/978-1-7998-9148-2.ch001

INTRODUCTION

Gastronomy is currently understood as one of the main motivations of tourists for their destination selection. The concept of gastronomic tourism involves a strong experiential dimension and a connection with the local culture of the regions. Indeed, gastronomy has become an indispensable element in knowing and understanding the culture and lifestyle of a territory, so it encompasses a set of traditional values associated with new trends in tourism, such as respect for culture and tradition, a healthy living style, authenticity, sustainability, and, above all, experience (UNWTO, 2012). In fact, food can be "considered as a reflection of the culture of a country and its people" (du Rand & Heath, 2006, p. 207).

Tourists look for connections and experiences that are rooted in their destinations (Boyle, 2004). They are increasingly sophisticated, making the choice of the destination they visited part of their lifestyle and a way of expressing their identity (Morgan et al., 2004). Regional products and gastronomy stand out among the multiple characteristics that provide authenticity to a destination. "It is a destination's food, people and places which make up its heritage and its character and which thereby provide a destination with its own authenticity" (Yeoman et al., 2007, p. 1135).

Inserted within gastronomic tourism, olive oil tourism is a relatively recent tourist segment, but it has been growing in the last decades all over the world, namely in Greece, Cyprus, Australia, Italy, and Spain, which translates to strong economic development in the rural regions where olive groves and mills are predominant. These structures, which in the past were only used to produce olive oil, have now taken on touristic functions, making these rural areas more dynamic. Indeed "the important role of olive oil mills as drivers of social development and structural change in municipalities (Mozas-Moral 2019), implies that the spaces occupied by agricultural cooperatives can be understood as a collective good" (Alonso & Gonzáles, 2021, p.176).

The practice of this type of tourism is strongly linked to the sustainable development of the rural areas and contributes to the enhancement and conservation of the unique identity of these places. This is accomplished through the promotion of traditional activities to tourists from urban areas who seek to enjoy unique and memorable experiences there (Elias & Barbero, 2017).

Olive tourism has the potential to generate a considerable number of experiences contextualized in the territory, from tastings and harvesting to massages with olive oils. The profiles of tourists looking for these varied experiences may differ. When looking at the motivations for choosing olive oil tourism activities, Fernández et al. (2022), in a study conducted at the Spanish regions of Andalusia and Extremadura, segmented the olive oil tourists into four main groups. The first group was the Healthy Olive Oil Tourists, whose "main motivation [in] choosing olive oil tourism is to engage in activities that are beneficial to their health [and] they do not tend to spend the night away from home during their trip," Fernández et al. (2022, p. 3). The second group identified by the authors was the Olive Oil Flavours Hunter Tourists, who were mainly motivated to learn about the different varieties of olive oil. This group also showed a propensity to spend a night way during the visit to the olive oil region. The third segment, which the authors labeled as Designation of Origin Olive Oil Tourists, were motivated to explore the area of a Designation of Origin. "These are tourists who spend at least one night away from home (72%) and has *[sic]* the highest level of income in excess of €1800," Fernández et al. (2022, p. 3). Finally, the fourth segment, labeled Extra Virgin Olive Oil Tourists, is "motivated by a desire to learn about the taste properties of EVOO (69.2%) and spend at least one night away from home, but in this case, they do so in the same city where the company offering the olive oil tourism activities that they have carried out is located (70%)," Fernández et al. (2022, p. 4).

The development stage of different markets may also influence the motivations of the olive oil tourists because their knowledge of olive oil is also different. Hwang and Felitti (2021), in their study based on the experience economic framework, conclude that for the U.S. olive oil tourist, the educational experience is more related to the motivation and satisfaction of the tourists than aesthetics, entertainment, and escapism.

The experiences to be enjoyed by this different segment of tourists usually are offered by different small, regional actors and need to be integrated (Murgado, 2013). Although some of these experiences have already been identified (Bezerra & Correia, 2018), it is yet to be understood how they could be articulated and integrated to generate a more cohesive and stronger view of a specific territory as an olive tourism destination.

The aim of this chapter is to understand the relevance of an actor to integrate and contextualize the diverse experiences that can be generated by olive oil tourism in a specific region. The selected region was Mirandela (a municipality in the Northest of Portugal), and the actor under analysis will be the Olive Tree and Olive Oil Museum (OOM).

In addition to this introduction, the chapter is divided in five sections. Firstly, is the theoretical background, mainly focused in the experiences in tourism as well as in the olive oil tourism as a source of touristic experiences. It follows a description of the methodology used in this paper. After that, the case of the OOM is presented, followed by an analysis and reflection. The last sections are devoted to the conclusions as well as to suggestions for further research.

THEORETICAL BACKGROUND

Experiences and Tourism

The central assumption of the experience economy is based on the consumers' demand for new concepts for products and services. Customers challenge producers to offer experiences with added value through additional services to the core product or service, differentiating them from the competition and creating unique values and authentic experiences, for which customers are willing to pay more (Denmark Government, 2003; Mehmetoglu & Engen, 2011; Pine & Gilmore, 1998; Richards, 2013). Pine and Gilmore (1998) represented, in a simplified way, the process, both for the consumer's experience and the producer's benefit, which translates into the added value incorporating the experiences.

According to Ilincic (2014), in a generic perspective, all forms of tourism, in essence, come from the experience economy, pointing to the shift from delivering goods and products to producing engaging and memorable experiences. Tourism as an industry has also undergone major transformations. The essence of tourism in today's world is based on the development and delivery of travel experiences and visits to a range of individuals and groups who wish to see, understand, and experience the nature of different destinations and the way people in these places live, work, and enjoy life (Ritchie et al., 2010).

The experiences are the main product of the tourism sector, with direct impact on the satisfaction of tourists. This is a challenge for the Destination Management Organizations to analyze, as they evaluate which experience to offer and how to improve it and make it more positive (Buhalis & Amaranggana, 2015). Thus, destinations are expected to offer personalized services to each different type of tourist in order to exceed their expectations by enhancing their tourism experience (Buhalis & Amaranggana,

2015; Smith et al., 2016). This experience will enrich the way tourists value the trip, with the greatest benefit of personalization being an increased level of emotional and physical comfort.

Morgan et al. (2010) noted that experiences that allow participants to engage with the identity of a place should not be undervalued. In turn, Marujo (2016) highlighted that each tourist experience is unique, assuming different perceptions for each tourist and taking into account the experience's multidimensional and sensory nature. Every observation or experimentation that takes place in a given destination results in a unique experience for the tourist, whether positive or negative, which occurs at a given moment, depending on the state of mind and the way he or she relates to the destination.

Tourism embraces the expectation of finding new and diverse experiences that are commonly different from those present in everyday life (Morgan et al., 2010). Thus, the search, whenever possible, to deliberately create new cultural and tourist manifestations to stimulate the interest of tourists is incessant. It is assumed that tourists travel to consume experiences and tourism is qualified as the experience and consumption of the place. Oh et al. (2007) noted that this new demand for unique and memorable experiences, in the tourism sector, forces companies to structure a differentiated and valued offer for products and services. Here, lies authenticity as an important component of experience tourism, appreciated by tourists, as consumers, and to be considered by Destination Management Organizations, as responsible for structuring the offer (Couret, 2016).

According to Richards and Wilson (2006), the value of the product or service and the quality of the tourism experience can be increased by shifting from tourists' passive consumption of shows or creative spaces to tourists' active participation in the creative process, or creative tourism. These authors argued that the application of creativity affects the very nature of tourism, especially concepts associated with tourist contemplation and authenticity of the tourist experience. Authenticity, according to Cook (2010), exists externalized to the tourist, being a characteristic inherently found in an object, product, service, event, culture, or place, giving it measurable quality or characteristics, inseparable from a perception, a feeling, an experience, or a value resulting from social processes and cultural values.

In short, the current tourist is a mega-consumer, and tourism, which has always been an experience, is increasingly transformed into something experiential. The tourist wants to live sensory, affective, and imaginary experiences in a new emotional connection with the destination and tourism services (Serra, 2015).

Olive Oil Tourism

Olive oil tourism is a relatively recent segment, which combines tourism with agricultural activity and emerges mainly in Mediterranean countries with large olive groves, such as Spain, Greece, Turkey, Italy, Morocco, and Portugal, but is also beginning to gain some expression in countries like Australia, the United States, Argentina, and Japan (Fernández et al., 2019).

We can define olive oil tourism as "a form of domestic tourism (especially in rural areas) related to gastronomy, which allows for the essence of the culture encompassing the world of the olive to be captured while deepening knowledge about everything connected to olive oil" (Millán et al., 2014, p. 180). It is tourism that is related to the olive grove culture and that due to its specificities presents itself as an excellent complement to the profitability of a certain territory, also representing the guarantee of sustainable development. According to Torre et al. (2015) the olive grove landscapes and their diversity are cornerstones for the development of olive oil tourism, and all the activities that support these two variables are enhancers of job creation and population settlement, which translates into the development of rural areas.

Although olive oil tourism is a recent concept, it has already been proven that it can play a very important role in the creation of wealth of a certain geographical area. The greatest example of this is Italy, which saw, in January 2020, olive tourism being integrated into its state budget and becoming regulated like the wine tourism sector. This regulation brings together, in the definition of olive tourism, the cultivation of olive trees, the olive oil production processes, its marketing and tasting, and all activities related to visits to mills and olive groves and that supposes a recreational and informative action (Granitto, 2020).

The literature has analyzed olive tourism through different perspectives. It has addressed and analyzed olive tourism's demand and supply, reviewed concepts, studied its possible impact on rural development, and proceeded to categorize this segment in more consolidated tourism typologies (Mogollón et al., 2019). Bezerra and Correia (2018) also pointed out the possibility that olive tourism can be analyzed in the light of creative tourism, since its essence is the participation in tourist experiences that generate memories, through the immersion in traditions and local culture, thus being an interactive, participated tourism, an experience tourism.

González et al. (2020) included olive oil tourism as a type of special interest tourism (SIT), finding "this form of tourism meets the needs of specific markets by focusing on diverse experiences and activities that fall outside the scope of general interest tourism," and they conclude "that the two key dimensions in the characterization of olive oil tourism as SIT are experience and sustainability." For these authors, the partnerships between multiple actors, both public and private, are relevant in order to consolidate the global experience of the destination.

In this context, it is important to note that the tourist experience has changed its object and abandoned the attachment to tangible aspects of service quality, to be dedicated to issues of emotional involvement of the tourist and the creation of memorable experiences, and this fact cannot be neglected by destination managers. However, there is still much to be discovered regarding the determinants of a memorable experience, and one of the paths pointed out is the involvement of the tourist in the action, the creation of interactive and creative tourism (Coelho et al., 2018).

In fact, the current tourist is no longer satisfied with just contemplating the places they visit, nor with the superficial knowledge that can be transmitted to them. They want to go beyond what they can observe; they want to be protagonist of his trip story (Pezzi & Vianna, 2015). In this sense, olive and olive oil tourism can be an extraordinary tool if it allows its practitioners to obtain memorable experiences through a deep involvement with traditions and local culture. This involvement can be achieved through participation in the various experiences that olive tourism offers and that are now listed: visits to mills, tastings of olive oil and other olive products, visits to olive oil museums and interpretive centers, participation in fairs and events on the theme, and participation in routes that include the exploration of the olive grove, with the possibility of participating in the process of harvesting the olives, among others (Bezerra & Correia, 2018; Millán et al., 2014; Morales et al., 2015; Torre et al., 2015).

Small organizations and tourist destinations usually offer multiple possibilities for tourists. However, from the tourist's point of view, these activities are merely a partial experience that will only contribute to the formation of an integrated experience, of a global tourist product (Lehtolainen, 2003). The dynamics of a regional tourist destination results mainly from the collective thought and the need for cooperation to create a structure between multiple partners. This led to a coherent and integrated product that became attractive to the tourist and produced a value-add for the territory. As such, these movements must be also understood as sustained policies for territorial development (Correia, 2006). Olive oil tourism is essentially a product that emerges from the territory, and the experiences that emerge from it should be liked and potentiated in order to create a holistic and global experience more relevant for the tourist and with more value for the region.

Millán et al. (2017) suggested that integration and coordination between a multitude of local agents is simultaneously "the main attractiveness and the difficulty of the development of tourism proposals linked to olive oil production" (p. 16). Also, Murgado (2013), analyzing the olive tourism in the Spanish context, noted that "there are no partnerships, cooperative agreements, or philosophy of working together between the actors involved in olive oil routes so each tourism provider operates independently" (p. 101). In line with this idea, Hidalgo et al. (2021) developed a study that surveyed the managers of companies located in Andalusia-Spain, and were among the first implementing the diversification of their activity towards olive oil tourism. The results show "that the priority of companies to collaborate is low, with the public sector being the preferred actor and the predisposition to invest economic capital in collaboration networks is minimal" (Hidalgo et al., 2021, p. 20). The authors also concluded that these leading companies need information, training, and incentives to develop the cooperation. Without this integration and cooperation, the quality of the touristic product is constrained. It looks clear that the olive oil tourism should combine, integrate, and articulate multiple actors and different territorial dimensions in order to generate valuable experiences.

Olive oil tourism and the sustainability of rural areas are interconnected; however, and despite the potential of this segment, for there to be effective gain for the territory, it is necessary that concrete measures are taken. With this in mind, Tregua et al. (2018) developed a study for which the objective was to understand the role of local agents in this sustainability. Through this study it was possible to confirm what was already emphasized by the literature: sustainability depends on the degree of cooperation between stakeholders, and the establishment of rules for the performance of agents within the oleotourism is fundamental for potential tourists to realize the offer. The objectives of local actors must be aligned to withdraw the desired profits for the territory.

Olive Oil Tourism in Portugal

The culture of the olive tree in Portugal is lost in time. According to history, it was the Phoenicians and the Greeks who were responsible for the introduction of the olive tree in the Iberian Peninsula, but the Romans and the Arabs were responsible for the production of olive oil (Reis, 2014). Portugal is currently the fourth-largest European country in relation to olive grove extension (Eurostat, 2019), and the ninth largest olive-oil producer and fourth biggest exporter of olive oil (6% of global exportations) worldwide (IOC, 2020). Simultaneously, it is the fourth largest consumer of olive oil per capita worldwide (IOC, 2020). There are six olive-oil Protected Designation of Origin (PDO) regions in Portugal (Trás-os-Montes, Beira Interior, Ribatejo, Norte Alentejano, Moura, and Alentejo Interior).

Regarding olive tourism, similar to what happens all over the world, it is still in its initial phase, with some expression of its existence a little everywhere where olive groves predominate. However, even in Alentejo, where olive tourism is starting to gain more dimension, its existence is associated with wine tourism, working as a complement to this activity, in rural tourism farms (Vallone, 2018). Thus, and according to what is described by Guerra et al. (2018), wine often appears as the main attraction and olive oil as the resource that completes the offer.

In a recent study about olive oil tourism in Portugal, Almeida and Silveira (2021) contextualized the supply in that country and concluded that there is a "lack of promotion of the segment as such: activities are promoted as part of or in addition to larger products; Lack of cooperation/synergies among stakeholders for the promotion and development of the segment; Lack of available information on the current supply, especially in the case of visitable production spaces and of the Olive-Oil Route of Trás-os-Montes; Part of the resources of olive-oil tourism is not yet exploited in a touristic way." (p. 162)

METHODOLOGY

The aim of this chapter is to understand the relevance of a specific actor to integrate and contextualize the diverse experiences that might be generated by olive tourism in a specifc region.

Mirandela was the destination chosen because it meets the conditions for the development of olive oil tourism. These include its geographical location (highly ruralized area); the climatic conditions that promote the predominance of olive groves and its effective exploitation, proven by the existence of 17 registered mills; the existence of endogenous products of reference, such as the PDO olive oils from "Trás-os-Montes," which have been awarded prizes worldwide; the olives and their different forms, including capers; and an indirect product that cannot exist without the presence of olive oil PDO, like the "alheira," the most famous smoked sausage of Portugal.

The Olive and Olive Oil Museum (OOM) was chosen in the municipality of Mirandela because it aims to be a living space, where experiences and knowledge are shared, having a public dimension with an objective of enhancing the tradition associated with the olive oil in the region.

In methodological terms, the work is based on a qualitative approach of exploratory nature, using the case study of OOM. The study authors' main goal was understanding the importance of the OOM as a potential olive-tourism-experiences integrating agent; thus, a semi-structured interview was conducted in order to collect the opinion of diverse stakeholders.

According to Garrod (2003), stakeholders play a key role gathering information for decision-making, in the development and management associated with tourism. For this reason and taking into account the intention of this investigation, it was important that the interviewees had knowledge in the area of tourism, olive oil/olive groves, the management of museum spaces, knowledge in territory management and knowledge in marketing and communication. We obtained the sample according a convenience approach (Table 1).

The interviews were carried out between July 2020 and August 2020; the testimonies were recorded and later transcribed and analyzed.

THE OOM CASE

The New Museology, founded on participation and the principles of citizenship, saw museums as unfinished works that continuously intervene in the community, so for Teixeira (2005), collective memory has become the heritage par excellence of these museums and interdisciplinarity the guideline of programs developed with the active participation of populations. In the context of the New Museology, innovative experiences emerged in different countries that established the objective of safeguarding and valuing the identity of each region, and it is in this context that, in 2017, opens the OOM in the municipality of

Table 1. Respondents and positions held

Name	Function
Vera Preto	Councilor for the Health, Tourism, Culture, Historical Heritage and Communication and Events Office of Mirandela Municipality
Fátima Fontoura	Head of the Development, Cultural, Tourism and Health Division of Mirandela Municipality
Armando Graça	Museologist and Coordinator of Armindo Teixeira Lopes Museum and OOM
Joana Portela	Tourism Technician at Mirandela Municipality
Susana Ribeiro	Director at the Operational Department of Tourism of Porto and North of Portugal
Helena Chéu	Professor at Instituto Piaget Viseu
Francisco Pavão	President of APPITAD
Domingos Pires	Portugal NTN Manager

Mirandela, whose main objective is to preserve, promote, and disseminate the memory of the role of the olive tree and its products in local and regional culture.

The proposal that this museological space presents is a journey through time, which begins at the end of the 19th century and takes place in the middle of the 20th century. The vehicle for the trip is the visits made to the rooms and corridors that house the estate donated by the various parishes that make up the municipality of Mirandela (see Figure 1).

In the entrance space there is a "terrace" and a shop; the next space, the central room of the former "Mirandelense" milling, houses the room "the oil mill" where there is a traditional press more than 100 years old. Crossing the mill, you reach the "the field and the fruit" room; in parallel is the "olive tree courtyard," then the room "the work and transformation of olives." The internal stairs lead to the last one exhibition room with the theme "oil and its uses." Still on this floor is the auditorium with 48 seats, which completes the exhibition route with the viewing of videos, debate sessions, and/or conferences (Dias & Viegas, 2005).

Figure 1. OOM showrooms
Source: OOM

Interview Analysis

One of the phases of the proposed methodology was based on the application of a semi-structured interview. For this purpose, a script was created, using only four questions, based on the specific objectives of this study. The questions asked were:

1. How do you describe the olive oil tourism segment in the municipality of Mirandela?
2. In your opinion, what strategic line should be followed to boost this segment in the municipality?
3. Do you consider that the OOM can play a decisive role in this boost? To what extent?
4. In your opinion, which olive oil tourism experiences could be leveraged through partnerships between OOM and other local agents?

Olive Oil Tourism Segment Description

Regarding the first question, the data collected in the interview refer to the emergence of this segment, and the potential for its development is enunciated by most respondents. Words such as "incipient or embryonic segment" are used to describe the moment that oil tourism is experiencing in the municipality of Mirandela. Table 2 presents some of these positions.

Table 2. Segment diagnosis

Domingos Pires	"…the potential we believe it exists… it exists, it's here, we have quality in the product …"
Armando Graça	"…oil tourism in the municipality of Mirandela is at an almost embryonic stage, taking its first steps."
Francisco Pavão	"…the municipality of Mirandela has a huge potential for the practice of oil tourism,…"
Joana Portela	"I would say that the oil tourism segment in the municipality of Mirandela is incipient"
Vera Preto	"This is a segment that exists, we've felt it for some years now, but …in my opinion it's a segment that has a lot to explore and a lot to work on …"

Strategic Line to be Followed

Regarding the second question, the interviews point in general to the value chain creation need recommended by the AMA (2017). In this context, the responses obtained point to a strategic line of action where public-private partnerships should be considered. This means involving all of the stakeholders in the sector, as well as sensitizing producers and entrepreneurs to this opportunity. The training of the diverse actors involved is also pointed out as one of the premises of the strategy, as well as the dissemination and promotion of the segment, namely the promotion of internationally awarded olive oils. The creation of a major event and the association with another identity product (namely the "alheira," the most famous sausage in Portugal that includes olive oil on its ingredients) are also referred to as elementary points of the strategy. The promotion of tourist experiences and the communication of the segment are also considered of paramount importance for the development of a dynamic strategy. All of this must be coordinated with the goal to obtain a sustainable development of the territory, as emphasized by all the literature review. Table 3 transcribes two opinions that condense most of the information collected regarding the recommended strategy to be followed.

Table 3. Recommended strategy

Helena Chéu	"The strategy to be followed will always be the valorization of History and Culture; increased awareness of the region/county; more efficient and targeted promotion; improve coordination between different tourist regions and offers; intensify the interaction between the various sectors of activity, increase qualified human resources in the olive and olive growing areas."
Francisco Pavão	"… it is necessary to create targeted policies and strategies, which encompass all agents in the sector, olive growers, mills, restaurants, hotels,… establishing a "Letter of Principles" containing all the rules/obligations of the members. Then it will be necessary to inventory all the heritage linked to olive growing and olive oil, later creating paths and routes, which enhance the direct contact of visitors with the municipal olive oil reality. The training of agents will still be a necessity, especially with regard to the sensory analysis of olive oil, providing them with the minimum knowledge that allows to carry out an Olive Oil Tasting."

The OOM Role

Regarding the third question, and according to the interviews, OOM plays an extremely important role in the promotion and boosting of this segment.

As the central theme of this museum space is the olive tree, olives, and olive oil, which is an identity theme, not only in the municipality of Mirandela, but in the entire region, the museum's supreme objective should be the promotion and development of the segment in conjunction with all agents of the olive oil industry.

In this sense, OOM is indicated in the interviews as the arrival point or the starting point for any type of oil tourism experience. It is also seen as the welcoming place, where tourists are received and given the opportunity to know intimately the history and tradition associated with the olive tree and olive oil; it is the place where the visit is guided, and it is the anchorage point. Table 4 presents some excerpts from interviews that support and contextualize what was mentioned.

Table 4. OOM and the development of the segment

Fátima Fontoura	"OOM plays a key role, bearing in mind that the primary objective of creating this equipment is to leverage agro-food based tourism, so here we have the ideal conditions for this segment to be explored from here on."
Francisco Pavão	"OOM should be the link of this entire Oleotourism initiative in the municipality of Mirandela, both as a starting point for the different routes and visits, as well as a place of return."
Helena Chéu	"OOM is a key element in the oil tourism sector in the region. It is a very interactive space, which offers the possibility of getting to know the secrets of Olive Oil, its history, tradition, customs, curiosities, old and current technological processes, among other knowledge... it should be part of all visits, oil tourism routes and all activities related to the sector… it must always be a point of arrival!"
Joana Portela	"At this moment, OOM, is the only or one of the few symbols of the oil tourism segment in the municipality of Mirandela, fortunately we have the OOM… it is very important, I think it is perhaps the anchor, it could be the anchor for the development of the oil tourism sector in Mirandela… "
Susana Ribeiro	"Visits to the museum are essential for tourists to emerge and learn about the entire collection that constitutes the culture of olive trees and olive oil. Without anchorage points and interpretation centers, tourists have more difficulty in internalizing this culture and tradition, which they often ignore in its deepest essence."

Olive Oil Experiences

Finally, regarding the last question, the interviews suggest that the oil tourism experiences that can be proposed by the OOM, made available in conjunction with the local agents, are (see Table 5): visits to centenary olive groves, to old mills and to mills with normal operation, olive oil tasting protocols and/or training courses, promotion of "Olive Oil Tourist Guides," oleo therapy, promotion of the olive oil product in restaurants, with commented olive oil tastings, the design of an Olive Oil charter, guided tours to the OOM, participation in sensory analysis, and promotion of oil tourism routes. These experiences, leveraged through partnerships between the museum, restaurants, tourism companies, producers, educational establishments, travel agencies and accommodation, will create a production chain and a very interesting value proposition for the diverse local agents and for tourists. Additionally, this type of offer will contribute to avoiding seasonality.

Table 5. Olive oil tourism experiences

Fátima Fontoura	"Following the city's strategic objectives, visits to this museum, guided tours, experiences that can be enhanced there, such as tastings, olive oil tastings. Eventually, the experiences that we are, now within the OOM's strategy, trying to plan with stakeholders, with partners, that we are contacting are visits to olive groves, namely to centuries-old olive groves, old mills and even... mills with normal operation and all this can enhance our restaurants and also all the tourism, tourism entertainment and accommodation facilities that are in the county."
Helena Chéu	The oleotourism experiences to be promoted can be guided visits to olive groves, a didactic test and/or a training course, promotion of "Tourist Guides in Olive Oil", OleoTherapy, the promotion of the olive oil product in restaurants, with commented olive oil tastings, for example, conception of an Olive Oil Charter, a guided visit to the Olive Oil Museum, participation in an interactive sensorial analysis room, promotion of oil tourism routes, all of this has to be done through partnerships between Museum, catering, tourism companies, producers, establishments teaching, travel agencies, etc…"
Domingos Pires	"In our specific case, experiences for the whole year, essentially olive oil tastings and actions related to olive oil awareness that the OOM already does today. This for us, for the whole year is perfect. For the kind of experience we set up that is what we need. However, some experiences related to seasonality in partnership with the producer and linked to olive oil are also important."

ANALYSIS AND REFLECTION

In addition to the objective to preserve, promote, and disseminate the memory of the role of the olive tree and its products in local and regional culture, the OOM intends to be a living space, where experiences and knowledge are shared. However, the tourist offer of this equipment from the Municipality of Mirandela is constrained within its walls, when it could be strategically worked as a dynamic pole of oil tourism, fulfilling what is intended in the New Museology, carrying out its activities in articulation with the community and with other local actors.

It becomes clear the added value for the development of oil tourism in the territory if OOM would extend action beyond its doors. In this context, it is necessary to work on the product, adjusting it to the needs of the segment intended to work with and within the previously established tourist attractions. The OOM must be the link of the entire oil tourism initiative, both as a point of arrival and as a starting point, articulating the entire activity with the various agents and visitors. Figure 2 presents a proposal for connections and interactions that should be established between OOM and the diverse local actors.

The OOM should articulate the following oleotourism experiences:

- Guided tours to its interior with olive oil tasting protocol and/or the offer of olive oil toast, symbol of the "tiborna" that was eaten in the old mills. Through this experience, in an interactive environment, and visitors can learn about the secrets of olive oil, its history, tradition, customs, curiosities, the technological processes of olive oil production and its evolution over time, among other knowledge. The sensory analysis offered at the end of the visit will certainly be the content that will prolong the memory of the event experienced. The olive oil used in this experience should always be the PDO olive oil produced in the municipality of Mirandela.
- Guided tours to centennial olive groves. The landscape of the municipality of Mirandela is characterized by the abundance of olive tree culture for several centuries, traveling through the county is still possible to find magnificent examples of century-old olive trees, perfectly preserved and that still today play an important role in the local economy. The potential of the olive grove can be explored through contemplation and/or through co-creation, making the visits coincide with the productive processes. The proposal that we present is the creation of a package "in the day for a day," where the tourist is given the possibility, with professional accompaniment, and according to the time, to know the varieties of trees, to participate in the pruning processes, to harvest the olives, and to take them to the olive press where later they will also participate in the extraction of olive oil. The closing of the proposal will always be the tasting of olive oil and the possibility of purchasing it.

Figure 2. The OOM as an olive oil industry integrative actor

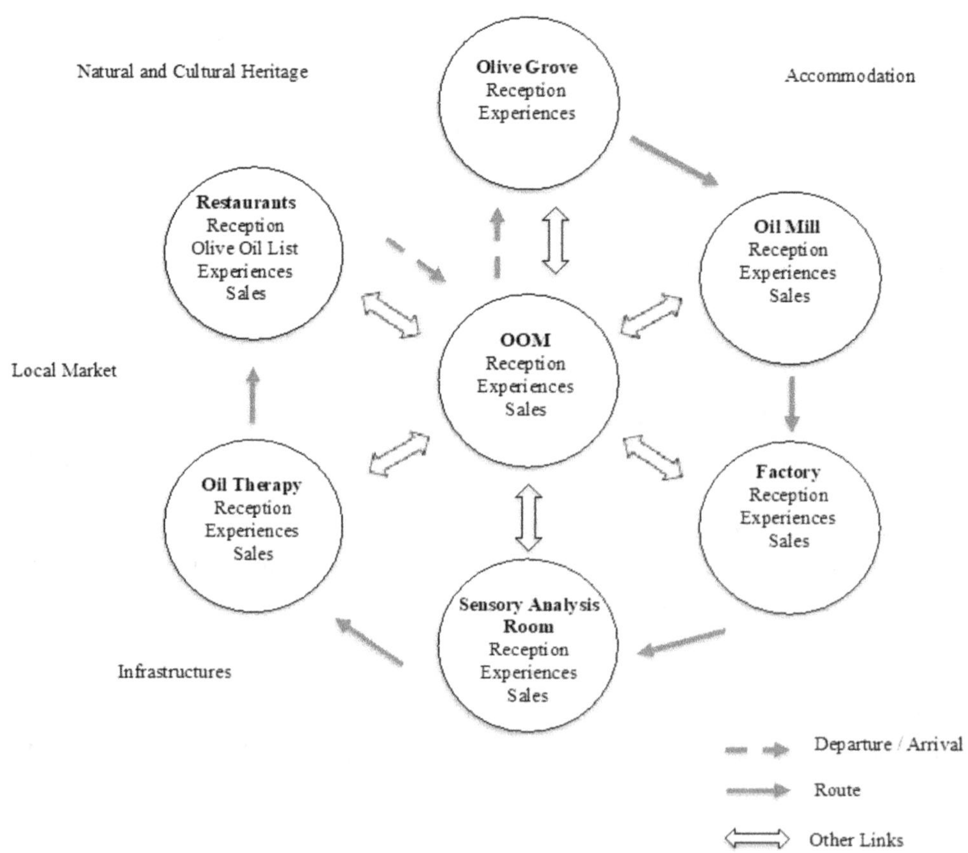

- Olive Oil Tasting Protocol or Training Course, followed by a guided visit to an historical and/or in-normal-operation olive oil mill. This experience will enable visitors to learn about the theme "extra virgin olive oil" (EVA), from the organoleptic characteristics, to the olfactory-gustatory sensations associated with it. Visitors will also be able to learn more about the extraction methods of EVA. It is proposed to articulate this experience with the Professional School of Agriculture and Rural Development of Carvalhais, for having the appropriate facilities for this service and its own mill and know-how.
- Oleotherapy. Massages with essential oils are already widespread on the market for some time; however, it is recommended to dilute this oil in another of vegetable base, which can damage the skin by direct application. The benefits of pure oil, applied directly to the skin, have been explored since the classical age, and since then its moisturizing, healing, and protective effects have been proven. The expression of this activity is almost none in Mirandela, there are only two aesthetic centers working in this area, but it is something very recent. OOM should seek to articulate with the employment and vocational training institute, training activities, aimed at people linked to physiotherapy or aesthetic branches, in order to create a sustainable offer. These facilities should also provide health and wellness products and cosmetics based on olive oil.
- Interactive room for sensory analysis. The sensory analysis uses the five senses to evaluate characteristics of a given product, in this case olive products: olives, olive oil, olive paste, toast, innovative products based on olive oil. It would be an excellent opportunity to demonstrate the differentiating character of endogenous products of the municipality of Mirandela, in a single location. This room should include an area where the products used for sensory analysis will be available for sale. This experience should be arranged with all producers of olive oil and olives for preservation.
- Workshops. These are yet another experience to be proposed to the local community, with a view to bringing them closer to the museum and to develop complementary products, so the proposal is to hold a workshop on chocolates, a workshop on bakery, a workshop on olive oil soap, and other areas with strong connections to olive oil.
- Geo-referenced oil tourist routes. The routes are made using the points of interest referenced in the territory, so it is proposed that, starting from the MOA, to discover the landscape, the buildings, the crafts, the canning industry, gastronomy, sensory analysis and massages; thus, it is available to the visitor all the olive oil potential of the municipality of Mirandela in a single offer. Although it was referred that the MOA would be the starting point, it can also be the arrival point, reinforcing in both cases the welcoming image.

CONCLUSION

In the context of olive oil tourism, it is of utmost importance that strategic alliances are established between the various local actors, public and private, which allow the creation of composite tourism products, which articulate with each other to combat seasonality (Millán et al., 2014). These alliances should also involve residents and stimulate their cooperation, in order to minimize issues that may derive in irregularities and/or unfair competition, harmful to the image of the destination and its progress.

OOM has the potential to trigger the process for the development of olive oil tourism in Mirandela Municipality through a vision that goes far beyond its walls. Together with identified networks of partners, different experiences are possible to be made available, which will evolve the costumer in a holistic experience around the olive oil.

The philosophy behind it is should be related to the quality of the partners, rather than their quantity, in order to structure a composite product that arouses interest and visits to this extensive region, whose richness is based on a nuance of flavour, stories, and diversity. From this perspective, it is understood that from a conceptual point of view, it is imperative to present a value proposition, in which the benefits for the territory and for the members of this OOM network are remarkable, using essentially the attributes of olive oil as a regional driver. However the continuity of OOM action, mainly focused inside of its walls, will inevitably represent the postponement of this potential.

LIMITATIONS AND FURTHER LINES OF INVESTIGATION

The main limitation of this research is related to the few academic studies devoted to this theme, mainly on the supply side. The analysis is subjective and based on a scarce number of interviews and because of that should also be considered as a limitation of the study. The study would benefit from complementary information, for example, the number of tourists that visited Mirandela with olive oil tourism motivation. In fact, the impact of the insights of the chapter could be higher if it would include the perspective of the tourists who actually visit Mirandela.

Thus, as future lines of research, it is suggested to study the actual demand of this touristic product in the region, interviewing the customers who actually visit Mirandela to understand their profile and main motivations that is: why they choose olive oil tourism in the first place and then decide to look for that kind of experience in Mirandela. On the other hand, it is also proposed to investigate the situation of the link between olive oil tourism and other important touristic products in the region, namely wine tourism. As the olive oil tourism is still in its infancy, a further line of research suggests the understanding about how it has been promoted so far and the selling points that the touristic actors have been using in order to attract tourists. Finally, it is suggested to research the role that technology might have in olive oil tourism, specifically to potentiate the integration of different products offered by a network of actors.

REFERENCES

Almeida, I., & Silveira, L. (2021). Olive-oil tourism in Portugal: Contextualization of the current supply. *Journal of Tourism & Development*, *36*(2), 153–165. doi:10.34624/rtd.v36i2.23289

Alonso, D., & Gonzáles, J. (2021). Museumization of olive oil mills: Instruments for conservation of the heritage of the olive grove culture. *Ge-conservacíon*, *20*, 173–183. doi:10.37558/gec.v20i1.1036

American Marketing Association. (2017). *What is marketing?* https://www.ama.org/the-definition-of-marketing/

Bezerra, R., & Correia, A. (2018). The potential of olive oil for creative tourism experiences in the northern region of Portugal. *Revista Portuguesa de Estudos Regionais*, 56-72.

Boyle, D. (2004). *Authenticity: Brands, fakes, spin and the lust for real life*. Harper Perennial.

Buhalis, D., & Amaranggana, A. (2015). Smart tourism destinations enhancing tourism experience through personalisation of services. In I. Tussyadiah & A. Inversini (Eds.), *Information and Communication Technologies in Tourism* (pp. 377–389). Springer International Publishing. doi:10.1007/978-3-319-14343-9_28

Coelho, M. F., Goslingb, M. S., & Almeida, A. S. (2018). Tourism experiences: Core processes of memorable trips. *Journal of Hospitality and Tourism Management*, •••, 11–22. doi:10.1016/j.jhtm.2018.08.004

Cook, P. S. (2010). Constructions and experiences of authenticity in medical tourism: The performances of places, spaces, practices, objects and bodies. *Tourist Studies*, *10*(2), 135–153. doi:10.1177/1468797611403048

Correia, R. (2006). *The relevance of collective action in tourism* [Conference session]. Cutting Edge Research in Tourism Conference, University of Surrey. http://hdl.handle.net/10198/5837

Couret, C. (2016). Managing experiences within the field of creative tourism: Best practices and guidelines. In M. Sotiriadis & D. Gursoy (Eds.), *The Handbook of Managing and Marketing Tourism Experiences* (pp. 235–259). Emerald Group Publishing Limited. doi:10.1108/978-1-78635-290-320161034

Denmark Government. (2003). *Culture and experience economy-5 new steps The Danish Growth Strategy*. Author.

Dias, M. G., & Viegas, E. J. (2005). *Projecto do Museu da Oliveira e do Azeite, Mirandela* [Olive and Olive Oil Museum Project, Mirandela]. Ouro Virgem - Revista do Museu da Oliveira e do Azeite - Mirandela.

du Rand, G. E., & Heath, E. (2006). Towards a framework for food tourism as an element of destination marketing. *Current Issues in Tourism*, *9*(3), 206–234.

Elias, S. R., & Barbero, A. C. (2017). Situación del oleoturismo y lineamientos para su desarrollo en la región del sudoeste bonaerense, Argentina [Situation of oleotourism and guidelines for its development in the southwestern region of Buenos Aires, Argentina]. *Revista Interamericana de Ambiente y Turismo*, 91-104. https://riat.utalca.cl/index.php/test/article/view/338/pdf

Eurostat. (2019, March 1). Olive trees cover 4.6 million hectares in the EU. *Eurostat*. https://ec.europa.eu/eurostat/web/products-eurostat-news/-/DDN-20190301-1?inheritRedirect=true

Fernández, J., Montilla, J., Hidalgo, I., & Fernández, M. (2022). Evaluating olive oil tourism experiences based on the segmentation of demand. *International Journal of Gastronomy and Food Science*, *27*. Advance online publication. doi:10.1016/j.ijgfs.2021.100461

Fernández, J. I., Montilla, J. C., & Hidalgo, I. C. (2019). Introducing olive-oil tourism as a special interest tourism. *Heliyon*, 1–8. doi:10.1016/j.heliyon.2019.e02975 PMID:31872132

Garrod, B. (2003). Local participation in the planning and management of ecotourism: A revised model approach. *Journal of Ecotourism*, 33–53. doi:10.1080/14724040308668132

González, J. A., Armenteros, E. M., & Ruiz, F. J. (2020). Characterization of olive oil tourism as a type of Special Interest Tourism: An analysis from the tourist experience perspective. *Sustainability*, 1–10. doi:10.3390u12156008

Granitto, Y. (2020, January 31). *O negócio* [The business.] OliveOilTimes: https://www.oliveoiltimes.com/pt/briefs/oleotourism-becomes-law-in-italy/79917

Guerra, I. R., Molina, V., & Quesada, J. M. (2018). Multidimensional research about oleotourism attraction from the demand point of view. *Journal of Tourism Analysis: Revista de Análisis Turístico*, *25*(2), 114–128. doi:10.1108/JTA-06-2018-0017

Hidalgo, I., Montilla, J., & Fernández, J. (2021). Desarrollo sostenible del oleoturismo: Un análisis de la colaboración entre actores [Sustainable development of oleotourism: An analysis of collaboration between actors]. *Revista de Ciencias Sociales*, *27*(2), 19–38. doi:10.31876/rcs.v27i2.35896

Hwang, Y., & Felitti, D. (2021). A quest for the healthy lifestyle and learning experiences in olive oil tourism. *Journal of Tourism and Cultural Change*. Advance online publication. doi:10.1080/14766825.2021.1917586

Ilincic, M. (2014). Benefits of creative tourism - The tourist perspective | Tourism | Creativity. *Academia.Eu*, 99–113. https://pt.scribd.com/document/309796523/Benefits-of-Creative-Tourism-The-Tourist-Perspective

International Olive Council. (2020). www.internationaloliveoil.org/

Lehtolainen, M. (2003). Public infrastructure investments and their role in tourism development in the Finnish Lake Region. In *Proceedings of Lake Tourism Conference 2003*. Savonlinna Institute for Regional Development and Research International.

Marujo, N. (2016). Turismo, turistas e experiências: abordagens teóricas [Tourism, tourists and experiences: theoretical approaches]. *Revista: Turydes Revista Turismo y Desarrollo*. http://hdl.handle.net/10174/20116

Mehmetoglu, M., & Engen, M. (2011). Pine and Gilmore's concept of experience economy and its dimensions: An empirical examination in tourism. *Journal of Quality Assurance in Hospitality & Tourism*, *12*(4), 237–255. doi:10.1080/1528008X.2011.541847

Millán, G., Arjona, J., & Amador, L. (2014). A new market segment for olive oil: Olive oil tourism in the South of Spain. *Agricultural Sciences*, *5*(3), 179–185. doi:10.4236/as.2014.53022

Millán, G., Arjona, J., & Amador, L. (2017). Olive oil tourism: Promoting rural development in Andalusia. *Tourism Management Perspectives*, *21*, 100–108. doi:10.1016/j.tmp.2016.12.003

Mogollón, J. M., Clemente, E. D., Fernández, J. A., & Cerro, A. M. (2019). Olive oil tourism: State of the art. *Tourism and Hospitality Management*, 179–208. doi:10.20867/thm.25.1.5

Morales, P. M., Guzmán, T. J.-G., Cuadra, S. M., & Agüera, F. O. (2015). *Análisis de la demanda del oleoturismo en Andalucía* [Analysis of the demand for oleotourism in Andalusia.]. Revista de Estudios Regionales.

Morgan, M., Lugosi, P., & Ritchie, J. R. B. (2010). *The tourism and leisure experience: Consumer and managerial perspectives*. Channel View Publications.

Morgan, N., Pritchard, A., & Pride, R. (2004). *Destination branding: Creating the unique destination proposition*. Butterworth Heinemann.

Mozas-Moral, A. (2019). *Contribución de las Cooperativas Agrarias al Cumplimiento de los Objetivos de Desarrollo Sostenible. Especial Referencia al Sector Oleícola* [Contribution of Agricultural Cooperatives to the Fulfillment of the Sustainable Development Goals. Special Reference to the Olive Sector]. Centro internacional de investigación e información sobre la economía pública, social y cooperativa.

Murgado, E. M. (2013). Turning food into a gastronomic experience: Olive oil tourism. In N. Arcas, F. N. Arroyo López, J. Caballero, R. D'Andria, M. Fernández, R. Fernandez Escobar, A. Garrido, J. López-Miranda, M. Msallem, M. Parras, L. Rallo, & R. Zanoli (Eds.), *Present and future of the Mediterranean olive sector* (pp. 97-109). CIHEAM. https://om.ciheam.org/om/pdf/a106/00006809.pdf

Oh, H., Fiore, A. M., & Jeoung, M. (2007). Measuring experience economy concepts: Tourism applications. *Journal of Travel Research*, *46*(2), 119–132. doi:10.1177/0047287507304039

Pezzi, E., & Vianna, S. L. (2015). Experiência turística e o turismo de experiência: Um estudo sobre as dimensões da experiência memorável [Tourist experience and experience tourism: a study on the dimensions of the memorable experience]. *Revista Turismo em Análise*, *26*(1), 165–187. doi:10.11606/issn.1984-4867.v26i1p165-187

Pine, J., & Gilmore, J. H. (1998). Welcome to the experience economy. *Harvard Business Review*, *76*(4), 97–105. PMID:10181589

Reis, P. (2014). *O olival em Portugal dinâmicas, tecnologias e relação com o desenvolvimento rural* [The olive grove in Portugal dynamics, technologies and relationship with rural development]. Animar - Associação Portuguesa para o Desenvolvimento Local. https://www.animar-dl.pt/site/assets/files/1273/o_olival_em_portugal.pdf

Richards, G. (2013). Tourism development trajectories - From culture to creativity? In G. Richards & J. Wilson (Eds.), *Encontros Científicos - Tourism & Management Studies, 297–303*. Routledge. doi:10.4324/9780203933695

Richards, G., & Wilson, J. (2006). Developing creativity in tourist experiences: A solution to the serial reproduction of culture? *Tourism Management*, *27*, 1209–1223. doi:10.1016/j.tourman.2005.06.002

Ritchie, J., Tung, S., & Ritchie, R. J. B. (2010). Tourism experience management research: Emergence, evolution and future directions. *International Journal of Contemporary Hospitality Management*, *23*(4), 419–438. doi:10.1108/09596111111129968

Serra, F. (2015). *Turismo de experiência - Tempos de gestão* [Experience tourism - Management times]. http://www.temposdegestao.com/turismo/turismo-de-experiencia

Smith, M. K., Ferrari, S., & László, P. (2016). Service innovations and experience creation in spas, wellness and medical tourism. In D. G. Marios Sotiriadis (Ed.), *Tourism Experiences* (pp. 299–319). Emerald Group Publishing Limited.

Teixeira, D. J. (2005). *O Ecomuseu de Barroso. A Nova Museologia ao serviço do desenvolvimento local* [The Ecomuseum of Barroso. The New Museology at the service of local development] [Master Dissertation]. Universidade do Minho.

Torre, M. M., Hidalgo, L. A., & Fuentes, A. J. (2015). El oleoturismo: Una alternativa para preservar los paisajes del olivar y promover el desarrollo rural y regional de Andalucía [Oleotourism: An alternative to preserve the landscapes of the olive grove and promote rural and regional development of Andalusia]. *Revista de Geografía Norte Grande*, *60*, 195–214. doi:10.4067/S0718-34022015000100011

Tregua, M., D'Auria, A., & Marano-Marcolini, C. (2018). Oleotourism: Local actors for local tourism development. *Sustainability*, 1–20. doi:10.3390u10051492

United Nation World Tourism Organization. (2012). *Global report on food tourism*. https://www.e-unwto.org/doi/epdf/10.18111/9789284414819

Vallone, G. (2018). *Turismo do azeite ganha adeptos no sul de Portugal* [Olive oil tourism wins fans in southern Portugal]. Folha de S. Paulo: https://www1.folha.uol.com.br

Yeoman, I., Brass, D., & McMahon-Beattie, U. (2007). Current issue in tourism: The authentic tourist. *Tourism Management*, *28*, 1128–1138. doi:10.1016/j.tourman.2006.09.012

Chapter 2
The Growing Importance of Gastronomy Tourism in China

Poshan Yu
https://orcid.org/0000-0003-1069-3675
Soochow University, China & Krirk University, Thailand

Heng Tang
The Intellectual Property School, Jiangsu University, China

Ding Zuo
Independent Researcher, China

Ramya Mahendran
https://orcid.org/0000-0001-9585-9077
Independent Researcher, India

ABSTRACT

With improving standards of living in China, people can afford travelling for leisure, and gastronomy tourism (also known as food tourism and culinary tourism) has been gaining popularity. Through this tourism, the travelers get to explore different regional foods, cuisines, and understand the local culture. After the COVID-19 outbreak, the tourism industry has taken a hard hit. But we are slowly returning to normalcy, and the demand for international travel will recover. Food, being one of six elements of tourism alongside accommodation, transportation, travel, shopping, and entertainment, cannot be ignored. This chapter aims to study how and whether the Chinese government has issued various policies to support gastronomy tourism and promote the regional food culture in context of tourism for visitors and travelers. Four Chinese cities will be used for discussion due to their featured international popularity and distinctive gastronomy. Finally, recommendations will be given to stakeholders.

INTRODUCTION

China's Tourism Industry

DOI: 10.4018/978-1-7998-9148-2.ch002

Figure 1. Number of domestic tourists in China from 2010 to 2019 (in billions)
Source: NBSC

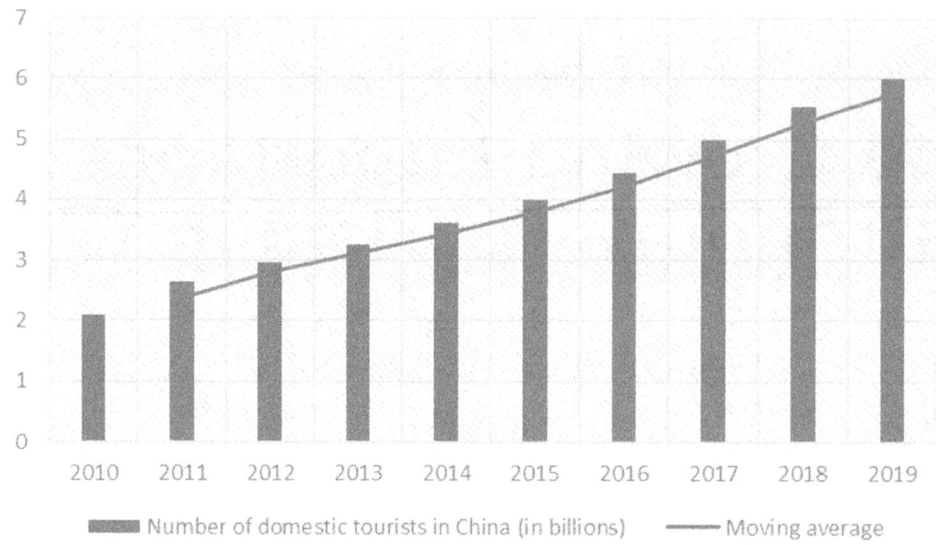

Post-World War, modern tourism became an emerging industry. China's modern tourism journey began a little later, whose formative years lie between the founding of new China and the early stages of reform and opening-up (Fu, 2014). In 1978, the number of visitors from foreign countries was 1.81 million, grossing 263 million dollars as foreign exchange earnings from international tourism. It set a new high in China's tourism history, as it exceeded the sum of earnings of the previous 20 years, and the dawn of China's modern tourism development began.

With the deepening of reform and opening-up, the tourism industry has developed rapidly and has become an important engine of economic growth. It has bought socio-economic benefit to the communities through employment and income earning opportunities.

According to the data from the National Bureau of Statistics of China [NBSC], the domestic tourism market grew steadily from 2010 to 2019. In 2010, the number of domestic tourists was just over 2.1 billion and by end of 2017 it had exceeded 5 billion, reaching a new high. In the next two years this number increased by 1 billion, rapidly reaching 6.006 billion by the end of 2019, with a growth rate of 8.43% compared with the same period in 2018. Consequently, the total income from domestic tourism also showed a significant year by year growth in the decade starting 2010. In 2010, the income was at 1,257.98 billion and by 2014 it increased by nearly 1.5 times, exceeding 3,031.19 billion. From 2015 to 2019 it maintained a growth rate of more than 10% and the total income was at 6,000 billion in 2019. The following figures show China's number of domestic tourists and the total income from domestic tourism from 2010 to 2019 (Figure 1 & 2).

An industry's contribution to GDP is an important indicator of economic development and industry structure. The contribution of tourism to China's GDP has risen from 10.39% in 2014 to 11.05% in 2019. The World Travel and Tourism Council [WTTC] (2020) predicts that the contribution rate will reach 12.9% in 2028. The chart below depicts contribution of China's tourism industry to GDP from 2014 to 2019 (Figure 3).

Figure 2. China's total income from domestic tourism 2010-2019 (in billion yuan)
Source: NBSC

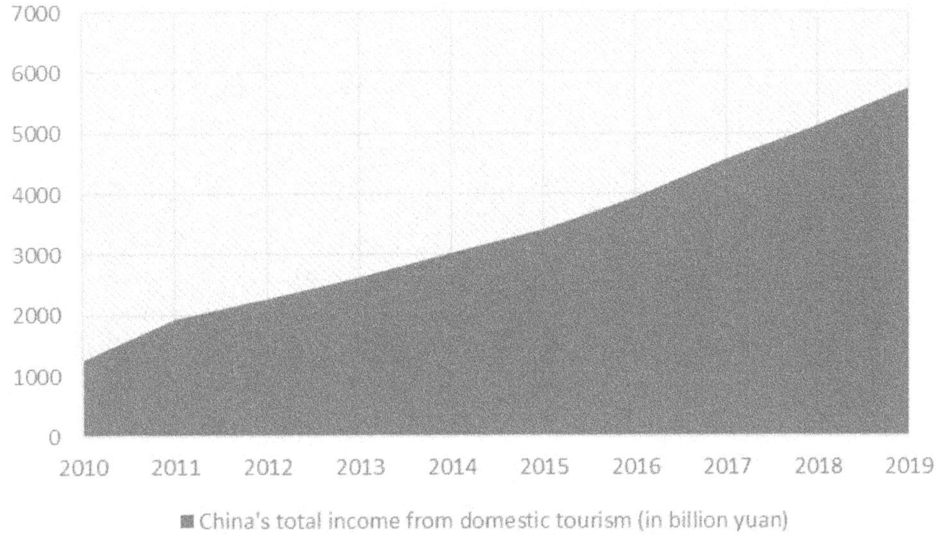

Figure 3. Contribution of China's tourism industry to GDP 2014-2019
Source: Ministry of Culture and Tourism of the People's Republic of China

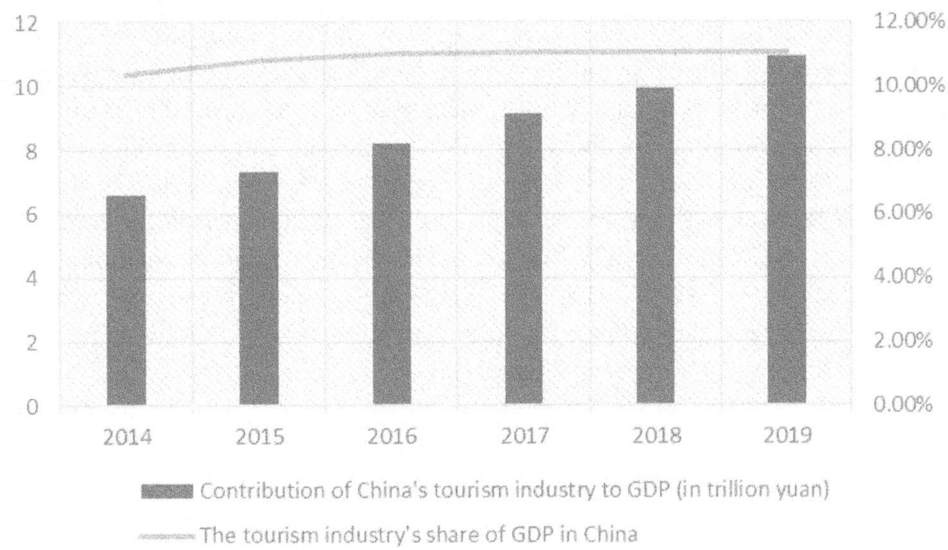

It is noticeable that tourism plays an important role in economic development and people's life. The previous deputy director of the China National Tourism Administration [CNTA], Sun (1994), once called tourism "A Sunrise Industry Forever".

Figure 4. Expenditure on tourism catering in China 2012-2017
Source: xinhuanet.com; NBSC

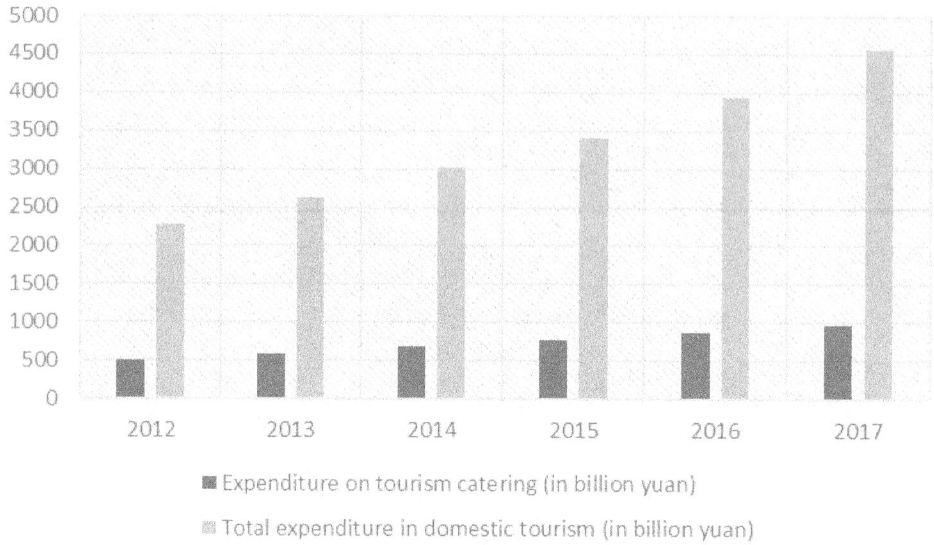

THE BACKGROUND OF CHINA'S GASTRONOMY TOURISM

In 1980, He, president, and chief editor of China Tourism News, first proposed the three elements of tourism: restaurant, transportation, and service, which had practical significance for the development of China's tourism industry at that time (Fu, 2014). Subsequently, in 1991, Sun presided over the publication of China's Tourism Economy Development Strategy Report, which for the first time clearly stated the six elements of tourism, namely "accommodation, transportation, food, travel, shopping, entertainment" and since then has become widely recognized. Both the original three elements and the later expanded six elements include "food". Balderas-Cejudo et al, support that food and tourism are two industries with important synergies (as cited in Martín et al., 2020). The white paper on the consumption power of China's cuisine tourism (XINHUANET, 2018) released at the 2nd Boao international cuisine culture forum pointed out that, from 2012 to 2017, the expenditure on tourism catering continued to grow, reaching 972.671 billion by the end of those six years, and always accounted for above one fifth of the total expenditure in domestic tourism years (Figure 4).

The links between tourism and food are multifaceted and deep. "Food" is not only an important resource to tourism, but also an embodiment of society's consumption demand (Zhong et al., 2021b). According to Reynolds, food is a "non-optional" component of the tourism experience when it comes to satisfying physiological needs (as cited in Boyne et al., 2002) and "for many tourists, food and eating are important social and experiential elements of their holiday" (Boyne et al., 2002, p. 91).

Food is one of the motivating factors for visiting tourist destinations. Hence, the type of tourism related to food, namely, gastronomy tourism, has been on the rise. It was first put forth by Long as "culinary tourism" in 1998 (as cited in Zhong et al., 2021a, p. 463), then Long (2004) further defined it as the experience and participation in the food life of people in other regions, not just limited to the introduction or consumption of food. The Committee on Tourism and Competitiveness [CTC] of the

World Tourism Organization [UNWTO] defined gastronomy tourism as "a type of tourism activity which is characterized by the visitor's experience linked with food and related products and activities while travelling" (2021b). UNWTO suggested that, "Gastronomy is about much more than food. It reflects the culture, heritage, traditions, and sense of community of different people. It is a way of promoting understanding among different cultures, and of bringing people and traditions closer together" (2021a).

The origin of China's gastronomy tourism began with the popular "tour banquet" or "boat banquet" during the ancient times. It was not until the first Guangzhou International Gourmet Festival, which was held in 1987 that China's gastronomy tourism developed in the form of food festival (Zhong et al., 2021a). Chinese scholars have different definitions of gastronomy tourism. For example, Shang (2015) thinks that gastronomy tourism includes activities with attributes of leisure and culture, where the main motivation is to taste food and experience the local food culture which is enriched by the natural and cultural landscape. Chen, Lu and Li D. L. (2018), Li X., He and Liu (2019) believe that gastronomy tourism emphasizes the attraction of food related tourism, as a way that provides tourists with a unique and cultural experience that makes a trip unforgettable.

In general, whatever be the definition, there's no doubt that today's tourists have set higher standards and expectations for food and related services during their trip, and it takes more than filling the belly to satisfy their demands. While going through different touristic activities, people start paying more attention to the quality, characteristics, and cultural background of food (Zhong et al., 2021a). It's high time we gained deeper insight into gastronomy tourism.

CHARACTERISTICS OF CHINESE GASTRONOMY TOURISM POST-COVID-19

In 2020, the COVID-19 pandemic triggered a global economic crisis that could last for years. The tourism economy, which combines the experience of traveling, socializing with the local culture and staying in different places, has fallen victim to the pandemic and is facing unprecedented challenges and negative impact.

According to the Ministry of Culture and Tourism of the People's Republic of China, the number of domestic tourists was 2.88 billion in 2020, a decrease of 3.02 billion from 2019, dropping by 52.1% (2021, February 18). The domestic tourism revenue also decreased significantly, making only 2.23 trillion yuan in 2020, a decrease of 3.5 trillion yuan from 2019, dropping down by 61.1% (Ministry of Culture and Tourism of the People's Republic of China, 2021, February 18). The China Tourism Academy predicted that, the number of inbound tourists and international tourism revenue were 27.2 million and 17 billion dollars, with a year-on-year decrease of 81.3% and 87.1% respectively (2021).

However, iResearch (2020) pointed out that, "Tourism consultation has become a rigid demand and it is prospective in the long run. The annual compound growth rates of the catering and tourism markets in the past decade are 11.4% and 18.3%, respectively. In 2013, the revenue of the tourism market reached 2.63 trillion yuan, surpassing the catering market to become a driving force of the tertiary industry (Figure 5). In recent years, the 80s and 90s born (Gen Y) have gradually become the main consumer groups. They have contributed to the boom in family vacations, leisure and holiday travel. Based on this long-term perspective of population structure and consumption habits, the market trend of Chinese public travel will be optimistic and will not change significantly in the long term" (p. 3). The China Tourism Academy also stated that, tourism consumption is rigid, because the rate of travel depends on one's income and leisure time, which are both growing (2021, February 22). The continuing COVID-19

Figure 5. Revenue of China's catering and tourism markets from 2010 to 2019
Source: www.iresearchchina.com

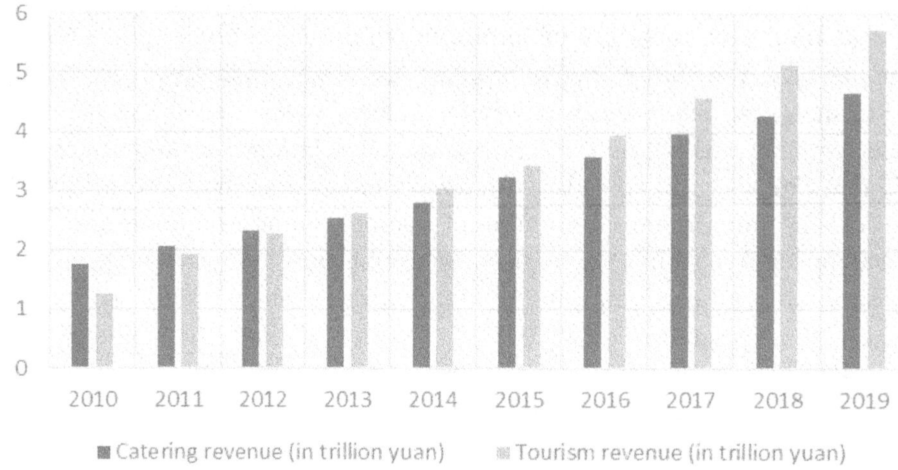

pandemic has globally decreased the willingness to travel abroad, according to a survey conducted by Baidu, about 87 percent of the demand for traveling in China centered on domestic travel (2020).

Data from the Ministry of Culture and Tourism of the People's Republic of China show that domestic tourism is gradually improving. According to the quarterly data of 2020, the year-on-year decline rates of the number of domestic tourists were 83.4%, 51%, 34.3% and 32.9% in four quarters, indicating that the rate of travel has started showing an upward trend (Ministry of Culture and Tourism of the People's Republic of China, 2021, February 18). In the first half of 2021, the number of domestic tourists totaled 1.871 billion, going up by 100.8% from the same period last year and the domestic tourism revenue was 1.63 trillion yuan, with a year-on-year growth of 157.9% (Ministry of Culture and Tourism of the People's Republic of China, 2021, July 29).

The tourism industry is facing the problem of supply-demand mismatch, product structure (also known as BOM – bill of materials) imbalance and inefficient investments. Therefore, from another perspective, COVID-19 provides a buffer period for removing ineffective production capacity, adjusting product structure, and improving the quality of development for the whole industry to adapt to the new normal (Huang & Liang, 2020). Modern tourism tends to be diversified and personalized, highlighting the leisure needs of the tourists. To transform the crisis into opportunities, it is crucial to establish an effective tourism supply system, which is demand-oriented (Ibid).

Food, as one of the six elements of tourism, remains a major focus in the post-COVID-19 recovery. UNWTO and the Food and Agriculture Organization of the United Nations [FAO] stated that, "As governments seek to revitalize their hard-hit tourism industry, gastronomy will have a key role to play enriching the tourism offerings and stimulating economic development in destinations all around the world" (2021). Since the outbreak of COVID-19, most of large-scale chain restaurants, snack stands, fast food stalls and food streets have responded to the government's call to cease or restrict operations and temporarily closed their business, resulting in a revenue loss of about 90% (XINHUANET, 2020). According to a survey report from Deloitte and China Cuisine Association [CCA], as of March 2020, around 95 percent of the surveyed catering enterprises in China had already experienced or were expecting

Figure 6. Expected fund shortage among restaurants due to coronavirus in China 2020
Source: Deloitte & CCA

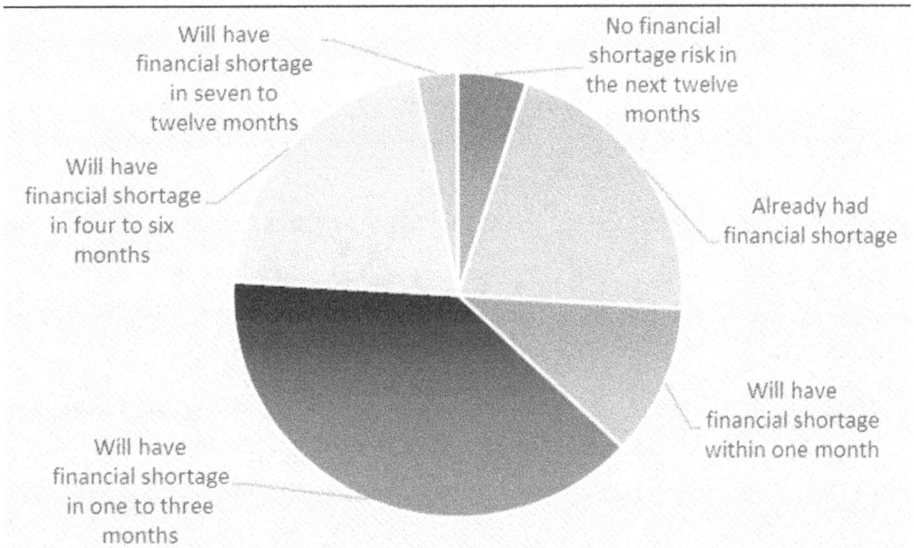

a financial shortage to occur in the next twelve months, and about 70 percent of the catering enterprises expected a year-on-year decrease of more than 70 percent (2020) (Figure 6). The lack of cash flow was a reason for the financial shortfall. The figure 6 below shows the expected fund shortage. However terrible though the situation is, based on personal consumption factors, catering will be the first to recover after COVID-19, and then travel with accommodation (iResearch, 2020).

As we recover from the blow of COVID-19 and people start reintegrating back to their societies there may remain some concerns of hygiene and safety of food and related services during domestic and international travel. Catering enterprises in tourist destinations are seeking new methods to meet said needs. The traditional catering industry is characterized by in-store dining and in-person payment. However, in the aftermath of COVID-19, contactless-services and minimalized personal interactions have come into picture, thereby limiting the traditional dining methods. As of March 2020, about 69% of the catering enterprises in China reported a loss of more than 80% of their dine-in customers (Deloitte & CCA, 2020) (Figure 7).

The outbreak of COVID-19 in 2020 shifted the business strategies of many restaurants in China. Around 78% of restaurants mainly focused on online food delivery after outbreak, in comparison to only 15% who did so before the outbreak, and only 2.3% of the restaurants said their main business was still offline as of March 2020 (iiMedia Research, 2020) (Figure 8). XINHUANET finds that, online takeout and delivery has become the prime choice, providing not only ready to eat food, but also grocery and semi-prepared, ready-to-cook food for consumers who are willing to prepare it by themselves (2020).

Different from the common practice of simply employing online delivery to increase turnover, some merchants have expanded their restaurant business through collaboration with supermarkets (Li, 2021a). In Li's interview with Wang, who is the owner of a halal restaurant in Beijing, the customer flow of supermarkets far exceeds that of ordinary restaurants during the pandemic, in consequence, takeout food often sells better in supermarkets than online ordering (2021). In case of in-store dining, intelligent ordering system is widely used. It means that customers will scan the QR code (Quick Response code)

Figure 7. Impact of coronavirus on dine-in customer number in China 2020
Source: Deloitte & CCA

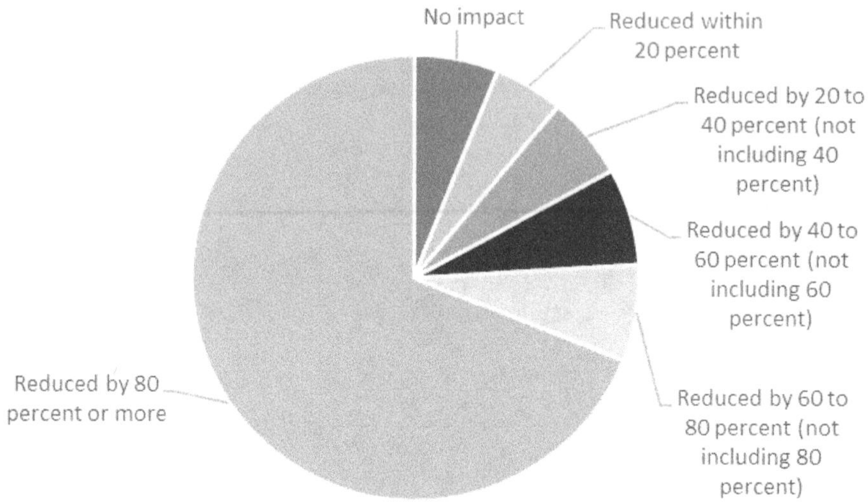

to complete the ordering process. After the order is placed, it is automatically transmitted to the back kitchen and service desk. The whole process reduces the number of serving staff and also improves staffing efficiency. At the same time, it can also shorten the waiting and dining time of guests and improve the turnover rate (Shi & Li, 2021). The structure of personnel has also been adjusted.

Robots are now being used to provide contactless delivery and services thereby shifting hospitality sector from "high-touch" to "high-tech" (as cited in Fuste-Forne, 2020). Some enterprises are cooperating with each other and sharing employees who are used to complete some relatively simple tasks,

Figure 8. Coronavirus impact on operational models of restaurants in China 2020
Source: www.iimedia.cn

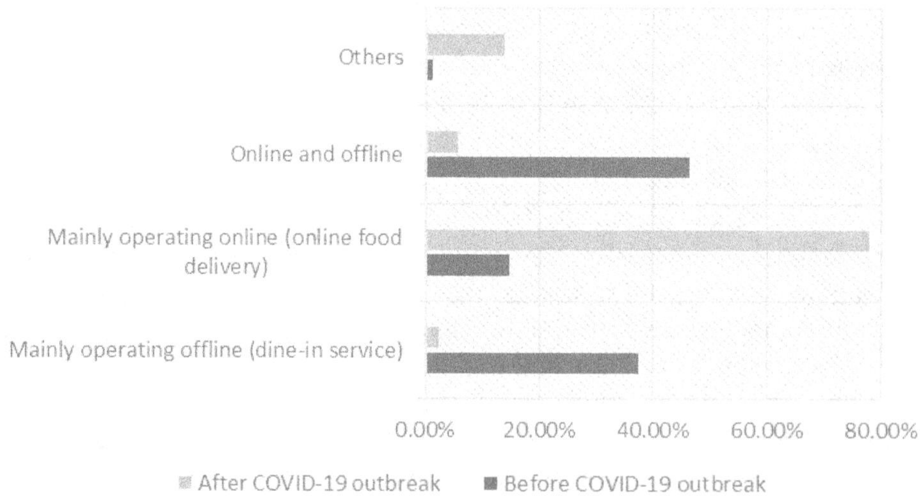

which reduces the labor cost (XINHUANET, 2020). Additionally, because street-stall economy is now advocated by the state, several catering enterprises take advantage of this opportunity to create mobile advertising (Shi & Li, 2021). From the research by the China Tourism Academy, there have been growing innovations in smart tourism and its digitally driven transformation. The "cloud tourism" (also known as "virtual tourism". It is a kind of "at-home" tourism which includes a whole host of experiences. Tuning into livestreams of short and long trips, even giving museum and art exhibition tours) spawned by live broadcasting, enriches the residents' daily tourism and leisure activities, parallelly meeting the content demand of tourism consumption (China Tourism Academy, 2021, February 22). Some catering enterprises also publicize food through live broadcasting, creating a new form of combining internet and ordering (Shi & Li, 2021).

Gastronomy tourism deserves attention in the post-pandemic era. The China Tourism Academy pointed out that the potential of gastronomy tourism has rapidly unleashed after COVID-19 (2021, February 22). Driven by the demand of both tourists and residents, city tours with gastronomy as the theme is also more frequent (China Tourism Academy, 2021, July 13).

The Roles of Chinese Government in Assisting Hospitality and Tourism Sector in Promoting Gastronomy Tourism

Institution is a series of established rules, law-abiding procedures, and moral and ethical norms of behavior (North, 1994). It includes formal institutions, informal institutions, and their implementation, which define the incentive structure of economy (Ibid). According to North and Thomas (2009), an institution that can provide effective incentives to individuals is the decisive factor to ensure economic growth.

To promote healthy and sustainable development of the tourism industry, the Chinese government has introduced a series of policies and measures in recent years, many of which emphasize the importance of gastronomy.

The 13th Five-Year Plan for tourism development was released on December 7, 2016. In this document, the State Council put forth the recommendations for improving the quality of catering industry to promote the level of tourism elements and upgrade the industrial structure (2016). It specifically underlines combining Chinese culture and traditional food, the in-depth exploration of folk snacks (traditional snacks), and the conscious establishment of featured food brands, to expand their influence and competitiveness thereby piquing attracting tourist interests (Ibid). In the second quarter of the following year, 2017, the government work report proposed the concept of all-for-one tourism, whose purpose is to develop tourism by integrating various industries (State Council, 2017). As one of the categories of intangible cultural heritage, traditional food has created competitive advantages in gastronomy tourism (Bertan, 2020). The number of travels organized with the purpose of seeing the cultural values of a place has been increasing in the recent years (as cited in Bertan, 2020). On June 12, 2017, China National Tourism Administration published the guidelines for the establishment of all-for-one tourism demonstration areas. It suggested tapping into traditional folk snacks and enhancing brand effect, along with the construction of characteristic catering blocks and cultural themed hotels (China National Tourism Administration, 2017). On September 10, 2018, in response to the proposal by a deputy to the National People's Congress, the Ministry of Culture and Tourism of the People's Republic of China said it would insist on integrating culture into tourism, treating gastronomy as an important resource of intangible cultural heritage (2018, September 10).

Gastronomy tourism helps create opportunities, including jobs, most notably in rural destinations (UNWTO, 2021a). The 2019 Central Document No.1 clearly stated that leisure tourism and B&B (Break and Breakfast) should be attached importance to in rural areas (Central People's Government of the People's Republic of China, 2019). According to the Ministry of Culture and Tourism of the People's Republic of China, the Agricultural Bank of China [ABC] was going to provide 100-billion-yuan credit line to rural areas focusing on tourism development, supporting the development and promotion of tourist offerings like local cuisine and creative cultural goods (2019, July 26). In post-pandemic, on the premise of health and safety, local cuisine was given importance in the recovery of rural tourism, including related food culture and cooking skills (Ministry of Culture and Tourism of the People's Republic of China, 2020, July 17).

When planning regional tourism, the Chinese government also has paid attention to the role of gastronomy. In 2018, the Ministry of Culture and Tourism of the People's Republic of China proposed to develop all-region tourism in both Manzhouli in Inner Mongolia and Fangchenggang in Guangxi on the border, to build characteristic catering blocks and support residents to open restaurants (2018, April 16). The China Tourism Academy has pointed out that, the tourism industry plays an important role in the development of Guangdong-Hong Kong-Macao Greater Bay Area [GBA] (2021, February 22). The Ministry of Culture and Tourism of the People's Republic of China has suggested gastronomy tourism in GBA and advocated the integration of culture and tourism (2021, January 5).

After COVID-19, the state has also issued several supporting policies to help the tourism and catering industry tide over the difficulties. The Minister of Culture and Tourism of the People's Republic of China decided to temporarily refund part of the tourism service quality deposit to support travel agencies in coping with operational difficulties (2020, February 5). The State Taxation Administration (2020) proposed that the longest carrying forward period of losses of enterprises in difficult industries greatly affected by COVID-19 can be extended to 8 years, including the tourism and catering industry. The China Banking and Insurance Regulatory Commission [CBIRC] emphasized that the financial institutions should not blindly withdraw loans, cut off loans and suppress loans in hard hit industries, such as accommodation, and catering (2020).

Now the pandemic at home is taking a turn for the better and China has officially entered the 14th Five-Year Plan period in 2021. Tourism development in the new period focuses on the integration of culture and tourism. On March 8, 2021, the Minister of Culture and Tourism of the People's Republic of China recommended to fuse public cultural services with tourism and innovate culture-related catering business (Minister of Culture and Tourism of the People's Republic of China, 2021, March 8). In April, the 14th Five-Year Plan for culture and tourism development was issued. It paid attention to the construction of tourism destinations and underlined the pattern of all-for-one tourism (Minister of Culture and Tourism of the People's Republic of China, 2021, April 29). Besides, the 14th Five-Year Plan for tourism development made clear requests for improving the quality of tourism catering, promoting the combining of tourism catering and culture services, and developing gastronomy tourism (Minister of Culture and Tourism of the People's Republic of China, 2021, April 29). To vigorously develop the tourism economy and meet the people's growing needs for a better life, the Ministry of Culture and Tourism of the People's Republic of China has decided to build the first batch of state-level night culture and tourism consumption agglomeration zones which concentrate on the development of tourism, with rich cultural connotation and prominent regional characteristics (2021, July 5). In particular, the ministry said that food streets with no cultural messages, which allows tourists to appreciate the locality, would not be taken into consideration (Ministry of Culture and Tourism of the People's Republic of China, 2021,

July 5). Meanwhile, the ministry has also decided on organizing the 2021 national culture and tourism consumption season. With the theme of stimulating the potential of culture and tourism consumption and promoting the high-quality development of industries including catering and accommodation. The activity provides consumption subsidies, preferential promotions, and other measures to benefit the people, and increases publicity efforts to strive to convert online traffic into offline traffic (Ministry of Culture and Tourism of the People's Republic of China, 2021, July 19).

In these connections, the Chinese authorities proposed to use intellectual property (IP) rights to enhance culture and tourism economy. IPs are exclusive rights to the fruits of intellectual labor created by human beings in social practice. In 2015, the Ministry of Culture and Tourism of the People's Republic of China proposed to use cultural tourism IP to promote industrial development (December 31, 2015). Cultural tourism IP refers to an IP system with cultural characteristics, brand core and unique values under the integration of cultural and tourism elements (Yi & Hu, 2019). These IPs help increase the added value of products, bridge the gap between the catering and tourism industries, and meet the needs of cross-border development of gastronomy tourism (Wang, 2020).

DISCUSSION OF CHINESE CASES

"City of Gastronomy" is one of the projects of Creative Cities Network [UCCN] by United Nations Educational, Scientific and Cultural Organization [UNESCO]. If a city is given this title, it means that the city's gastronomy has gained international recognition and is attractive to tourists. Up to now, four cities in China have won the title, namely Chengdu, Shunde, Macao and Yangzhou. They are distributed in different regions, and their gastronomy shows different characteristics. So, they are chosen as successful cases for discussion of China's gastronomy tourism.

Given China's enormous size, every region in China has its own distinct cuisine, referred to as "the 8 Great Cuisines of China" (as cited in Jackie, 2018) (Figure 9). These include the cuisines of Shandong, Jiangsu, Anhui, Zhejiang, Fujian, Canton, Hunan, and Sichuan, the latter being one of the most popular in the country and abroad. Chengdu, Sichuan, which is the first Chinese city to win "City of Gastronomy" in 2010, is the birthplace and principal area of the Sichuan cuisine (Figure 10). One important feature of the Sichuan cuisine is the diversity of tastes, which artistically integrates spiciness, sweetness, bitterness, sourness, and saltiness. However, cuisine from Chengdu and Sichuan provinces are set apart from the rest by their hot and spicy taste and the numbing flavor, which comes from the Sichuan pepper that is rare to find in other Chinese cuisines (as cited in Jackie, 2018). Hot pot is closely related to such taste, so, almost all tourists who come to Chengdu give hot pot a try, and some even view this as the main purpose of their visit. According to statistics, there are more than 300 varieties of snacks in Chengdu (Fang et al., 2020), of which the most famous ones are Mapo Tofu (tofu set in a spicy sauce), Lai tang yuan (a kind of sticky rice dumplings), Zhong dumplings, sliced beef and ox tongue in chili sauce and Kung Pao Chicken. In addition, rabbit head and brains, chicken gizzard and duck intestines also leave a deep impression on tourists and stimulate people to explore the charm of Chengdu cuisine, because these raw materials are usually discarded as waste in other places (Yang et al., 2019). There are also famous food streets in Chengdu, like the Broad and Narrow Alley and the Jinli Street, which have attracted numerous tourists. However, to find cheaper but authentic food, street venders and sidewalk snack booths present themselves as good options. A teahouse is a place where people drink tea and relax, and it also sells some local snacks. Popular in Chengdu, teahouses can be found almost everywhere by visiting tourists. From Yang et al.'s research, tourists frequenting these teahouses are more interested in the culture and the peaceful lifestyle behind the food, rather than being drawn by the spicy food that is being served (2019). These indicate that characteristic gastronomy really appeals to tourists. Moreover, the public have a quantity of opportunities to participate in gastronomy activities in Chengdu, such as China (Chengdu) International Food & Tourism Festival, which is held every year. Due to its popular food culture and abundant tourism resources, Chengdu has been rated as one of the best tourism cities in China by UNWTO (UNESCO, n.d. a).

Shunde, Canton, won "City of Gastronomy" in 2014. Being the birthplace of the Cantonese cuisine (Figure 11), Shunde was rated as the first city worth visiting for gastronomy by the Chinese website Zhihu. The Cantonese cuisine is one of the most representative and most influential food cultures at home and abroad. Since 60% of Chinese overseas settlers are from Guangdong (formerly Canton), most Chinese restaurants around the world are based on the Cantonese cuisine (Qi, 2019). With a long history and deep roots, the Cantonese cuisine has widely adopted cooking techniques from other cuisines, and even absorbed some advantages of western food to form its own style. This reflects their innovative spirit of maintaining tradition while constantly evolving (Hu, 2010). The Shunde cuisine retains and inherits traditional Cantonese cooking methods, and a lot of tourists are attracted by Shunde's authentic taste of the Cantonese cuisine. The city's chefs pay great attention to maintaining the original flavors of the ingredients and try their best to make dishes focusing on looks, smell and taste: visually attractive, inviting flavors and appealing to the palette. In terms of cooking techniques, the cooks use a wide range of comprehensive cooking methods like boiling, braising, frying, marinating, roasting, simmering, smoking, and steaming. The representative dishes of Shunde cuisine like steamed fish mouth and fried fish cake, are all prepared with locally sourced fishes, which also is ingrained in the local fishing culture. In today's industrialized world, they still adhere to the traditional local cooking methods and food materials, maintaining the freshness and localization to the greatest degree (Qi & Chen, 2018). Shunde has other advantages in developing gastronomy tourism. The characteristic Shunde cuisine is rich in cultural connotations, and the local government and non-governmental organizations are all committed to spread the food culture. As early as 2009, the local government had set up special funds for creative industrial design, supporting the integration of culture and different industries (Xie, 2015). Non-government organizations like Shunde Dietetic Association and Shunde Cook Association plays a significant role in promoting the development of the catering industry (Qi & Chen, 2018). Moreover, as hometown to many Chinese residents in foreign countries, Shunde is always on their mind. These migrated residents can rediscover their childhood memories and a sense of belonging in Shunde's traditional food, which is also conducive to the spread and development of Shunde's gastronomy (He et al., 2020). Meanwhile, its superior natural environment, rich cultural landscape, prosperous economy, and famous culinary art colleges also provide excellent conditions for local gastronomy tourism (UNESCO, n.d. b).

Macao, where different cultures meet, won "City of Gastronomy" in 2017. For over 500 years, Macao has been a well-developed trading port. It was because of these maritime commercial activities that people from different places carrying different cultural backgrounds, living habits and food cultures gathered in Macao. They introduced some new varieties of food to the Macao people, thus changing the local diet structure and promoting the communication between Chinese and Western food culture (Li, 2009). Combining the cooking techniques of Portugal, Southeast Asia, India and China, the Macanese cuisine has gradually evolved its unique flavors and are more suitable for Oriental tastes. Jackie (2018) once stated, "From Portuguese style egg custard tarts to fried pork chop sandwiches, crab congee, and Serradura (sawdust pudding), the Macanese cuisine is a great and insanely scrumptious fusion of Chinese

Figure 9. The 8 great cuisines of China
Source: Ministry of Natural Resources of the People's Republic of China

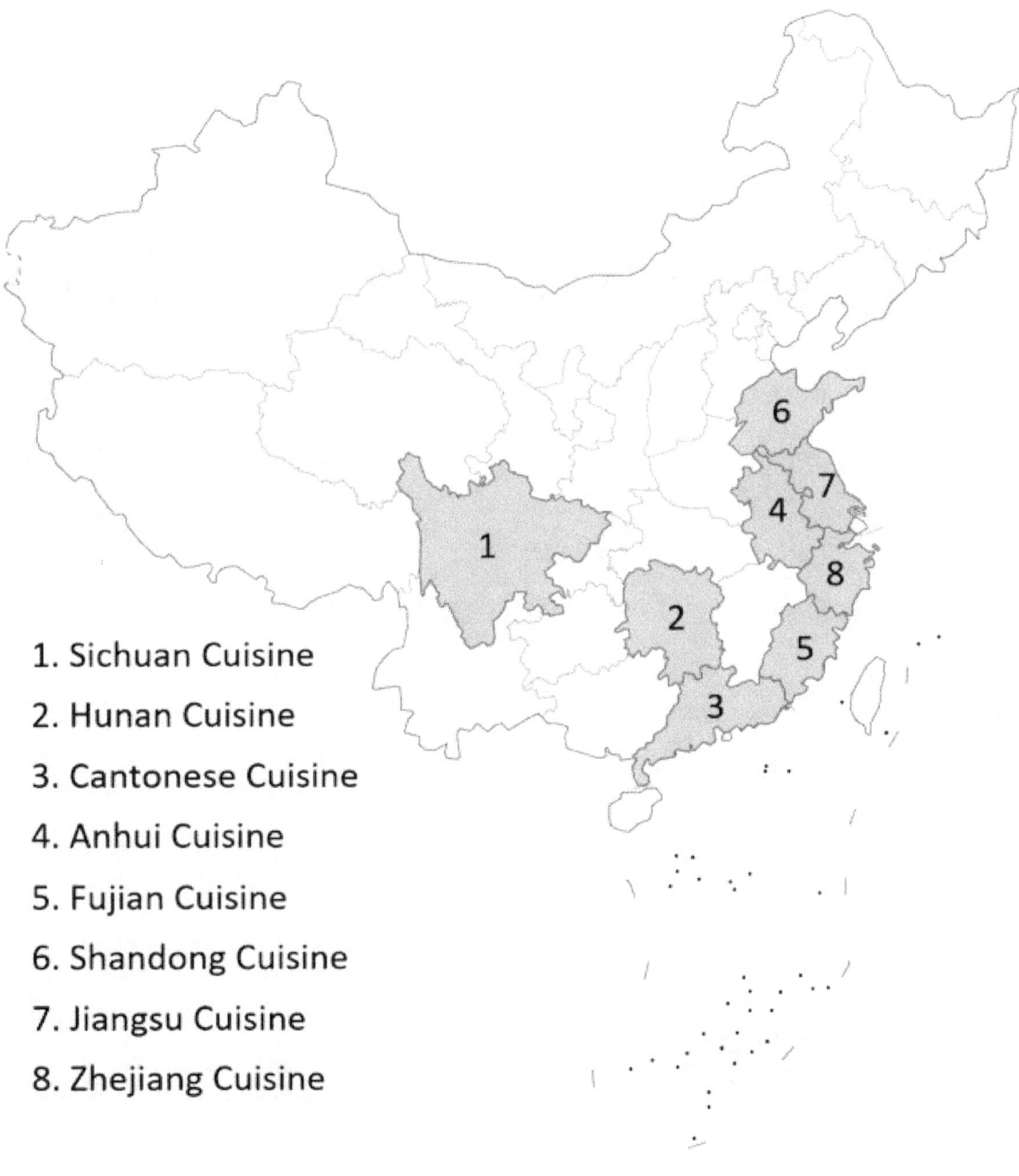

with Portuguese". It can be said that the Macanese cuisine is really a class apart. In the past, several people from India or Southeast Asian countries visited Macao to conduct business or for seeking employment, and they brought authentic Indian or Southeast Asian cuisine. Since most residents in Macao are Cantonese, the Cantonese cuisine is popular locally as well. Coupled with Macao's geographical advantages and low tax rates, foreign food can be imported from all over the world, it is no exaggeration to say

Figure 10. Word cloud for the Sichuan cuisine
Source: wordart.com

Figure 11. Word cloud for the Cantonese cuisine
Source: wordart.com

that Macao has brought together international cuisines. In Qu and Chen's interview with Feng, director of the United Association of Food and Beverage Merchants of Macao, he mentions the characteristic snacks that are omnipresent in Macao, the catering enterprises usually operate as small and medium-sized businesses. It becomes relatively easier to preserve the traditions given many of these enterprises have family-owned and family-run business model (2009). The Macao Special Administrative Region [SAR] Government also makes a difference in local gastronomy tourism. For example, it authorizes the United Association of Food and Beverage Merchants of Macao to help time-honored brands which are moribund (Qu & Chen, 2009) and holds Macao Food Festival at the same time as Grand Prix Macao to attract more tourists (Tang, 2020). An organizing official pointed out that, "The reputation is also inseparable from the fact that the Macao SAR (Special Administrative Regions) government has always put food hygiene, food safety and food legislation in the first place" (Xin, 2020).

Jiangsu cuisine (Figure 12), also called Huaiyang cuisine, is popular in the lower reach of the Yangtze River. Yangzhou, Jiangsu, which practices Huaiyang cuisine, won "City of Gastronomy" in 2019. As one of the four traditional Chinese cuisines, the Huaiyang cuisine has a long history and with a moderate taste it is suitable for all kinds of people. It holds taste as its core and nutrition as its purpose. In today's rich material life, people are no longer satisfied with just ample portion sizes and varieties, but also hope to eat healthily. The Huaiyang cuisine tastes light and often uses seasonal ingredients, befitting people's dietetic needs for healthier food choices. Besides, its detailed production process and material selection are also consistent with people's pursuit of exquisite lifestyle (Miao & Wu, 2020). Being the birthplace of the Huaiyang cuisine, Yangzhou is home to over 52 food intangible cultural heritage projects such as Fuchun tea making skills, many time-honored restaurants such as Yechun and Gonghechun, many food gathering areas like the Dongguan Street and the Slender West Lake night market, and more than 60 Chinese Master Chefs (Yang, 2021). The Huaiyang cuisine contains rich cultural connotations. It can be found in famous works of literature, such as Cao Xueqin's "The Story of the Stone" and Li Dou's "Yangzhou Pleasure-boat", so it is also known as literati dish. Associating the aesthetic taste and artistic conception of literary works, tourists can gain deeper insight into the Huaiyang cuisine and get a more profound tourism experience (Yang, 2021). In addition, because the Beijing-Hangzhou Grand Canal passes through Yangzhou, the city also carries a unique canal culture. Huaiyang cuisine not only has lot of river foods and delicacies which are made from raw material sourced from and through the river, but also the scenery around the Grand Canal itself acts as a tourism resource and showcases the cultural history of Yangzhou city (Xu, 2021). It's fitting to say that the cultural experience is a key point of Yangzhou's gastronomy tourism. UNESCO (n.d. c) posted, "Every year, Yangzhou hosts over 100 food-themed conferences, exhibitions, festivals and contests. These events are cross-cutting platforms organized to promote sector development and international exchange and cooperation, enabling Chinese and foreign citizens to share the delights of creative gastronomy". In terms of policy creation and strategic planning, the Yangzhou government has developed the 13th Five-Year Plan on the Development of Yangzhou's Trade and Movement of Goods Sector. This plan plays an important role in the development of gastronomy and creative sectors. In addition to these long-term plans, Yangzhou which has over 20,000 catering enterprises and creates over 245,000 jobs, has established a Creative Gastronomy Development Fund, worth 2 million US dollars per year, to support heritage, talent exploitation, innovation, and enterprises in the field of traditional gastronomic culture (UNESCO, n.d. c).

Figure 12. Word cloud for the Jiangsu cuisine
Source: wordart.com

OPPORTUNITIES & CHALLENGES FOR DEVELOPING REGIONAL GASTRONOMY TOURISM

Gastronomy tourism was defined as participation in the special or unique food experiences related to a tourist destination (Horng et al., 2012). Güzel and Apaydin (2016) stated, "Food is one of the assets of a culture that stayed local which makes a destination different than the other". To experience the culture, tourists may go for a "local food" (Sims, 2009).

Considering the theoretical discussion developed previously, the following hypothesis is proposed: Local specialties has a positive impact on tourists' choices on tourism destinations. Questionnaire by wenjuan.com was issued to the public following a random statistical design from July 28th to August 16th, 2021. Finally, 576 respondents from 30 provincial-level administrative regions (identified using Regional Internet Registry of Internet protocol) IP addresses finished the questionnaires. Respondents who said they would choose to travel somewhere due to local specialties continued to answer questions about channels by which they got to know about local specialties, reasons why local specialties were attractive, and the places they chose to go and try the local specialties and factors that were discouraging gastronomy tourism (Figure 13). But respondents who were unwilling to go traveling on account of local specialties only answered the question about factors discouraging gastronomy tourism after they expressed their unwillingness in the first question (Figure 13). The questionnaire's results were analyzed by spssau.com.

Figure 13. Flow chart for the questionnaire

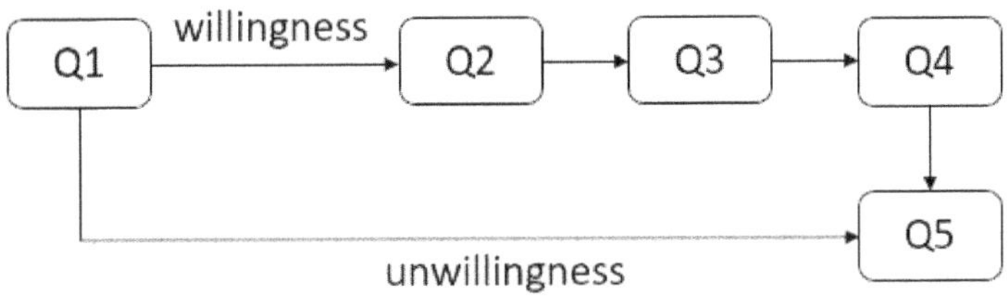

Q1: Willingness/Unwillingness to travel somewhere due to local specialties
Q2: Channels by which tourists get to know local specialties
Q3: Reasons why local specialties are attractive
Q4: Places tourists choose to go for local specialties
Q5: Factors discouraging gastronomy tourism

Among the respondents, 87.33% said they were willing to travel somewhere due to local specialties (Table 1). Local specialties really play a role in tourists' choice in tourism destinations. Montarini and Staniscia (2009) proposed that, locality of food should be strengthened to help tourism destination obtain sustainability, promote regional development, and compete with global competitors.

Table 1. Willingness/unwillingness to travel somewhere due to local specialties

Categories	Frequency	Percent	Cumulative Percent
Willingness	503	87.33%	87.33%
Unwillingness	73	12.67%	100.00%
Total	576	100.0%	

About the channels by which tourists get to know the food in tourism destinations, most respondents said they got to know through social networking sites and their friends who have previously been to said destination (Table 2). The results are similar to that of Yang's research, which finds that it is the most common for travelers to share their travel experiences on social networking sites, including the local food there and in essence acts as an electronic word of mouth (2020). In recent years, China's social media has flourished. As the data of iiMedia Research showed, the scale of mobile social users in China is exceed 900 million in 2020 which is an increase of 7.1% from 2019. The popularity of the short video and live broadcast industry has transformed the way people communicate from simple text chat to picture sharing, to voice communication, and finally to video social interaction (iiMedia Research, 2021). It's noteworthy that some respondents especially pointed out that their information came from social video websites, such as Bilibili. With free creative environment, diversified high-quality content and good interaction, these websites have been well received by the public (Lu & Li, 2021). About one third of the respondents mentioned that their information came from tourism websites or apps. In the popular domestic tourism websites or apps like Ctrip and Mafengwo, the food sector is always in a visible position, they have become important platforms for tourism destination advertising.

Table 2. Channels by which tourists get to know local specialties

Categories	Response		Popularity rate(n=503)
	n	Response rate	
Tourists' friends who have been to	348	32.49%	69.18%
Social networking sites	443	41.36%	88.07%
Tourism websites	149	13.91%	29.62%
Search engines	118	11.02%	23.46%
Others	13	1.21%	2.58%
Total		100%	
Goodness of fit: χ^2=580.013 p=0.000			

When it comes to reasons why local specialties were attractive, 7/8th of respondents said special and authentic tastes, and there were high number of respondents who wanted to enjoy the culture behind the food, accounting for above 60% (Table 3). The era of experience economy has arrived (Pine & Gilmore, 1999). In this era, consumers' consumption demands have changed. They no longer simply consume products or services, but pay more attention to the overall consumption experience, and the psychological feelings when consumption occurs (He, 2020). Tourism is one industry which majorly relies on creating and repeatedly making these experiences possible. Meneguel et al. (2018) stated that restaurants are providers of cultural elements. For tourists, tasting local specialties is a way to enrich their tourism experiences (Lu & Li, 2021) and the importance of gastronomy tourism lies in the physical and sensual experience that is a result of the desire to engage with local food (Ellis et al., 2018). China has a vast territory, and the diets in different regions vary widely. The development of gastronomy in different regions is closely related to regional culture, which refers to cultural characteristics and holds significant difference with other regions that have been imbibed into production activities and everyday living, by virtue of the regional geography, climate, nation, religion, politics, and economy (Liu, 2012). It's the combined pattern of behaviors and thoughts of people in a certain space range (Zhang, 2008). Such differences are particularly obvious in dietary preferences. For example, Northeast China is at a high latitude, and the cold climate causes a shortage of fresh vegetables during winter, so pickle making is popular, making it easier to store and use for longer periods of time (Yu, 2005). When it comes to Southwest China, spicy food can't be ignored. The local climate is always wet and cold from winter through spring, so eating spicy food is a good way to keep warm, treat cold and dampness. On the contrary, the residents of Guangdong and southern Fujian prefer light flavors due to the warm climate, which has spawned a unique "herbal tea" culture (Ma & Liu, 2021). Concluded from the questionnaire results, the novel tastes and culture of those local specialties really attract tourists.

According to the questionnaire, local food streets, night markets and time-honored restaurants were chosen as the most possible places tourists would go for local specialties, whose cumulative ratio was nearly 80% (Figure 14). Food streets and night markets are both gathering places of a city's food resources. Owing to the optimistic expectation for the interests of agglomeration, there is generally no shortage of other relevant facilities for tourist activities around food streets (Hou & Chen, 2008). Thus, food streets are good choices for gastronomy tourism. The night market is a product of the development of night economy. According to a survey by the China Tourism Academy, more than 80% of respondents have a

Table 3. Reasons why local specialties are attractive

Categories	Response		Popularity rate(n=503)
	n	Response rate	
Bringing better insight into local culture	306	22.53%	60.83%
Special and authentic tastes	440	32.40%	87.48%
High popularity	278	20.47%	55.27%
Food focused events	90	6.63%	17.89%
Manufacturing technology	88	6.48%	17.50%
Connotation of food	152	11.19%	30.22%
Others	4	0.29%	0.80%
Total		100%	
Goodness of fit: χ^2=721.814 p=0.000			

strong desire to go around at night. By October 1, 2020, China had issued 197 policies highly related to the night economy, 82 of which directly mentioned the night economy (China Tourism Academy, 2020, October 25). When it comes to time-honored brands, the rational purchasing theory points out that brand cognition is the first link for consumers in the purchasing process (Feng et al., 2006). In Yuan's (1997) opinion, brand cognition consists of two parts: one is product cognition, that is, consumers' perception of the value the product brings and their loyalty to it, the other is cultural cognition, that is, consumers' perception of corporate image, operation mode, corporate culture, and other elements. As time goes by, some local popular restaurants become time-honored brands. It means apart from being recognized they also carry local culture and historical significance.

Figure 14. Pareto chart for places tourists choose to go for local specialties

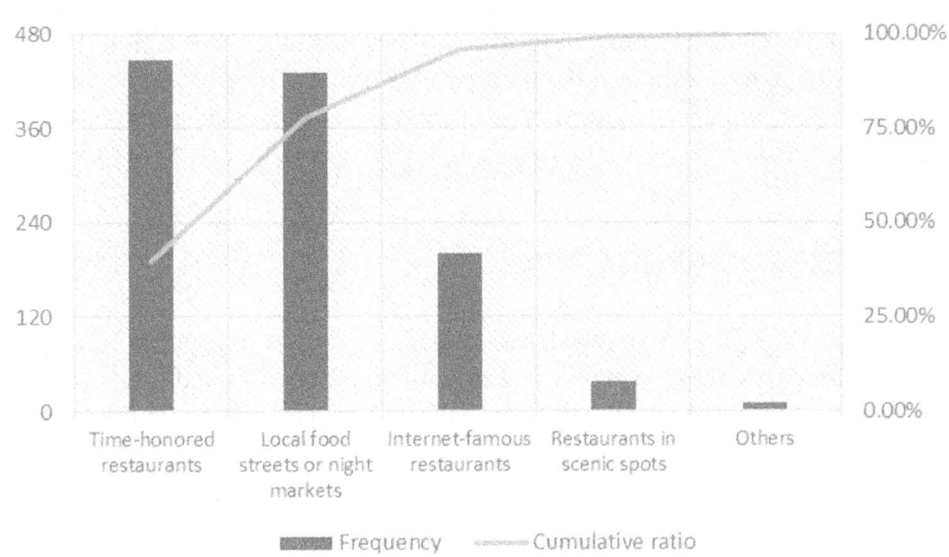

Table 4. Factors discouraging gastronomy tourism

Categories	Willingness/Unwillingness to travel somewhere due to local specialties (%)		Total(n=576)
	Willingness(n=503)	Unwillingness(n=73)	
Lack of distinctions	274(54.47)	28(38.36)	302(52.43)
Insufficient publicity	228(45.33)	19(26.03)	247(42.88)
Monotonous form of food	247(49.11)	24(32.88)	271(47.05)
Worries about food hygiene and safety	278(55.27)	17(23.29)	295(51.22)
Unattractive food culture	237(47.12)	44(60.27)	281(48.78)
Others	16(3.18)	3(4.11)	19(3.30)
Chi-Square Test: χ^2=19.074 p=0.002			

Jacobsen and Haukeland (2002) found that the tourists' restaurant selection criteria depend on the physical conditions and the ambiance of the restaurant, the quality of the food and the service quality and the friendliness of the service staff. Sánchez-Cañizares and López-Guzmán (2012) suggested that the image of the tourist destination might get affected negatively if the quality of local food is poor. Thus, a destination interested in developing gastronomy tourism has to consider factors that affect the success of it (Chaney & Ryan, 2012). Five main factors discouraging gastronomy tourism were mentioned in the questionnaire (Table 4). As P value is equal to 0.002, less than 0.05, there're no big differences between people who were willing or unwilling to travel somewhere due to local specialties when talking about discouraging factors. Distinctions, publicity, diversity, safety and cultural aspects of food should be thought about in tourism destination marketing management.

However, it's worth noting that the questionnaire results have limitations. The results from the questionnaire cover only 30 of 34 provincial-level administrative regions in China, lacking data from Tibet, Ningxia, Yunnan and Taiwan. Figure 15 shows the source of respondents' IP addresses. The darker the color, the more the respondents from that provincial-level administrative region (Figure 15). It can be seen that the color of Jiangsu Province is relatively dark, but the color of some areas is very light, indicating that respondents are unevenly distributed across the country, which has a certain impact on the results of the questionnaire.

CONCLUSION & RECOMMENDATIONS

With the improvement of people's living standards, food no longer just satiates one's hunger in the tourism experience but also paves a way to experience the culture around it. Food has always been a major point of consideration when deciding what to eat and where travel to, because it gradually becomes an important way for tourists to experience the local food and culture. Additionally, it exerts other influences on tourism destinations, such as the designing of tourism products and marketing (Henderson, 2009).

Figure 15. The source of respondents' IP addresses

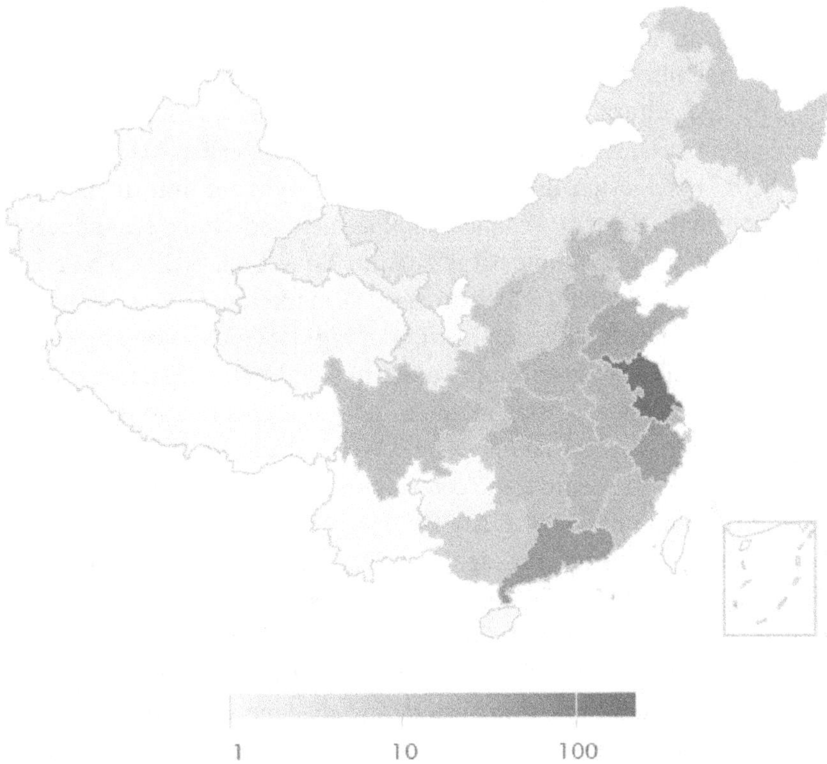

China's gastronomy tourism has shown good prospects, with some regions having become famous gastronomy tourism destinations and gaining popularity all over the world. Severe as the impact by COVID-19 on tourism, the food element is estimated to make a difference in the recovery of tourism industry.

In order to promote sustainable gastronomy tourism, IP can't be ignored. With unique attributes, IP always can condense the flow of fans and has high value realization ability (as cited in Wang, 2020). The core value of IP is content. In essence, an object, a scene, a story, a cartoon character, a movie, etc., can all be derived into an IP symbol after incubation of unique ideas. IP has connection function, location function and aggregation function (Hong, 2018). In the Internet era, wemedia develops rapidly, and IP attracts people's attention quickly because of its uniqueness. Under the role of big data, it can connect information with target groups at low cost, accurately select and actively gather the target group, which is commonly referred to as "fans", so the other end of IP is traffic (Guan, 2020). Today's tourism industry has entered the era where content is the king, when scarce and personalized tourism products mean customer base and large traffic. IP has become a feature that highlights the differences between one's own products and similar products (Cao & Zhu, 2018). For the tourism industry, food is not only a carrier of destination culture, but also one of the powerful tools for destination brand building (Wang, 2020). The differentiation of food constitutes the characteristics of differentiated tourism destinations, which form an effective market attraction (Ibid).

When a brand appears, it can't be called IP. IP is the advanced stage of brand evolution, which is conducive to the renewal, development and iteration of the existing brand system (Wang, 2020). Fields described the gastronomy experience at the destination as a chance to participate in another culture, try new recipes and visit places with a strong identity with certain local foods (as cited in Hernandez-Rojas et al., 2021). Gastronomy IP should not be limited to tasting and introducing food (Long, 2004), enriching and extending brand content is the necessary way to build IP. It can integrate culture and education, such as launching food related cultural products and cooking courses which enable tourists to participate in producing, making and processing food, so as to enrich their experience (Li et al., 2019).

Buonincotri et al. (2017) suggested that, tourists rely on dynamic information like social media, online reviews and blogs prior to the trip. Pesonen and Pasanen underlined that advertisments on social media may have influence on people's choices (as cited in Yu & Sun, 2019). The construction of IP is inseparable from the Internet and big data, which are used as marketing tools to accurately push products to tourists (Guan, 2020). With the effective application of big data technology and deep mining of useful information (Cao, 2021), sellers in gastronomy tourism should identify a target market and discover a marketing segment's wants and needs, thus maintaining long-term relationships with customers and establishing authority and expertise within the industry (Ward, 2021). Such Internet marketing also helps in creating Internet celebrities who promote the catering trend and tourism hot spots and have become a new demand of tourism consumption (Lu & Xu, 2021).

Additionally, the government's tourism management departments need to formulate relevant measures and industrial development plans, highlighting the role of gastronomy in tourism (Li, 2021b). Cui and Zheng (2019) suggested, "The lack of tourism food safety guarantee system in China has not effectively cured the problem of tourism food safety. The construction of tourism food safety guarantee system is based on the coordination of system, function system and management system. China's tourism food safety guarantee system is deficient in the construction of laws and regulations, the function of the main body of supervision, the construction of social supervision mechanism, the competition of enterprises in tourist attractions and the construction of social credit system. A number of measures should be taken to promote the construction of tourism food safety guarantee system and solve the problem of tourism food safety". In addition to promulgating relevant preferential policies, the government should also pay attention to food hygiene, food safety and food legislation, in order to ensure the rights and interests of consumers and improve the country's tourism image.

REFERENCES

Baidu. (2020, June 10). *New travel, new experience: Baidu tourism industry study 2020*. https://yingxiao.baidu.com/new/home/product/details?id=7206&ly=article_next

Bertan, S. (2020). Impact of restaurants in the development of gastronomic tourism. *International Journal of Gastronomy and Food Science*, *21*, 100232. doi:10.1016/j.ijgfs.2020.100232

Boyne, S., Williams, F., & Hall, D. R. (2002). On the trail of regional success: tourism, food production and the Isle of Arran Taste Trail. In A. M. Hjalager & G. Richards (Eds.), *Tourism and Gastronomy* (pp. 91–114). Routledge.

Buonincontri, P., Morvillo, A., Okumus, F., & van Niekerk, M. (2017). Managing the experience co-creation process in tourism destinations: Empirical findings from Naples. *Tourism Management*, *62*, 264–277. doi:10.1016/j.tourman.2017.04.014

Cao, M. T., & Zhu, X. H. (2018). Research on the current situation of tourism IP. *China Market*, (35), 70–78. doi:10.13939/j.cnki.zgsc.2018.35.070

Cao, Z. H. (2021). Research on the marketing strategy of food e-commerce under the guidance *of E-Commerce and Network Marketing. Food Industries, 512*(4).

Central People's Government of the People's Republic of China. (2019, February 19). *Suggestions on giving priority to the issues of agriculture, rural areas and farmers.* http://www.moa.gov.cn/ztzl/jj2019zyyhwj/2019zyyhwj/201902/t20190220_6172154.htm

Chaney, S., & Ryan, C. (2012). Analyzing the evolution of singapore's world gourmet summit: An example of gastronomic tourism. *International Journal of Hospitality Management, 31*(2), 309–318. doi:10.1016/j.ijhm.2011.04.002

Chen, L., Lu, Y. H., Li, D. L., & Jiang, X. T. (2018). An empirical study on culinary tourism destination brand personality and its impact in the context of Confucian culture. *Luyou Xuekan, 1*, 25–41. https://www.cabdirect.org/cabdirect/abstract/20183093032

China Banking and Insurance Regulatory Commission. (2020, April 9). *Notice on doing well the key work of banking and insurance services in the field of agriculture, rural areas and farmers in 2020.* http://www.cbirc.gov.cn/cn/view/pages/ItemDetail.html?docId=899357&itemId=878&generaltype=1

China National Tourism Administration. (2015, December 17). *China tourism statistics bulletin 2014.* http://www.xinhuanet.com/travel/2015-12/17/c_128541973.htm

China National Tourism Administration. (2016, October 18). *China tourism statistics bulletin 2015.* http://zwgk.mct.gov.cn/zfxxgkml/tjxx/202012/t20201204_906456.html

China National Tourism Administration. (2017, June 12). *Guidelines for the establishment of all-for-one tourism demonstration areas.* http://zwgk.mct.gov.cn/zfxxgkml/zykf/202012/t20201213_919347.html

China National Tourism Administration. (2017, November 8). *China tourism statistics bulletin 2016.* http://zwgk.mct.gov.cn/zfxxgkml/tjxx/202012/t20201215_919598.html

China National Tourism Administration. (2018, February 6). *2017 Annual tourism market and comprehensive contribution data report.* http://zwgk.mct.gov.cn/zfxxgkml/tjxx/202012/t20201204_906468.html

China Tourism Academy. (2020, October 25). *Report on China's night economy 2020.* http://www.ctaweb.org.cn/cta/xsjl/202103/c730a94cae6a4847b63679756926f851.shtml

China Tourism Academy. (2021, February 22). *From overall recovery to high-quality development: Analysis of tourism economy in 2020 and its development forecast in 2021.* http://www.ctaweb.org.cn/cta/ztyj/202103/80532de9677f4c8685428b5a362e5ed9.shtml

China Tourism Academy. (2021, July 13). *Survey report on the recovery of urban leisure market in the first half of 2021.* http://www.ctaweb.org.cn/cta/ztyj/202107/08e09c680cfe4172ae635cdeb05dcbdd.shtml

Cui, M., & Zheng, H. D. (2019). The construction and countermeasure of China's tourism food safety guarantee system. *Food Industries, 40*(10), 275–279. https://www.cabdirect.org/cabdirect/abstract/20183205065

Deloitte & China Cuisine Association. (2020, March). *Impact of COVID-19 on finance and operations in the Chinese catering industry*. https://www2.deloitte.com/content/dam/Deloitte/global/Documents/About-Deloitte/COVID-19/gx-covid-19-impact-on-finance-operations-in-the-chinese-catering-industry.pdf

Ellis, A., Park, E., Kim, S., & Yeoman, I. (2018). What is food tourism. *Tourism Management, 68*, 250–263. doi:10.1016/j.tourman.2018.03.025

Fang, Y., Li, L. L., & Xu, P. (2020). Chengdu tourism and characteristic snack industry fusion development status analysis based on SWOT. *Shanxi Agricultural Economy*, *13*, 28–29. doi:10.16675/j.cnki.cn14-1065/f.2020.13.017

Feng, J. Y., Mu, W. S., & Fu, Z. T. (2006). A review of consumer purchase intention. *Modern Management Science*, *11*, 7-9.

Fu, W. Y. (2014). Review of the "Six Elements of Tourism". *Journal of Zhongzhou University*, *31*(5), 14–17. doi:10.13783/j.cnki.cn41-1275/g4.2014.05.005

Fuste-Forne, F. (2021). Robot chefs in gastronomy tourism: What's on the menu. *Tourism Management Perspectives*, *37*, 100774. doi:10.1016/j.tmp.2020.100774

Guan, P. (2020). IP construction of rural food tourism from the cultural output of Li Ziqi. *Journal of Jincheng Institute of Technology*, *13*(4), 33–36. doi:10.1016/j.tmp.2020.100774

Guzel, B., & Apaydin, M. (2016). Gastronomy tourism, motivations and destinations. In C. Avcıkurt, M. S. Dinu, N. Hacıoğlu, R. Efe, A. Soykan, & N. Tetik (Eds.), *Global Issues and Trends in Tourism* (pp. 394–404). St. Kliment Ohridski University Press.

He, J. T., Zhu, N., Feng, W. L., & Huang, Y. Y. (2020). Research on the economic development platform of Shunde food culture. *China Journal of Commerce*, *9*, 7–8. doi:10.19699/j.cnki.issn2096-0298.2020.09.007

He, M. L. (2020). *Research on the construction of experience value evaluation system of food tourism* [Master's dissertation]. Soochow University. https://kns.cnki.net/KCMS/detail/detail.aspx?dbname=CMFD202101&filename=1020152831.nh

Henderson, J. C. (2009). Food tourism reviewed. *British Food Journal*, *111*(4), 317–326.

Hernandez-Rojas, R. D., Folgado-Fernandez, J. A., & Palos-Sanchez, P. R. (2021). Influence of the restaurant brand and gastronomy on tourist loyalty: A study in Cordoba (Spain). *International Journal of Gastronomy and Food Science*, *23*, 100305. doi:10.1016/j.ijgfs.2021.100305

Hong, Q. H. (2018). *Taking the IP, command the world.* China Travel & Tourism Press.

Horng, J. S., Liu, C. H., Chou, H. Y., & Tsai, C. Y. (2012). Understanding the impact of culinary brand equity and destination familiarity on travel intentions. *Tourism Management*, *33*(4), 815–824. doi:10.1016/j.tourman.2011.09.004

Hou, B., & Chen, Z. Y. (2008). Study on the development pattern of catering street on the basis of urban tourism. *Culinary Science Journal of Yangzhou University*, *4*, 26-31.

Hu, X. F. (2010). A deep reflection over Guangzhou's loss of "Gourmet Capital". *Journal of Guangdong University of Finance & Economics*, *25*(6), 77–80.

Huang, X. Y., & Liang, R. Q. (2020, March 3). How can the tourism industry turn "crisis" into "opportunity" in the post-pandemic? *China Travel News*. https://www.traveldaily.cn/article/136134

iiMedia Research. (2020, April 12). *China's catering and food delivery merchants during COVID-19 report*. https://www.iimedia.cn/c400/70742.html

iiMedia Research. (2021, January 4). *2020-2021 China mobile social industry report*. https://www.iimedia.cn/c400/76205.html

iResearch. (2020, April 20). *2020 outlook for the recovery of hospitality and tourism industries in China after COVID-19 pandemic*. http://www.iresearchchina.com/content/details8_61202.html

Jackie. (2018, December 16). *Top 12 UNESCO creative cities of gastronomy foodie destinations to visit*. https://lifeofdoing.com/unesco-creative-cities-of-gastronomy-foodie-destinations/

Jacobsen, J. K. S., & Haukeland, J. V. (2002). A lunch with a view: Motor tourists' choices and assessments of eating-places. *Scandinavian Journal of Hospitality and Tourism*, *2*(1), 4–16.

Li, H. N. (2021, August 18). Not afraid of the epidemic, the catering industry actively seeks change and looks forward to a bright future. *China Economic Times*. http://jjsb.cet.com.cn/

Li, J. (2021). Current situation and countermeasures of tourism food industry in Yunnan, *Food. Research for Development*, *42*(6), 229–230. https://doi.org/CNKI:SUN:SPYK.0.2021-06-041

Li, S. B. (2009). *Research on the Development of the Tourism of Macao* [Master's dissertation]. Huazhong University of Science & Technology. https://kns.cnki.net/KCMS/detail/detail.aspx?dbname=CMFD2012&filename=1011276171.nh

Li, X., He, X. D., & Liu, S. Y. (2019). The development trend of culinary tourism at home and abroad. *Tourism Research*, *11*(4), 5–9.

Liu, G. (2012). Landscape design of sanhe quezhu culture city based on regional culture. *Anhui Architecture*, 2. doi:10.16330/j.cnki.1007-7359.2012.02.060

Long, L. (2004). *Culinary tourism (material worlds)*. The University Press of Kentucky.

Lu, C. B., & Xu, T. R. (2021). Dimensions and focus of perceived value of Internet celebrity restaurant visitors —— Empirical research based on network text analysis. *Journal of Researches on Dietetic Science and Culture*, *1*, 33–42. https://doi.org/10.19913/j.cnki.2095-8730msyj.2021.0003

Lu, L., & Li, P. L. (2021). Research on the development of food culture tourism under the guidance of Food Culture and Tourism. *Food Industries*, *3*, 374–375. https://doi.org/CNKI:SUN:SPGY.0.2021-03-110

Ma, L. J., & Liu, C. (2021). Study on the preference and influencing factors of potential customers for city food in China — based on the perspective of network attentions. *Resource Development & Market*, *37*(3), 365-371. doi:10.3969/j.issn.1005-8141.2021.03.016

Martín, J. C., Roman, C., Lopez-Guzman, T., & Moral-Cuadra, S. (2020). A fuzzy segmentation study of gastronomical experience. *International Journal of Gastronomy and Food Science*, *22*, 100248. https://doi.org/10.1016/j.ijgfs.2020.100248

Meneguel, C. R. D. A., Mundet, L., & Aulet, S. (2019). The role of a high-quality restaurant in stimulating the creation and development of gastronomy tourism. *International Journal of Hospitality Management, 83*, 220–228. https://doi.org/10.1016/j.ijhm.2018.10.018

Miao, Y. D., & Wu, C. Z. (2020). Practice and suggestions on the construction of "City of Gastronomy" Yangzhou from the perspective of Creative Cities Network. *Humanities World*, (13), 2-6. doi:10.16737/j.cnki.rwtx81281190.2020.13.001

Ministry of Culture and Tourism of the People's Republic of China. (2015, December 31). *Looking back to 2015: "Culture+" has become a new economic engine*. https://www.mct.gov.cn/whzx/bnsj/whcys/201512/t20151231_760240.htm

Ministry of Culture and Tourism of the People's Republic of China. (2018, April 16). *Implementation plan for the construction of Manzhouli in Inner Mongolia and Fangchenggang border tourism pilot zone in Guangxi*. http://zwgk.mct.gov.cn/zfxxgkml/scgl/202012/t20201213_919404.html

Ministry of Culture and Tourism of the People's Republic of China. (2018, September 10). *Reply to recommendation 3240 of the first session of the 13th National People's Congress*. http://zwgk.mct.gov.cn/zfxxgkml/fwzwhyc/202012/t20201206_916874.html

Ministry of Culture and Tourism of the People's Republic of China. (2019, February 12). *Basic situation of tourism market in 2018*. http://zwgk.mct.gov.cn/zfxxgkml/tjxx/202012/t20201204_906481.html

Ministry of Culture and Tourism of the People's Republic of China. (2019, July 26). *Notice on Agricultural Bank of China's financial support for the construction of national rural tourism key villages*. http://zwgk.mct.gov.cn/zfxxgkml/zykf/202012/t20201213_919373.html

Ministry of Culture and Tourism of The People's Republic of China. (2020, February 5). *Notice on temporarily refunding part of the tourism service quality deposit to support travel agencies in coping with operational difficulties*. http://www.gov.cn/zhengce/zhengceku/2020-02/07/content_5475676.htm

Ministry of Culture and Tourism of the People's Republic of China. (2020, March 10). *Basic situation of tourism market in 2019*. https://www.mct.gov.cn/whzx/whyw/202003/t20200310_851786.htm

Ministry of Culture and Tourism of the People's Republic of China. (2020, July 17). *Notice on normalizing epidemic prevention in rural tourism and accelerating market recovery*. http://zwgk.mct.gov.cn/zfxxgkml/zykf/202012/t20201213_919386.html

Ministry of Culture and Tourism of the People's Republic of China. (2021, February 22). *To build an internationally influential cultural and leisure Guangdong-Hong Kong-Macao Greater Bay Area*. http://zwgk.mct.gov.cn/zfxxgkml/zcfg/zcjd/202101/t20210106_920503.html

Ministry of Culture and Tourism of the People's Republic of China. (2021, February 18). *Domestic tourism data in 2020*. http://zwgk.mct.gov.cn/zfxxgkml/tjxx/202102/t20210218_921658.html

Ministry of Culture and Tourism of the People's Republic of China. (2021, March 8). *Suggestions on promoting high-quality development of public cultural services*. http://zwgk.mct.gov.cn/zfxxgkml/ggfw/202103/t20210323_923230.html

Ministry of Culture and Tourism of the People's Republic of China. (2021, April 29). *14th Five-Year Plan for culture and tourism development*. http://zwgk.mct.gov.cn/zfxxgkml/ghjh/202106/t20210602_924956.html

Ministry of Culture and Tourism of the People's Republic of China. (2021, July 5). *Notice on building the first batch of state-level night culture and tourism consumption agglomeration zones*. http://zwgk.mct.gov.cn/zfxxgkml/cyfz/202107/t20210706_926242.html

Ministry of Culture and Tourism of the People's Republic of China. (2021, July 19). *Notice on organizing the 2021 national culture and tourism consumption season*. http://zwgk.mct.gov.cn/zfxxgkml/cyfz/202107/t20210726_926673.html

Ministry of Culture and Tourism of the People's Republic of China. (2021, July 19). *Domestic tourism data in the first half of 2021*. http://zwgk.mct.gov.cn/zfxxgkml/tjxx/202107/t20210729_926770.html

Montarini, A., & Staniscia, B. (2009). Culinary tourism as a tool for regional re-equilibrium. *European Planning Studies*, *17*(10), 1463–1483. https://doi.org/10.1080/09654310903141656

National Bureau of Statistics of China. (n.d.). *Number of domestic tourists in China from 2010 to 2019 (in billions)*. Author.

North, D. C. (1994). *Structure and change in economic history*. Shanghai People's Publishing House.

North, D. C., & Thomas, R. P. (2009). *The rise of the western world: A new economic history*. Huaxia Publishing House.

Pine, B. J., & Gilmore, J. H. (1999). *The experience economy*. Harvard Business Press.

Qi, E. Z., & Chen, J. (2018). Research on brand promotion strategy of "City of Gastronomy" Shunde. *Market Forum*, *2*, 37-41. doi:https://doi.org/CNKI:SUN:JHTS.0.2018-02-011

Qi, L. L. (2019). The construction of characteristic town of gastronomy of Leliu, Shunde: Based on the development of industry. *Journal of Guangdong Industry Polytechnic*, *18*(1), 27–30. https://doi.org/10.13285/j.cnki.gdqgxb.2019.0006

Qu, Q. K., & Chen, H. B. (2009, December 3). Food has become a cultural feature of Macao tourism. *China Trade News*, 6.

Sánchez-Cañizares, S. M., & López-Guzmán, T. (2012). Gastronomy as a tourism resource: Profile of the culinary tourist. *Current Issues in Tourism*, *15*(3), 229–245. https://doi.org/10.1080/13683500.2011.589895

Shang, W. W. (2015). *Research on tourist satisfaction and loyalty of time-honored catering brand from the perspective of food tourism motivation* [Master's dissertation]. Henan University. https://kns.cnki.net/KCMS/detail/detail.aspx?dbname=CMFD201601&filename=1015658405.nh

Shi, Y. Y., & Li, Y. L. (2021). The influence of COVID-19 on the development of China's catering industry and its countermeasures. *Modern Business*, (1), 30-32. doi:10.14097/j.cnki.5392/2021.01.009

Sims, R. (2009). Food, place and authenticity: Local food and the sustainable tourism experience. *Journal of Sustainable Tourism*, *17*(3), 321–336.

State Council. (2016, December 7). *13th Five-Year Plan for tourism development.* http://www.gov.cn/zhengce/content/2016-12/26/content_5152993.htm

State Council. (2017, March 5). *2017 Government Work Report.* http://www.gov.cn/guowuyuan/2017zfgzbg.htm

State Taxation Administration. (2020, March 15). *Guidelines on preferential tax policies for epidemic control and economy development.* http://www.gov.cn/xinwen/2020-03/15/content_5491493.htm

Sun, G. (1994). Tourism, a sunrise industry forever. *China Reform*, *7*, 17-20. doi:https://doi.org/CNKI:SUN:ZGGG.0.1994-07-005

Tang, J. Z. (2020, December 3). The Macanese cuisine is not repeated. *Chinese People's Political Consultative Conference Daily*, 8.

United Nations Educational, Scientific and Cultural Organization. (n.d.a). *Chengdu.* https://en.unesco.org/creative-cities/chengdu

United Nations Educational, Scientific and Cultural Organization. (n.d.b). *Shunde.* https://en.unesco.org/creative-cities/shunde

United Nations Educational, Scientific and Cultural Organization. (n.d.c). *Yangzhou.* https://en.unesco.org/creative-cities/yangzhou

Wang, Y. (2020). Research on IP commercialization operation of food tourism in Leshan city. *Journal of Sichuan Tourism University*, *2*, 52–56. https://doi.org/CNKI:SUN:SCPR.0.2020-02-013

Ward, S. (2021, March 5). *What is Internet marketing*? https://www.thebalancesmb.com/internet-marketing-2948348

World Tourism Organization. (2021a). *Gastronomy.* https://www.unwto.org/gastronomy

World Tourism Organization. (2021b). *Gastronomy and wine tourism.* https://www.unwto.org/gastronomy-wine-tourism#:~:text=The%20Committee%20on%20Tourism%20and%20Competitiveness%20%28CTC%29%20of,food%20and%20related%20products%20and%20activities%20while%20travelling

World Tourism Organization & Food and Agriculture Organization of the United Nations. (2020). *Gastronomy key to supporting hard-hit tourism industry.* https://unric.org/en/gastronomy-key-to-supporting-hard-hit-tourism-industry/

World Travel and Tourism Council. (2020). *Contribution of China's travel and tourism industry to GDP 2014-2028.* https://www.statista.com/statistics/249794/contribution-of-chinas-travel-and-tourism-industry-to-gdp/

Xie, J. S. (2015). Research on the development of Shunde cultural and creative industry. *Culture Journal*, *12*, 130-136. doi:https://doi.org/CNKI:SUN:WHXU.0.2015-12-041

Xin, W. (2020). The reasons why Macao has gained "City of Gastronomy". *China Food Industry*, *1*, 65-69. doi:https://doi.org/CNKI:SUN:SPZG.0.2020-01-025

XINHUANET. (2018, April 25). *The white paper on the consumption power of China's tourism cuisine 2018*. http://www.xinhuanet.com/food/2018-04/25/c_1122738918.htm

XINHUANET. (2020, March 2). *Report of the current situation and trend of China's catering industry under COVID-19*. http://www.xinhuanet.com/food/2020-03/02/c_1125652997.htm

Xu, S. Y. (2021). Study on the strategy of "three beauties" to boost the local economy of Yangzhou under the impact of new epidemic situation. *Journal of Hubei Open Vocational College, 34*(14), 110-111+116. doi:https://doi.org/CNKI:SUN:HBHS.0.2021-14-050

Yang, C. H., Feng, M. H., & Chen, Y. (2019). A study on Chengdu culinary image based on tourist perception. *Journal of Researches on Dietetic Science and Culture*, *36*(4), 15–22. https://doi.org/CNKI:SUN:YZPX.0.2019-04-004

Yang, L. (2021). *Research on the development of Yangzhou urban tourism brand based on symbiosis theory* [Master's dissertation]. Yangzhou University. https://kns.cnki.net/KCMS/detail/detail.aspx?dbname=CMFDTEMP&filename=1021050858.nh

Yi, K. G., & Hu, H. X. (2019, December 10). Construction path and value creation of cultural tourism IP. *China Tourism News*. http://www.ctnews.com.cn/

Yu, C.-E., & Sun, R. (2018). The role of Instagram in the UNESCO's creative city of gastronomy: A case study of Macau. *Tourism Management*, *75*, 257–268. doi:10.1016/j.tourman.2019.05.011

Yu, Z. T. (2005, January). *Taste formation from the perspective of regional food culture*. http://www.dili360.com/cng/article/p5350c3d7eaea252.htm

Yuan, D. H. (1997). Discussion on the cognitive value of a product. In Chinese Psychological Society (Ed.), *Abstracts from the 8th National Conference on Psychology* (p. 321). Chinese Psychological Society.

Zhang, F. Q. (2008). The concept and the study methods of regional culture. *Zhejiang Social Sciences*, *4*, 63-66+50+127. doi:10.14167/j.zjss.2008.04.008

Zhong, Z. J., Lin, J. P., Zhou, M. Q., Zhang, H., & Han, Y. J. (2021a). "Food" tourism at home and abroad: Research comparison on "Food Tourism" "Gourmet Tourism" "Diet Tourism". *Resource Development & Market, 37*(4), 463-471. doi:10.3969/j.issn.1005-8141.2021.04.013

Zhong, Z. J., Zhou, M. Q., & Lin, J. (2021b). Research process of "food" tourism at home and abroad. *Yunnan Geographic Environment Research, 33*(2), 69-78. doi:https://doi.org/CNKI:SUN:YNDL.0.2021-02-010

Chapter 3
Analysis of the Traditional Gastronomic Offer as an Engine for Local Tourism Development:
Guayas, Ecuador

Alejandra Ycaza
Ecotec University, Ecuador

Verónica Baquerizo
Ecotec University, Ecuador

Javier Vallejo
Ecotec University, Ecuador

Guido Valero
Ecotec University, Ecuador

ABSTRACT

Gastronomy is an important part of tourism, an element that by sharing one dish and its ingredients can bring culture, traditions, and flavors from one destination all together. One of the most attractive parts of the gastronomic offer from a territory is the history behind the dish, which also includes special ingredients and way of preparation passed from one generation to the other. This culinary experience is a key motivation for a tourist to visit a specific destination. This study analyzes the traditional gastronomic offer from Guayas, Ecuador and the relationship between gastronomy and the development of local tourism. The main result will show the importance of a well-structured offer to influence the motivation of a tourist to go to a specific destination. This will increase the demand of gastronomic tourism and create economic growth in a sustainable matter.

DOI: 10.4018/978-1-7998-9148-2.ch003

INTRODUCTION

One of the main examples of a country`s identity is its gastronomy, which provides cultural, historical, and traditional character to the destination, and can create great motivation among tourists to visit a specific place.

Due to the pandemic of 2020, international tourism has decreased, but local tourism can be the catapult to generate economic growth through visits by the residents of the same country. This is an opportunity for local destinations to display their traditional dishes, making sure visitors feel the need to travel to enjoy their local food, understanding the history and traditions behind it.

Local tourism development must be considered by the central government, as well as by local and sectional authorities, as a fundamental and transcendental part of the support needed for the development of national territory. Therefore, creating transparent policies is imperative for each destination to be sustainable throughout the years.

Gastronomy in Ecuador is known as one of the cultural patrimonies in Latin American, with historical identity. The variety of ingredients and flavors represent ancestral traditions.

People from Guayas (guayasenses) are well known for a variety of cultural and geographical characteristics, but one of the most distinctive marks is their gastronomy, where seafood, vegetables, and fruits are necessary for their unique flavors.

This analysis of the traditional gastronomic offer as an engine for local tourism development in Guayas, can provide numerous benefits, including:

- Understanding which communities are prepared
- Analyzing routes and different zones
- Identifying principal stakeholders within this productive chain
- Generating a better understanding of the processes that need to be in place to position gastronomy as a fundamental element to develop local tourism.

This research intends to highlight the actual offers from some of the main potential destinations for gastronomic tourism. It aims to analyze whether the communities are ready to welcome tourists, and which could be their insignia dishes to be able to attract local tourism to them. From the 25 cantons, the study will focus on the following: Milagro, Samborondòn, Playas, Balao, and Naranjal. These destinations can become part of a traditional gastronomic route, and some of them will be able to offer other activities that will complement gastronomy, such as hiking, trekking, traditional festivals, and agricultural activities, among others.

BACKGROUND

Regarding the importance of gastronomic tourism to develop any destination, there are different examples that confirm how much this activity can motivate a tourist or visitor to travel to a specific place, create a full culinary experience, increase local revenues, and even become a repeat customer. As Fuentes et al. (2015) indicate in their study, understanding that the tourist experience is a cumulative process, in stages, including the previous phase (before travelling to the destination), the destination resort, and

Figure 1. Nationalities of surveyed members

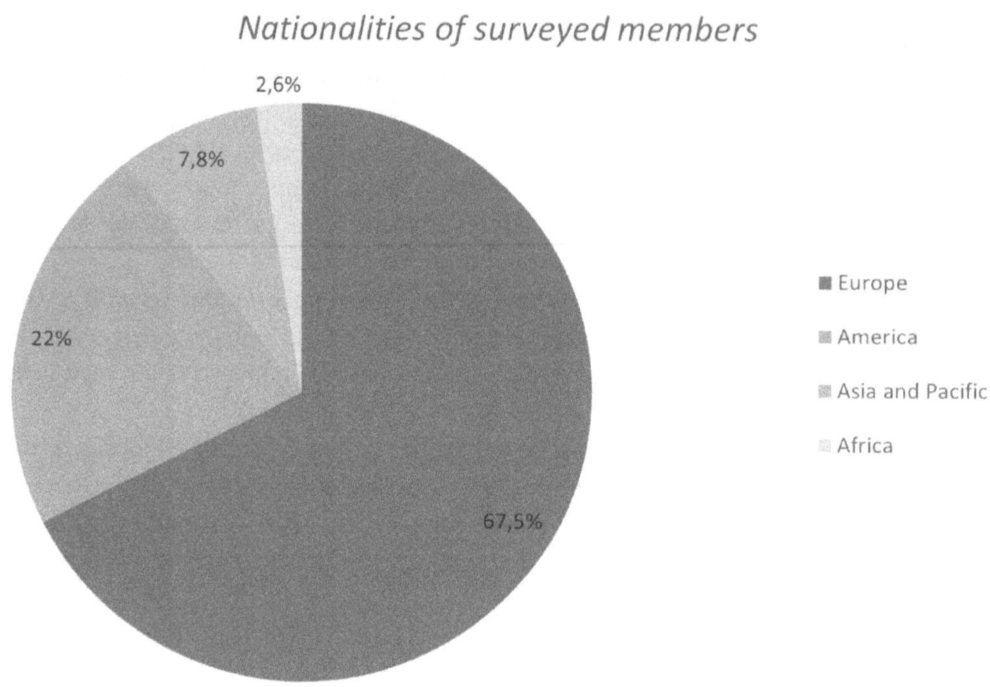

after the holidays, is one of the main factors to achieve success, by being prepared and create a complete offer that will fulfill the tourist desires and expectations.

A survey conducted by The World Tourism Organization (UNWTO) in collaboration with MADISON MK in 2016, brought very important results regarding the importance of Gastronomy tourism in different destinations, from the point of view of both the private and public sector. As Sorcaru (2019) indicated in his article, this survey was targeted to affiliate UNWTO members, including not limited to, Destination Management Organizations (DMOs), educational institutions, marketing and consultancy firms, accommodation providers, the meetings industry, food and beverage providers, and wineries. The goal was to better understand the perspective of the private and public sectors on gastronomy tourism and its importance to a destination, and to establish the importance of gastronomy as an engine for the development of the economy. The survey responded to the following objectives: the importance of gastronomy tourism, promotion and strategy, measurement of the tourism experience, impact on the community/destination, as well as opinion of gastronomy tourism (UNWTO, 2017, 2018).

The chart in Figure 1 presents the nationalities from the 77 members that responded to the survey. Of participants, 67.5% were from Europe, 22% were from the Americas, 7.8% were from Asia and the Pacific, and 2.6% were from Africa

The motivational factors for tourists to visit a destination are presented in Figure 2, analyzing Latin America placing gastronomy as one of the main factors.

As shown in the figure, tourists will travel for different factors like Cultural Tourism, shopping, wellness, Gastronomy, Natural Tourism, sports, religious, and medical reasons. But for Latin America the

Figure 2. Motivational factors Latin America
(Adapted from (UNWTO, 2017))

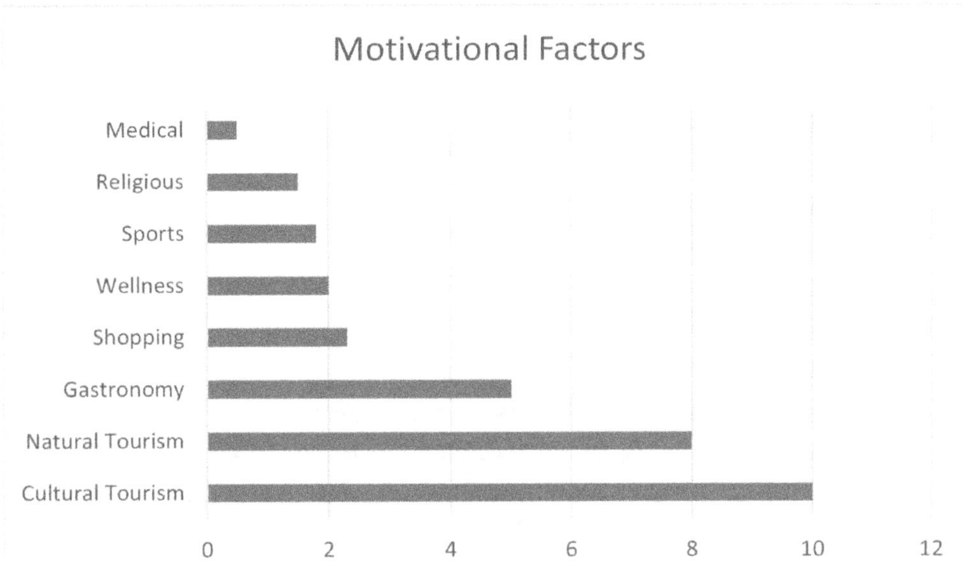

main factors are Cultural, Natural, and Gastronomy, this is a very important factor to consider specially because this study is made in Latin America.

The world's 10 most popular foodie destinations in 2018 are presented in Figure 3, and visitors want to experience the culture of each place through their food, that is a way to create a connection with the destination from the flavors and recipes offered to every tourist (Clark, 2018).

There are various real case scenarios, such as Popayan, México, and Andalucía that show, in diverse ways, how important gastronomy is for those destinations—from the ingredients to the geographical zones, the history and taste, and the growth of a new market segment that was not considered before, and which is now part of the tendencies that move travelers all over the world (Rodriguez, 2020; Rodríguez-Gutiérrez et al., 2020). The gastronomy of every country is the scenery of the historical evolution that represents each destination and conveys numerous factors to contribute and generate tendencies based on experiences, transforming them into an economic optimizer. (Entorno turístico, 2020)

Matta (2014) mentions that the gastronomic culture of a country is a cultural product that reflects the history and identity of its population, and that the country gets rich from the mixtures generated from colonial processes of acculturation or, in contrast, from the will to resist them.

According to UNWTO (2017, 2018), gastronomic tourism is a "Type of touristic activity that is characterized by the experience involving travelers during their trips related to the gastronomy of a place. Also, from the authentic culinary experiences, gastronomic tourism can also include other types of activities, such as visiting local producers, participating in gastronomic festivals, or attending a cooking class" (UNWTO, 2017, 2018).

UNWTO and the Basque Culinary Center (BCC, 2019) presented a new set of guidelines to help touristic destinations to benefit amid the recent interest for gastronomic tourism. The initiative was presented at the fifth World Forum for Gastronomic Tourism, celebrated in San Sebastian (Spain). Based on these guidelines, this study has considered the analysis of certain destinations in Latin America that

Figure 3. 10 Most Popular Foodie Destinations
(Adapted from Clark (2018))

have transcended in the past years as developers of the gastronomy of their countries, placing them on the world map due to their products and services, focused on what they call "insignia dishes."

According to a National Geographic (2018) study, the biggest gastronomies in the world stand side-by-side with the culture of that specific country. Peru is another good example of a developed gastronomic destination. The dishes combine pre-Inca and Inca history, the Spanish conquers, and finally the distinct cultures that have resided in this country, such as African, French, Cantonese, Japanese, and Italian, which have produced a fusion of new and original flavors. Perú has implemented excellent strategies, according to (Fuentes et al., 2015), who mentioned the importance of cultural identity, gastronomic fairs, implementation of a country brand, gastronomic routes, experiential marketing, among others, which have allowed Peru to achieve positive results generating internal tourism growth.

Since 2010, global interest for Mexican food has arisen due to the acknowledgment of Michoacán cuisine in the list of Intangible Cultural Heritage of Humanity, provided by (UNESCO, 2010b). One of the main descriptions stated in this article is that their knowledge and techniques express community identity, reinforce social bonds, and build stronger local, regional, and national identities. These efforts in Michoacán also underline the importance of traditional cuisine as a mean of sustainable development (Linares et al., 2018).

The creation of different gastronomic routes in Mexico is a leading example of how much gastronomy can improve and develop local tourism when implanted correctly. Some of the main gastronomic routes according to the Mexican touristic atlas (Atlas turístico de Mexico, n.d.) are: "Cooking from 2 worlds" (Cocina de dos mundos, Chihuahua); "From coffee to vanilla" (Del Café a la vainilla, Veracruz); "From sea to lagoon" (Del Mar a la laguna, Tamaulipas); and "The Altar from the day of the death," (El altar del dìa de los Muertos, Michoacán), among others.

APPROACH TO LOCAL AND TRADITIONAL GASTRONOMIC OFFERS

Analysis of the Actual Situation in Guayas

Guayas province has a diversity of traditional dishes that not only reflect the culture and main ingredients provided by the coastline and agricultural fields, but also the recipes passed from one generation to the other. These recipes include, in some cases, a specific way of making the dish to get special flavors from that part of the country.

The pandemic has brought real problems for tourists that like to explore unfamiliar places, to experience what they don´t have in their cities. Some of these destinations are driven by religious or cultural traditions that enable travelers to enjoy the gastronomy while in the area. However, there are only a few places that will motivate tourists to visit for the food. One of the main reasons for this small traffic of tourists is the lack of promotion of their gastronomic offers, and everything involved in the preparation of a dish, as part of the development of local tourism. Another reason is the empiric way of doing tourism, which is not entirely ready to receive tourists in an efficient and sustainable way.

Methodology

An important part of this study included territorial visits, surveys to residents and interviews with key actors to gather information from the main sources involved in this activity, as well as understanding the process that must be made within these destinations to be able to welcome tourists and give them a unique gastronomic experience. The surveys were conducted to a total of 130 residents, customers of the main restaurants in those locations, to gather information regarding market preferences. The interviews were conducted to fifteen owners of local restaurants in the five locations chosen in the study, to understand the history behind their most popular dishes, as well as the recipes and way of preparation.

The present research describes the characteristics and actions as objects of the study presented in the real environment. A literature review was also performed to create the design of the theoretical perspective. Observation was used as part of the scientific methods, which raised the knowledge from the abstract to the concrete. This design was applied from facts, phenomena, and several aspects that were related to the analysis of the traditional gastronomic offer as an engine for local tourism development, which will establish the extent of the benefits with the implementation of the proposal.

This investigation was qualitative, and information was obtained through surveys, and interviews that enabled the researchers to analyze the different perspectives of the actual and potential offers, establishing various ideas that will improve service quality to enhance the economy of the destinations. As a complement, throughout some real case scenarios, success examples were revised from documents and other investigations to enable comparisons and establish a real base of what would work in each case.

OBSERVATION AND SELECTION OF INSIGNIA DISHES FOR EACH DESTINATION

During the fieldwork, five destinations were visited to choose the insignia dishes with their recipes to be included in the proposal to develop this new opportunity of gastronomic tourism in Guayas.

Figure 4. Milagro's main popular dishes

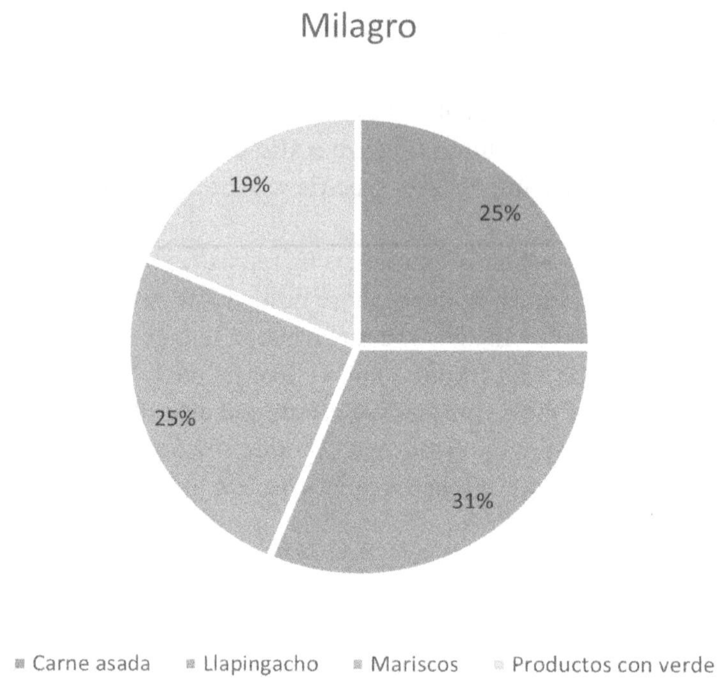

Milagro

The first town visited was Milagro, the second largest city in the Guayas province. Milagro is an outback canton full of charm, experiential tourism, and rural life. It was formerly known as the "Land of the Pineapple," but now it has grown for its agricultural activities and mixed gastronomy that combines flavors and recipes from different provinces and regions of Ecuador. Milagro has become a popular spot to enjoy food from the coastline, highlands, and even the Amazon region.

At kilometer 26, Cumana Estate Farm will offer cocoa plants, where a variety of improved national cocoa goods are made and exported annually. Visitors can experience and learn how to prepare their own artisan chocolate.

Other haciendas of fincas surround the area, such as Finca Chaparral, Sarita Miranda, Apio, and Perejil, with their special recipes made with duck. Milagro is a city of opportunities that create businesses combining agriculture, nature, and gastronomy.

The Doughnut chart in Figure 4 presents Milagro´s main popular dishes. They include roast beef (Carne asada), Llapingacho, seafood (Mariscos), and green plantain products (Productos con verde). According to this pie chart, LLapingacho is the most popular dish in the area.

One of the most popular dishes in this city did not originate from there. The Llapingacho consists of two potato cakes filled with cheese and white onion, cooked on a flat top grill, served with fresh avocado, red longaniza, salsa criolla, peanut sauce, and a fried egg. Llapingacho is served with chorizo, fried eggs, beet salad, and avocado. The dish is one consumed daily by locals and it is easily found in restaurants, food markets, and local fairs in the Province of Tungurahua.

Traditionally, these flavors and types of cooking come from the Tungurahua province. While the authors were doing their fieldwork, the residents recommended this dish from a local family restaurant. And from all the other dishes in the area, this was chosen as one of the most traditional and popular dishes in Milagro.

This type of preparation and cooking technique comes from pre-Hispanic and Creole origin. The name of this very traditional dish comes from the Kichua culture. In their language, "llapina" means to crush, and from the Spanish, "gacha," soft dough (Ministerio de Turismo, 2018).

Insignia Dish for Milagro: Llapingacho

(Yields 4 Servings)

Potato Cake:

- 1 kg. Potato (Russet), washed and peeled
- 3 tablespoons of Butter
- 2 teaspoons of achiote (oil)
- 1 stick of a type of green onion, called white onion (native of Ecuador), chopped
- Salt and white pepper

Filling:

- 2/4 of a pound, fresh cheese, crumbled or cut into little squares
-

Sides:

- 4 Chorizo Ambateño.
- 1 avocados, cut into strips
- 6 eggs, fried
- 4 layers of iceberg lettuce, chopped
- Salt and white pepper

Salsa Criolla (Creole Sauce):

- 4 red onions, julienned
- 5 lemons (sutil), juiced
- 4 tomatoes, julienned
- 6 sticks of a type of green onion, called white onion (native of Ecuador), chopped
- 2 cilantro stems with leaves, chopped
- 1 tablespoon of vegetable oil
- Salt

Peanut Sauce:

- 4 tablespoons Peanut butter, no sugar added

- 2 teaspoons garlic, paste
- 1/2 cup white onion (a type of green onion), chopped
- 1 teaspoon cilantro, fresh, chopped
- ½ cup milk
- 1 teaspoon flat parsley, fresh, chopped
- 1 teaspoon achiote oil
- Sal and white pepper

Preparation

1. Pre heat a flat top griddle.
2. Cook the potatoes with plenty of water. Add some sticks of white onion and salt. When they are cooked, drain them. Pass the potatoes through puree press to obtain smooth and creamy dough.
3. Knead gently and add the achiote oil. Season it with salt.
4. Form a ball, and using both thumbs, form a pocket at the middle of the ball.
5. Fill the pocket with cheese. Cover the filling with some potato dough and shape it into the shape of a crab cake.
6. Put some lard or oil on the flat top griddle.
7. Brown the potato cakes on both sides. Set aside.
8. Add more oil and brown the chorizo Ambateño. Cut into big chunks and set aside.

Salsa Criolla

1. Put the onions into a bowl and season with salt.
2. Add the lemon juice.
3. Then add the tomato, white onion, cilantro leaves, and the oil. Mix well
4. Correct the seasoning if needed and reserve in the fridge.

Peanut Sauce

1. Pre heat a saucepan.
2. Add the achiote oil.
3. Add the white onions and cook for a couple minutes. Add the garlic paste and cook for a couple of seconds.
4. Incorporate the peanut butter and stir.
5. Add the milk and keep stirring until the peanut butter dissolves.
6. Turn of the heat and add the fresh herbs.
7. Season with salt and pepper and set aside.

Serve: For the traditional presentation, place two Llapingachos and top with a fried egg. Beside the potato cakes, place some pieces of chorizos, then some salsa criolla and avocado. It is also often served with cooked beets and peanut sauce on top or on the side of the egg, which is optional.

Figure 5. Samborondon`s main popular dishes

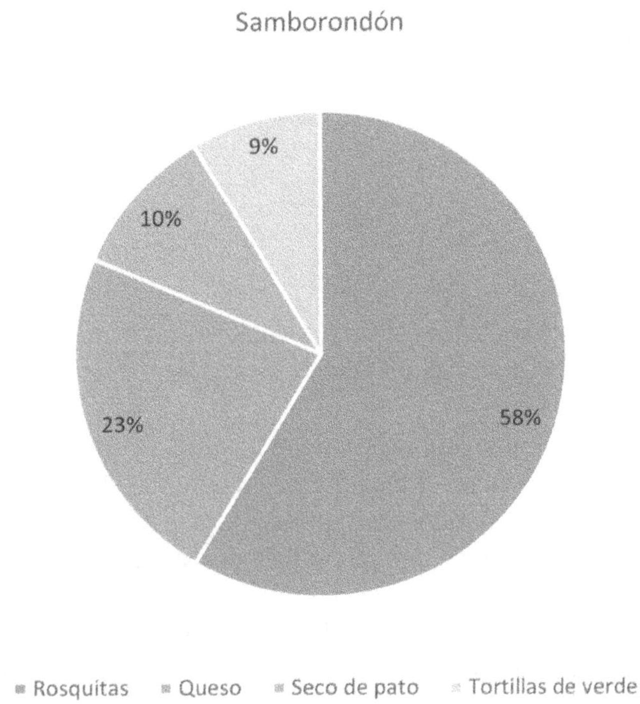

Samborondón

Samborondón is called "The equestrian Capital of Ecuador" for its horse-riding tradition and constant equine activity. Samborondón is a rice-growing canton that produces about 40,000 hectares of rice per year. It also offers clay artisanries and naval crafts (canoes). In the area, visitors can find rural and urban touristic spots such as The Buijo Racecourse, Guayaquil Historical Park, Pottery Workshops, Sánchez Aguilar Theater, Old Municipal Palace, and Playa El Encanto, among others. (Diario El Expreso, 2021)

The gastronomy that Samborondón offers, includes a variety of typical dishes, such as: corn tortillas, green plantain tortillas with queso criollo, green plantain tortillas with crawfish, corviches (like a kibbeh made of albacore and peanut butter), and seco of duck. Fish casseroles, fried sea bass, stuffed ball broth, sea bass bun, catfish broth, and the unique donuts that are the pride of Samborondeños (Universo, 2015, 2017).

The chart in Figure 5 presents Samborondón´s main popular dishes. They include rosquitas, cheese (Queso), duck (Seco de pato), and green plantain tortillas (Tortillas de verde). According to this chart, Rosquitas are the most popular dish in the area.

One of the most iconic gastronomy preparations is the Rosquitas. These are very well known by the local people. The most renowned Rosquitas are the ones form the Barroso Bakery, a traditional family business.

Insignia Dishes for Samborondón: Rosquitas

(Yields 4 Servings)

Ingredients

- 3 cups flour
- 4 ea. Egg yolks
- 4 tablespoons of butter, no salt
- 1 pinch of salt
- 1 teaspoon of baking powder
- 4 tablespoons of white sugar
- ½ cup milk

Preparation

1. Pre-heat oven to 180 C.
2. In a large bowl put the flour, salt, baking powder and sugar.
3. Mix well and let it rest for later.
4. In a separate bowl, add the butter and whip it until creamy.
5. Add the yolks and continue beating at medium speed until the ingredients are incorporated.
6. Add the flour mixture to the butter and egg yolk mixture. Knead with your hands inside the bowl and gradually add the milk, paying attention to the consistency of the dough. Stir the dough well until smooth and dense.
7. Place the dough on the work table and sprinkle with flour.
8. Divide the dough into small balls. Roll out the dough, forming strips 10 centimeters long.
9. Weave two strips together, form a donut (circle) and pinch both ends to close the circle.
10. Once all the donuts have been made, place them on a greased tray or covered with greaseproof paper. Bake at 180 C for 20 minutes until the rosquitas are golden brown.
11. Remove from oven and let cool on a rack. Serve the rosquitas with jam and as a side with tea or coffee.

The other insignia dish that was chosen for Samborondón was the Tortilla de Verde, usually served in the afternoon in a particular colorful and traditional spot in the city.

Tortilla de Verde con Camarones de Rio (green plantain tortilla with crawfish)
(Yields 2 -3 Servings)

Ingredients

- 4 ea. green plantains
- 1 ea. small red onion (diced)
- ¼ bunch Cilantro (chopped)
- 1 ea. egg
- 1- 2 teaspoons of annatto oil
- 1/2 teaspoon ground cumin
- 1 teaspoon garlic paste
- salt and ground white pepper, to taste
- 12 ea. small crawfish,

Preparation

1. Grate or blend (food processor) the green plantains.
2. In a bowl put the grated green plantain. Add the annatto oil, egg, onion, cilantro, ground cumin, garlic paste and season with salt and pepper.
3. In a pan with hot oil, add a couple of tablespoons of the green plantain mixture and fry.
4. On top add 4 crawfish. Pressing gently into the dough. Cook for a couple minutes.
5. Turn the tortilla and keep cooking, until golden brown.
6. Serve with some creole sauce or tabasco.

Playas

Playas is a canton with great beaches and several hotels and resorts to welcome tourists to the area. Beaches near Playas include: General Villamil Beach, Paseo Acuático Recinto Data de Villamil, El Arenal Beach, and Puerto Engabao, which is known for surfing and beach sports. Aside from tourism, Playas has a thriving fishing community and is known for its seafood cuisine.

Only a few minutes' drive east of town, visitors can find the fishing village of El Morro. One of its major attractions will include green mangroves, which are home to bottlenose dolphins. Local tourism is very active in El Morro and all of Playas. People from major cities visit these beaches on the weekends to relax and enjoy the wonderful climate and food that Playas has to offer.

The canton Playas has a wide variety of typical foods: sweat fish, mixed ceviches, seafood casserole, shrimp, fried corvina, grilled shrimp, fried fish, oyster ceviche, and breaded oysters, among others. During ancient times, many cantons all over Guayas used to cook in clay pots. At the present time, this tradition is over, but there is a dish called "Cazuela," which is served in little clay vessels. The process of cooking this dish is not using clay pots anymore. The dish is just assembled on the clay pot, then warmed on a direct fire, and served on a clay pot.

The chart in Figure 6 presents Playa´s main popular dishes. They include ceviche, seafood casserole (Cazuela), fried fish (Pescado frito), sweat fish (Sudado de pescado), and oysters (Ostras gratinadas), According to this chart, Fried fish is the most popular dish in the area.

Fishermen have a routine of fishing in the mornings, so there is an abundance of fresh fish. The restaurants fry the fish and serve it with rice, patacones, and creole sauce. This is a staple dish from the Playas.

Insignia Dish for Playas: Fried Fish
(Yields 1 Serving)
Ingredients

- 1 whole fish corvina, carita, smooth, clean and without scales
- 4 garlic cloves
- Cumin
- Salt and pepper

Salsa Criolla (Creole Sauce)

- 4 red onions, julienned
- 5 lemons (sutil), juiced
- 4 tomatoes, julienned
- 6 sticks of a type of green onion, called white onion (native of Ecuador), chopped

Figure 6. Playa`s main popular dishes

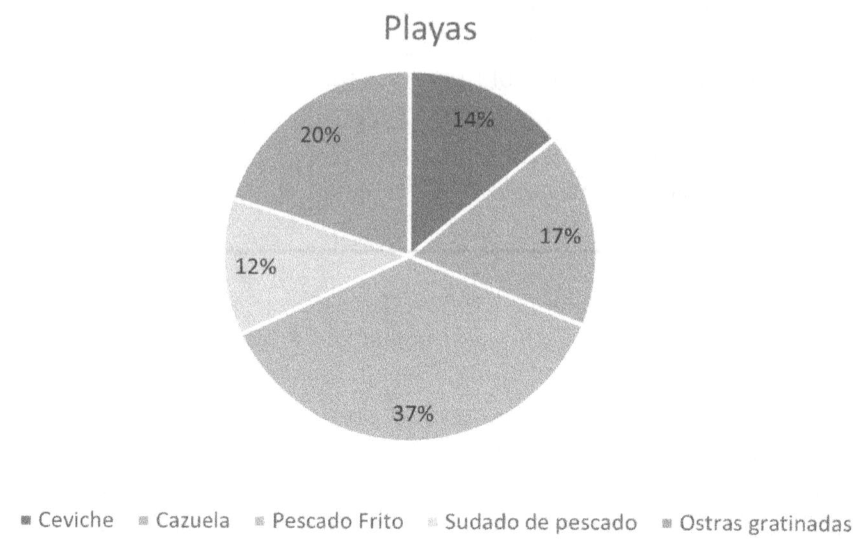

- 2 cilantro stems with leaves, chopped
- 1 tablespoon of vegetable oil

Preparation

1. Make 3 to 4 cuts on each side of the fish. They will look like tiger stripes.
2. Add lemon juice to each stripe cut, on both sides and inside the fish.
3. In a bowl, mix crushed garlic, salt, pepper, and ground cumin. You can also add some mustard.
4. Spread the dressing on each cut, on both sides and in the inside of the fish.
5. Let marinate for 30 minutes.
6. In a pan, put in the oil and heat it to fry the whole fish.

Balao

Balao canton is a potential touristic destination, because of its location and variety of flora, fauna, and bio aquatic activities. Balao is part of the Cocoa tourist route. It also offers La Paloma Beach, El Playòn, and El Recreo Hacienda, among others. The name of this canton was inspired by a native tree from the area that was found at the riverside (GoRaymi, 2020).

The most popular dish in Balao is Crab Salad. At this canton, the tourists can enjoy a very wide gastronomy based on seafood, with dishes such as: crabs with garlic sauce, crabs with creole sauce, ceviche, seafood rice, and many more.

The chart in Figure 7 presents Balao´s main popular dishes. They include crab salad (Ensalada de cangrejo), crab with creole sauce (Cangrejo criollo), seafood rice (Arroz marinero). According to this chart, Crab Salad is the most popular dish in the area.

Figure 7. Balao`s main popular dishes

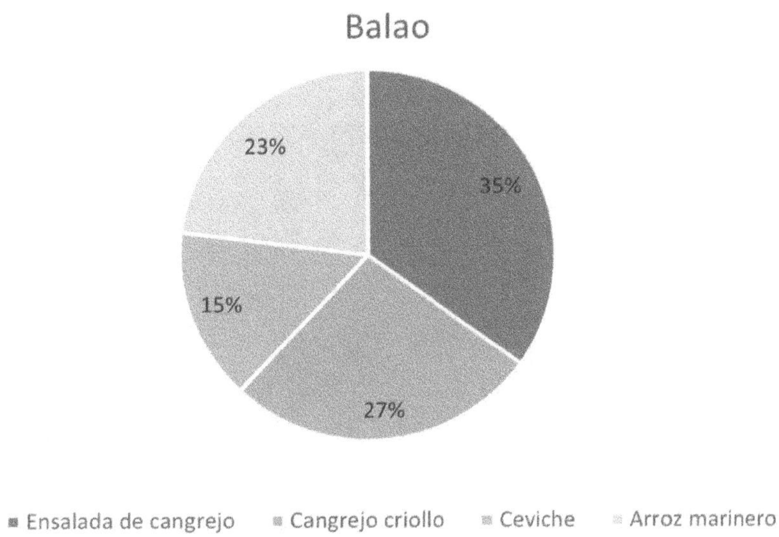

Crab salad is the most emblematic dish from Balao. This preparation was so important to them that on November 17, 2017, to celebrate the cantonización festivities, they organized the biggest crab salad of Ecuador. According to the information published by El Universo newspaper, the participants used more than 3,500 fresh crabs. (Universo, 2015, 2017).

Insignia Dish for Balao: Crab Salad
(Yield: 4–6 Servings)

Ingredients

- 1 lb. Potatoes, peeled and cut into big squares
- 1 red onion, cut into fine strips
- 2 tomatoes, seeded and cut in small squares
- 4 lemons, juiced
- ½ Cup green peas, boiled
- 1 lb. Crab meat, fresh
- 1 tablespoon Cilantro, fresh, chopped
- 6 layers of iceberg lettuce, chopped
- Salt and pepper

Preparation

1. Peel the potatoes and cut them into cubes.
2. Then put a pot with water and season it with some salt.
3. Put the potatoes and boil them for about 10–15 minutes, depending on the size. The potatoes should be just tender but not mushy. Let them cool and reserve.
4. In a bowl, add the red onion, potatoes, green peas, tomatoes, and the crabmeat.
5. Add the lemon juice. Season it with salt and pepper.

Figure 8. Naranjal`s main popular dishes

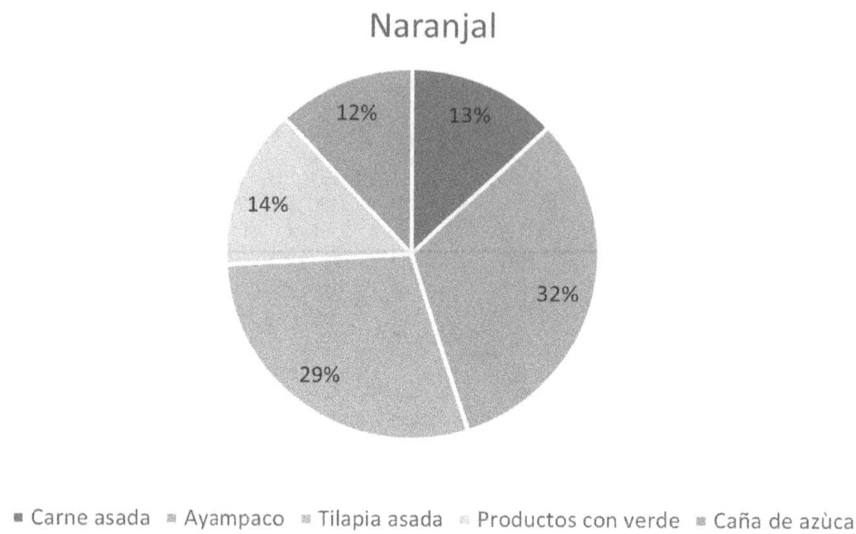

6. Mix well all the ingredients and rectify seasoning and the lemon juice, if needed. Reserve in the fridge.

Serve: Decorate with the crab claws and the chopped lettuce.

Naranjal

Naranjal canton is located southwest of the province of Guayas, one hour from Guayaquil, and two hours from Cuenca. It is an agricultural area with may products such as cocoa, sugar cane, rice, coffee, banana tree, and a wide variety of fruits (Pérez, 2017).

Due to its biodiversity, Naranjal offers a wonderful climate, with some natural reserves, landscapes, waterfalls, medicinal hot springs, and it is home to a vast variety of birds and animals. One of the main tourist attractions is the Churute Mangrove Ecological Reserve. Naranjal is also home of a Shuar Community, which opens its doors to visitors to display their traditions, handicrafts, and ancestral gastronomy.

The chart in Figure 8 presents Naranjal´s main popular dishes. They include roasted Tilapia (Tilapia asada), Ayampaco, roasted beef (carne asada), Sugar cane (Caña de azùcar), and green plantain products (Productos con verde). According to this chart, Ayampaco is the most popular dish in the area.

The Ayanpaco is a traditional and ancient recipe of the Shuar Community (province of Morona Santiago). The Shuar Community that lives in Naranjal in the mountain area migrated many years ago from the Amazon. History says that there was a fight between the leaders of the clans, and one of these groups decided to leave the Amazon and move to Naranjal, and the other to Bucay, also located in Guayas (Naranjal, 2020).

This recipe consists of a river fish wrapped in a Bijao leaf (a native leaf of the Amazon, typically used to wrap foods for cooking). The "Ayanpaco" traditionally can have several types of proteins not only the river fish. The recipe could also be made with some type of wild meat, chicken, duck, giblets, or the very popular Mukint, also called Chontacuro (the larva of the palm tree). The traditional side dishes are hearts of palms and boiled yuca (Shuarnum, 2020).

There is a very similar dish called "Maito" from Puyo and Coca in the Ecuadorian Amazon. The most relevant differences between "Ayampaco" and "Maito" is that Maito is seasoned with grain salt and wrapped in a Yaki Panga leaf (Amazonian Plant). Both preparations are accompanied by yuca, banana, or palm hearts, and served with a hot Guayusa drink. The Maito traditionally uses as protein or main element, the Cachama fish or the Carachama, a variety of Catfish that is very typical in the area.

The Guayusa is a traditional herb used by the indigenous people for centuries. There are stories that say that warriors used this herb to increase their focus and stamina. This traditional herb nowadays is known for its energizing and antioxidant properties.

Insignia Dish for Naranjal: Ayanpaco

(Yields 1 Serving)

Ingredients

- 1 Cachama or Tilapia, drawn (no intestines)
- 1 cup Fresh Palmito, chopped
- 3-4 Bijao leaves, whole
- 1 cup Yuca or green plantain, boiled (water)
- Salt

Salsa Criolla (Creole Sauce)

- 4 red onions, julienned
- 5 lemons (sutil), juiced
- 4 tomatoes, julienned
- 6 sticks of a type of green onion, called white onion (native of Ecuador), chopped
- 2 cilantro stems with leaves, chopped
- 1 tablespoon of vegetable oil
- Salt

Preparation

1. Place 3 Bijao leaves, one on top of the other.
2. Score the meat of the fish in a crosshatch pattern, cutting about 1/2 inch deep, to help the seasoning penetrate the meat. Season it with salt.
3. Then put the fish in the middle of the top leave.
4. Wrap the leaves from the outer rim to the center. Then fold the leaves lengthwise over the fish until you reach the stem.
5. Take another Bijao leave and cut a thin strip.
6. Use the strip as a string and tie the Bijao wrap.
7. Make a wood fire or with coal and put a wire rack on top.
8. Put the wrap on top of the wire rack. Cook it for approximately 15–20 min., turning the Ayunpaco, every few minutes.

Salsa Criolla

1. Put the onions into a bowl and season with salt.
2. Add the lemon juice.
3. Add the tomato, white onion, cilantro leaves, and oil. Mix well
4. Correct the seasoning with salt and pepper if needed and reserve in the fridge.

Serve

1. After cooking, carefully open the leaves. Cut the length of the leaves so the preparation will fit into a dish.
2. Serve with the boiled yuca and salsa criolla.

CROSS-SITE SYNTHESIS AND CREATION OF THE FIRST PROPOSAL FOR THE TERRITORIES

Gastronomic routes represent a complete offer of specialized tourism that from the geography, culture, traditional products, and selective cuisine, become a tendency in tourism for a more defined visitor. They must combine their dishes with the experience involved in the cooking techniques, gathering of main ingredients from local farms and producers, as well as working with small entrepreneurs to create a real culinary experience. They must also attract tourists with other activities and attractions, restaurants, and gastronomic events to make it even more appealing for this new customer to accept this offer.

When creating a gastronomic route, it is important to acknowledge that while food may be the main motivation for travelers, there must be other essential elements such as historical sites, artisans' products, and natural attractions, among others, that must be part of the destination visited.

In the following lists for each destination, touristic attractions have been combined with the insignia dishes proposed in this research for a complete planification of a touristic gastronomic route. In some places, the creation of a gastronomic route will be easier than in others, since these are already considered touristic destinations. Therefore, adding an extra motivator for people to travel to these sites will be easier for promotion and implementation of the offer.

The gastronomic route itineraries in these five potential destinations must involve information for visitors to learn about the process and creation of a dish from the first stage to the final step, which is when it is ready to be served at the table. If possible, as part of the routes, activities involving gathering of these products will create memorable experiences, allowing tourists to enjoy the culture, traditions, history, and the actual preparation all in one activity, bringing an identity to each specific route.

When creating culinary stops within gastronomic routes, local restaurants and markets will be advised to first promote the insignia dish to fulfill the expectation of the tourist when choosing that specific route. Insignia dishes are essential for a destination to be known and remembered, and therefore promoted by the actual customers when leaving the place, because it will remain in their memories as part of the brand created for that destination.

Milagro

Milagro is considered a commercial city. People do not visit this destination for recreation or to relax, but the development of agro tourism is preparing the tourist to see Milagro as a potential touristic destination. The Haciendas and fincas open the door for a full culinary experience in this city that will allow the visitors to collect the products and convert them into final dishes or beverages while acknowledging the history and traditions behind them.

The recommendation for a touristic gastronomic route in Milagro should include the following:

- Local Haciendas and fincas like Jersualén, Finca Chaparral, Sarita Miranda, Apio, and Perejil
- Culinary stop at the local family restaurant, "La Costa"
- Ingenio Valdez for sugar cane production
- Julio Gamboa Museum for the Milagro-Quevedo culture ancestral objects and ceramics
- Cumana Estate Farm offering cocoa production
- Culinary stop in Rancho Vasija de Barro and Apio y Perejil Haciendas

Samborondón

Tourism in Samborondón has increased due to the creation of touristic routes involving traditional activities that attract local visitors to enjoy new experiences. A gastronomic route will be an excellent addition to the offers and will be greatly appreciated and accepted by the new demand that now visits this wonderful city.

In this specific case, the creation of two gastronomic routes would be an excellent offer, because two different types of tourists visit Samborondón. One group visits the main town, and another group visits the developed Via a la Puntilla, a high-end neighborhood full of restaurants with all types of cuisine and gastronomic offers (Diario El Universo, 2019).

The recommendation for the first touristic gastronomic route in Samborondón should include the following:

- Culinary stop by the Rosquitas kiosk at the city's Riverwalk (Malecón)
- Samborondón Riverwalk to see the river, City Hall, and public parks and pools
- El Encanto Beach for a culinary stop and recreation
- The ancestral route "Rio Babahoyo" to have a canoe experience on the river (El Telegrafo, 2020)
- Culinary stop at the local market
- City Hall Museum
- Culinary stop at a local traditional kiosk, "Doña Esperanza," for shrimp tortillas

The recommendation for the second touristic gastronomic route in Samborondón should include the following:

- Historical Park in Samborondón
- Culinary stop in Plaza Lagos
- El Buijo Histórico
- Culinary stop in Plaza gastronómica El Buijo

- Culinary stops in La Piazza, Alhambra, Plaza Navona, and Las Terrazas
- Samborondón Riverwalk to see the river, City Hall, and public parks and pools
- Sambo Trolley ride
- Samborondón vision tour bus
- Teatro Sanchez Aguilar

Playas

Playas is already considered a touristic destination, and a gastronomic route will increase traffic to the area for more than one day.

The recommendation for a touristic gastronomic route in Playas should include the following:

- Puerto El Morro to visit the mangroves and pink dolphins
- Cerro El Muerto to visit the main church and the sanctuary
- Horseback riding in General Villamil Playas and via a Data Beaches
- Culinary stop at the local market
- Culinary stop at two of the main touristic and traditional restaurants in Playas, "Juan Ostra" and "La Ostra que fuma"
- Visit to Engabao Beach for surfing and other beach sports, as well as local fishing
- Culinary stop in Engabao Beach

Naranjal and Balao

The recommendation for these two destinations is to create one touristic gastronomic route, due to their close geographical proximity. Naranjal offers other activities and natural attractions to complement Balao and its wonderful cuisine. (Lugares turísticos, 2020)

The gastronomic route for Naranjal and Balao should include the following:

- Shuar Community to visit ancestral traditions
- Culinary stop at the Shuar Community
- Thermal water complex in the Shuar Community and other parts of the mountain
- Churute Mangrove ecological reserve
- Culinary stop at a Gastronomic local market in Naranjal town
- Culinary stop at La hueca del cangrejo
- EL Mate Touristic Trail
- San Antonia de Padua Church in Balao
- Puerto Balao River
- Culinary stop at Balao´s local market
- Culinary stop at a traditional and gourmet restaurant, "La Cascada"
- Culinary stop at a traditional local restaurant, "Doña Esperanza"

SOLUTIONS AND RECOMMENDATIONS

The results of this research show that there is a potential increase in touristic traffic in the mentioned destinations due to their gastronomy. Satisfied tourists will be returning to Guayas influenced by its culture and history. This result confirms that culinary experience plays a fundamental role as a crucial factor in achieving a high-level productive chain in the place visited.

There is a demand for local traveling, and agricultural and cultural tourism in Guayas are the primary elements in the list. Therefore, to include gastronomic tourism in these destinations will become key to achieving an increase in local revenues and satisfaction of the demand. In this study, as a dependent variable, reaching satisfaction with these insignia dishes and recipes can lead to a list of repeat customers that will create a local movement driven by gastronomic motivation.

Local authorities, as well as tour operators, will need to be part of creating these gastronomic routes to help with the correct planning and promotion, as well as formulating a complete and well-coordinated package that will have experts and local tour guides in each area involved to fulfill the needs of a gastronomic and touristic experience.

Due to the importance of gastronomy for local tourism development, acknowledging the academic and public roles will allow for the creation of strategies in benefits of tourism, like touristic products and packages, and improvement in the infrastructure for tourism reactivation.

FUTURE RESEARCH DIRECTIONS

This study will open other variables to be observed and analyzed. The next step in this study should involve an actual analysis of the complete offer per destination, the local authorities desire and disposition to create these routes, and the actual state of the producers, restaurants, and recreation activities to ensure a real proposal can be created.

After an analysis, the creation of a pilot gastronomic route should be in place to ascertain whether or not the main objective was achieved, and the customers were satisfied with the offer.

This chapter will show the gastronomic offer in the chosen territories, but for future lines of research, this study could be conducted in other destinations within the same province or be extended to other provinces and regions in Ecuador.

CONCLUSION

Gastronomy plays a very important role in tourism as a whole. Gastronomic tourism is an increasing issue that must be taken into consideration when creating new opportunities for local tourism development. Ecuador has vast variety of microclimates and magical geographical sites that offer a wide variety of elements that can be used as an excellent offer.

Guayas Province and its cantons have a great gastronomic variety and are known for their seafood, agricultural products, and fruits, such as plantains and coconuts. Their local dishes are made with special recipes full of history and tradition passed from one generation to another, creating their own styles and special flavors. This allows gastronomic tourism to grow in these five destinations, all with great potential to become gastronomic ambassadors for others to follow.

This study highlights that there is still much to do concerning public policies, infrastructure, and education within the residents, but the main objective can be achieved if processes are put into place and followed as needed. The creation of an actual plan for Milagro, Samborondón, Playas, Balao, and Naranjal, as well as the insignia dishes as part of the destinations' brands is imperative for these gastronomic routes to achieve success.

Merging public and private organizations to develop high-quality promotional strategies are also one of the main factors to achieve great results. Following success cases and expert guidelines, while using the specific tools needed according to each necessity will create new job opportunities, with the cooperation of all actors in the productive chain to establish the bases needed to integrate all the existing resources, with the fulfillment of a new and thriving demand.

The results of this study confirm that there is potential for this specialized touristic activity, and because of Ecuador´s excellent location, these gastronomic routes and their products will be a unique offer, bringing great profits while creating a sustainable offer that will transcend time, conserving the history and culture and integrating the culinary experience as whole.

REFERENCES

Atlas turístico de Mexico. (n.d.). https://atlasturistico.sectur.gob.mx/AtlasTuristico/inicio.do

Clark, P. (2018). *The world's 10 most popular foodie destinations.* https://www.travelpulse.com/gallery/destinations/the-world-s-10-most-popular-foodie-destinations.html?image=2

Diario El Expreso. (2021, March 20). Samborondón crea una playa artificial de agua dulce, al pie del río Babahoyo [Samborondón creates an artificial freshwater beach, at the foot of the Babahoyo River]. *Diario El Expreso.* https://www.expreso.ec/guayaquil/samborondon-crea-playa-artificial-agua-dulce-pie-rio-babahoyo-100958.html

Diario El Universo. (2019, November 7). El malecón de Samborondón tiene más atractivos [The boardwalk of Samborondón has more attractions]. *Diario El Universo.* https://www.eluniverso.com/guayaquil/2019/11/07/nota/7591352/malecon-tiene-mas-atractivos/

El Telegrafo. (2020, September 27). La vuelta ancestral del río Babahoyo, una nueva ruta fluvial [The ancestral return of the Babahoyo River, a new river route]. *Diario El Telegrafo.* https://www.eltelegrafo.com.ec/noticias/sociedad/6/samborondon-ruta-fluvial-rio-babahoyo

Entorno turístico. (2020, February 18). Turismo gastronómico: Una forma diferente de hacer turismo [Gastronomic tourism: A different way of doing tourism]. *Entorno turístico.* https://www.entornoturistico.com/turismo-gastronomico-una-forma-diferente-turismo/

Fuentes, R. C., Moreno-Gil, S., González, C. L., & Ritchie, J. B. (2015). La creación y promoción de experiencias en un destino turístico. Un análisis de la investigación y necesidades de actuación [The creation and promotion of experiences in a tourist destination. An analysis of research and needs for action]. *Cuadernos de Turismo, 35,* 71–94. doi:10.6018/turismo.35.221511

GoRaymi. (2020). Balao. *Go Raymi.* https://www.goraymi.com/es-ec/guayas/balao/ ciudades/balao-a87d9782e

Linares, E. C., Cortés, V. S., Rodríguez, M. M., & Padrón, B. R. (2018). Patrimonio gastronómico y desarrollo local en dos Pueblos Mágicos de México [Gastronomic heritage and local development in two Magical Towns of Mexico]. *Revista Iberoamericana de Viticultura, Agroindustria y Ruralidad, 5*(15), 22–38. https://www.redalyc.org/jatsRepo/4695/ 469565683002/469565683002.pdf

Lugares turísticos. (2020). 3 Lugares Turísticos en Naranjal [Tourist Places in Naranjal]. *Lugares turísticos*. https://www.lugaresturisticos.org/naranjal/

Matta, R. (2014). Gastronomic republic and country of cooks: Food, politics, media and a new idea of nation for Peru. *Colombian Journal of Anthropology*, *50*(2), 15–40. http://www.scielo.org.co/ pdf/rcan/ v50n2/ v50n2a02.pdf

Ministerio de Turismo. (2018). *Mapa gastronómico del Ecuador* [Gastronomic map of Ecuador]. https:// files.goraymi.com/2020/04/01/60d71579ff1651d857a1a6c8f25af41c.pdf

Naranjal. (2020). Historia del Centro Shuar [History of the Shuar Center]. *Naranjal Turistico*. https:// naranjal-turistico.es.tl/Centro-Shuar.htm

National Geographic. (2018). 5 países que destacan por su gastronomía a nivel mundial [5 countries that stand out for their gastronomy worldwide]. *National Geographic*. https://www.ngenespanol.com/ gastronomia/5-paises-gastronomia-mundial/

Pérez, A. F. (2017). Análisis de la demanda turística del cantón Naranjal- Provincia del Guayas, para el diseño de un plan promocional turístico [Analysis of the tourist demand of the canton Naranjal- Province of Guayas, for the design of a tourist promotional plan]. *Repositorios UG*, 1-113. https://rraae.cedia.edu. ec/ Record/UG_cb87e2a3ebef54452b5940726579f9f9

Rodriguez, E. (2020). Turismo gastronómico ¿Qué es y cómo es el turista gastronómico? [Gastronomic tourism What is and how is the gastronomic tourist?]. *Gastromarketing*. https://eloyrodriguez.com/ turismo-gastronomico/

Rodríguez-Gutiérrez, P., Santa Cruz, F. G., Gallo, L. S. P., & López-Guzmán, T. (2020). Gastronomic satisfaction of the tourist: Empirical study in the creative city of Popayán, Colombia. *Journal of Ethnic Foods*, *7*(1), 1–12. doi:10.118642779-019-0044-0

Shuarnum. (2020). *Gastronomía típica* [Typical Gastronomy]. https://shuarnum.com/gastronomia/

Sorcaru, I. A. (2019). Gastronomy tourism - A sustainable alternative for local economic development. *Annals of the University Dunarea de Jos of Galati: Fascicle: I. Economics & Applied Informatics*, *25*(1). Advance online publication. doi:10.35219/eai1584040912

UNESCO. (2010b). *Traditional Mexican cuisine - ancestral, ongoing community culture, the Michoacán paradigm*. https://ich.unesco.org/en/RL/traditional-mexican-cuisine-ancestral-ongoing-community-culture-the-michoacn-paradigm-00400

Universo, E. (2015). *Comunidad* [Community]. El Universo. https://www.eluniverso.com/noticias/2015/ 09/27/nota/5148923/luis-aveiga-mete-camaron-rio-tortillas-verde/

Universo, E. (2017). *En Balao se preparó la ensalada de cangrejo más grande del Guayas* [In Balao the largest crab salad in Guayas was prepared]. https://www.eluniverso.com/guayaquil/2017/11/02/nota/6462787/balao-se-preparo-ensalada-cangrejo-mas-grande-guayas/

World Tourism Organization. (2017). *Second global report on gastronomy tourism*. https://www.e-unwto.org/doi/pdf/10.18111/9789284418701

World Tourism Organization. (2018). *Glosario de términos de turismo* [Glossary of tourism terms]. https://www.unwto.org/es/glosario-terminos-turisticos

World Tourism Organization & Basque Culinary Center. (2019). *UNWTO.ORG. Guidelines for the Development of Gastronomy Tourism*. https://www.e-unwto.org/doi/pdf/10.18111/9789284420957

Chapter 4
Importance of Mediterranean Nutrition in Tourism Businesses and Current Approaches

Seydi Yıkmış
https://orcid.org/0000-0001-8694-0658
Tekirdağ Namık Kemal University, Turkey

Aysen Coban Dincsoy
Istanbul Gedik University, Turkey

Melikenur Türkol
Tekirdag Namık Kemal University, Turkey

Vahide Gizem Aydın
Tekirdag Namık Kemal University, Turkey

ABSTRACT

Eating as a social entertainment causes dynamism in tourism destinations. New diet tendencies in healthy food production act as a bridge between life science and the tourism industry. It forms a basis for alternative tourism. Studies are indicating that the Mediterranean-type nutrition model may have important roles in reducing the incidence of some important diseases and the risk of death, in terms of easy perception, psychological relaxation, and longer life. Another feature that distinguishes this model from other diet models and makes it of particular importance for tourism businesses is that it has a socio-cultural structure. It is not just a dietary pattern, but a cultural model that includes the entire food chain.

INTRODUCTION

Tourism and the Historical Development of Tourism

According to the Turkish Language Association, tourism; “A trip made with the needs such as resting,

DOI: 10.4018/978-1-7998-9148-2.ch004

seeing and getting to know, is the economic, cultural, technical, etc. taken to attract tourists to a country or a region. measures are all of the work done.''

The concepts that the sources brought to the literature approach about tourism are as follows:

- For trips to be selected as tourism, they should go to a place outside the travel region.
- It is necessary to benefit from the tourism supply of the travel region.
- It is necessary not to stay in the travel area permanently (Taşdelen, et al., 2020).

Tourism, which has passed from the French word 'tourism' to Turkish, is derived from the word 'tour'. The world tour, on the other hand, is used to mean returning to the same position again. Tourism; can be defined as pleasure, rest, going out for vacation, visiting places that are worth visiting and seeing during the journey, and then returning (Kocaman, 2011).

According to World Tourism Organization (WTO) in 1991; tourism is defined as "the activity of a person who goes to a place outside of his/her usual environment to stay for less than a specified period and whose main purpose is to travel other than to try an activity that makes money in the place he/she visits" (Özgüç, 1996).

The history of tourism is as old as the history of humanity. The beginning of travel begins with the invention of the wheel in 4000 BC. While trade travels were shaped with the invention of money, overseas travels began to be made with the development of sea vehicles. During the time of the Egyptian Empire, new roads were built and travel was made easier. During the Greek Empire, the Olympic Games, during the Roman Empire, gladiatorial fights gathered the traveling masses and for this purpose, accommodation and entertainment centers began to be created. The Greeks had people called 'proxenos', which means today's tour guides, to accompany and assist the masses who came to the city on these occasions. With the emergence of the Renaissance after Anno domini, the purpose of travel was in the form of entertainment, art, curiosity to discover new places, while with the Industrial Revolution, it was shaped as capital owners and material assets. In 1841, an Englishman named Thomas Cook brought a tourist group of 571 people to the festival held in England, and the first tourism movement in the modern sense was realized (Taşdelen et al., 2020).

REASONS DRIVING PEOPLE TO TOURISM

From ancient times to the present, tourism movements have been carried out for various purposes such as health, religion, entertainment, pleasure and trade. The reasons why people turn to tourism have been shaped according to the conditions they are in and the important turning points in history. For example, in the Middle Ages, the reason for the journeys called the Crusades was religion. While these travels are more organized and massive, today's understanding of travel is different (İçöz, 2014).

Today, the reasons that lead people to tourism movements have diversified. It is possible to list these movements as follows: (Taşdelen et al., 2020).

- Innovation and curiosity
- Rest
- Culture and education
- Fun and adventure-seeking

- Politics and diplomacy
- Religion
- Sport
- Health
- Occupation and work
- Feeling of missing and visiting friends
- Searching for different tastes (Gastronomic Tourism)
- Imitation and flaunt
- Personal shopping

Motivation is effective in people's decision to go on vacation. In shaping this motivation, the push and pull factors are effective in the decision to go on a vacation. Therefore, motivating factors are classified as push and pull factors (Yoon & Uysal, 2005). Push factors have been defined by Oh et al. 1995 as 'intrinsic factors that motivate individuals to travel' (Oh et al., 1995). The motivating factors for people to travel to other places for their desire to get away from their place, to discover new places and to have new experiences can be defined as push factors (Dürdaroğlu, Barboroz, Mercan, & Avcıkurt, 2020). Pull factors, on the other hand, are mostly expressed as concrete reasons. Uysal and Xiangping Li (2008) defined attractive factors as 'natural and historical beauties, delicious food, places where people live, recreation areas and external factors related to the image of the destination' (Uysal & Xiangping Li, 2008). According to another definition, pull factors are; the destination has its characteristics such as sun, blue sea, golden beach, cultural and sports activities (Klenosky, 2003).

In recent years, the attractive elements of the destination and the motivating factors for tourists to travel have been tried to be determined by using appropriate research methods (Klenosky, 2003). In 2005, Yoon and Uysal conducted a study on 148 tourists in Northern Cyprus. According to this study, push factors are stated under eight headings, while pull factors are stated under nine headings. push factors; 'exciting', 'education', 'resting', 'success', 'being with family', 'escape from the environment', 'entertainment and getting away from home', 'sightseeing', while attractive factors are: 'modern atmosphere' ', 'activities', 'favorable weather conditions, 'natural scenery', 'different cultures', 'hygiene conditions', 'nightlife', 'interesting cities', 'water activities' (Yoon & Uysal, 2005).

In the study conducted by Jang and Cai in 2002, it was investigated whether there is an effect of push and pull factors when tourists travel. According to the results of the study, it was concluded that there is an effect of push and pull factors. It has been observed that the most important driving factors in the choice of destination are gathered under the headings of 'information seeking', 'escape from the current environment, 'being with family. It has been observed that the most important attractive factors in the selection of destinations are gathered under the headings of "hygiene and safety", "easy accessibility", "sun and exotic atmosphere" (Jang & Cai, 2009).

In 2008, Sangpikul's research on the travel reasons of tourists going to Thailand found that 'information seeking', 'resting', 'ego satisfaction' and 'cultural attractions', 'activities', 'shopping', 'security and hygiene' Attractive factors such as ' have been reported to be effective (Sangpikul, 2008).

In the study carried out by Yousefi and Marzuki in 2012, the travel reasons of the tourists coming to the Penang region of Malaysia were questioned. At the end of the study, it was concluded that the driving factors were 'information seeking', 'ego satisfaction', and 'resting', while the attractive factors were 'environment and security, 'cultural and historical attractions, and 'touristic opportunities' (Yousefi & Marzuki, 2012).

Nutrition; It is the use of nutrients by the body for growth, development, maintenance of life and protection of health. Nutrition is a conscious behavior to meet the nutritional elements that the body needs, rather than just filling the stomach. It is not possible to sustain life and protect human health without nutrition. The importance of nutrition, which can be expressed as a life activity, is increasing day by day. Nutritional behavior should be considered as a whole in terms of physiological, psychological and sociological aspects (Korinth et al., 2010). Health tourism, on the other hand, is the use of tourism and health opportunities in the place where people travel to another location to protect their health and treat the diseases that exist in people (Dürdaroğlu et al., 2020).

TYPES OF TOURISM

Types of tourism vary over time. The types of tourism offered in a certain period are formed by the needs of the society, the level of cultural and social opportunities and available transportation. The spatial characteristic of tourism is the type of tourism that results from mixing more types of the same regional unit but is also affected by the season. Each type of tourism moves towards areas where the purpose can be achieved due to the climate and the unique traditions of the region or through potential facilities (Tureac & Turtureanu, 2008).

In general, six types of tourism are examined:

1. Relaxing tourism
2. Health tourism
3. Visiting tourism
4. Transit tourism
5. Reduced distance tourism
6. Professional tourism

Tourism operates in a field of competing interests, where many economic, social and political forces influence policy and management, cross traditional organizational boundaries and create interdependence among a wide variety of stakeholders (Kernel, 2005).

Tourism types represent the way tourism is carried out depending on its causes and external factors. In terms of touristic destination selection, it can be examined in two parts. In voluntary tourism, the destination is chosen by the free will of the person benefiting from the tourism services. Compulsory tourism is carried out by others for various reasons such as medical advice. The classification of tourism types is given in Table 1 (Tureac & Turtureanu, 2008).

GASTRONOMY TOURISM

The word 'gastronomy' is defined as a discipline that explores the relationship between all consumable foods and beverages and culture, and the art of eating and drinking well (**TDK, 2021**). According to the definition of the Turkish Language Association, gastronomy; is in the form of 'health-friendly, well-arranged, pleasant and delicious cuisine, eating order and system' (Kemer, 2011). Regardless of the purpose of travel, people must be fed to survive during their journey or when they arrive at their

Table 1. Classification of tourism types

Starting Point and Destination Criteria	Number of Participant Criteria	Organization Criteria	Criterion Season	Time Criteria	Criteria of Transportation Type	Social Criteria	Criteria for Age and Occupation of Tourist	Criteria for Destination Type
Domestic Tourism	Individual Tourism	Organized Tourism	Continuous	Very Long Term Tourism	Tourism by Train	Private Tourism	Youth Tourism	Mountain tourism
International Tourism	Group Tourism	Semi Organized Tourism	Tourism	Long Term Tourism	Tourism by Car	Social Tourism	Specific to Adult Tourism	Season Tourism
		Unorganized Tourism	Non-continuous Tourism	Term Tourism	Air Tourism		Specific to Old Generation Tourism	

Source: (Tureac & Turtureanu, 2008)

destination. This situation has made the food and beverage industry a part of tourism. Apart from meeting their eating and drinking needs, some people travel just because they enjoy eating and drinking. It is possible to consider the travels made for this purpose in the context of 'gastronomy tourism'. (Şengül & Türkay, 2016). In an unusual culinary culture, the type of tourism that reacts to witnessing, presenting and tasting the preparation stages of the dishes, investigating the characteristics of this culinary culture, discovering the meal habits and eating styles is called 'gastronomy tourism' (Öney, 2013).

In the studies carried out, it has been seen that tourists give more importance to food and beverage experiences in other types of tourism apart from gastronomic tourism (Kim et al., 2009). According to the results of the research conducted by Hall and Sharpler in 2003, it was emphasized that the tourists traveling with a tour attach importance to their food and beverage experience, they eat the food of that region, then they buy the food of that region and visit the local markets (M. & L., 2003). The importance of gastronomic tourism is increasing especially in communities with rural tourism opportunities, and cultural heritage can be developed by preserving it in order to promote traditional foods (Green & Dougherty, 2009; Alonso & Liu, 2011).

TOURISM ENTERPRISES

Tourism enterprises appear as economic units that meet the needs of people during their touristic travels. The tourism business offers a service that we can synthesize for information/reservation, transportation, accommodation, food and entertainment as a tangible form of existence of the business. Therefore, the tourism company has a very large customer base (Cosmescu, 2005).

Due to the increase in the needs of tourists and their diversity, tourism enterprises have also increased in terms of number and variety. Within a classification based on basic needs, tourism businesses are examined under the following subheadings.

Travel Businesses

With the increase in people's financial power, the development of technology and the globalization of the world, the rate of people moving from one place to another has increased. This has helped travel businesses gain importance (Taş & Zengin, 2019).

Accommodation Businesses

Hotels are commonly in accommodation establishments. Hotels show various features according to their location. The concept that each reflects varies according to its location and the purpose it serves. It has various classifications within itself. Apart from hotels, motels, hostels, auberges, campings, holiday resorts, spas and similar health tourism businesses constitute accommodation businesses (İrek, 1991).

Food and Beverage Businesses

Food and beverage businesses are organizations that help people meet their desire for rest and entertainment while meeting their nutritional needs. In such places, people not only want to eat, but also to be respected and served, and to meet their social and psychological expectations. For this reason, such enterprises undertake a social and cultural cohesive role. Among tourism businesses, food and beverage businesses play a major role in this respect. It is known that food and beverage businesses help promote rural areas and the country in terms of the different experiences they bring to people, protect moral values and contribute to creating a regional image (Barış, 2015). To provide successful food and beverage service, it is very important to act in accordance with the demographic structure of the segment to be addressed. In this case, it is necessary to plan according to customer expectations. Due to the changing living conditions, people have turned to the search for healthier, cleaner and safer food. The rate of preference for businesses that meet these conditions is increasing (Andersson et al., 2017).

Transportation Businesses

Transportation is one of the most important issues for tourist travel. Transportation enterprises are examined in four groups as land, sea, air and railway (Barış, 2015).

Recreation and Animation Businesses

The economic and social conditions of modern life, where the stress factor has increased, have led to the growth of this form of tourism. Pollution in big city centers, sedentary life of the townspeople, diseases manifesting themselves in various forms leads to the gradual alienation of nutrition from the most natural and the need for tourism to regain health. The destination of this type of tourism are spas and health-based businesses that do mixed treatments following recreational functions. The value of these resources with prophylactic and therapeutic properties is provided by the arrangement of facilities to provide medical treatment. Since participation in this type of tourism is based on medical references in most cases, it is far from seasonal fluctuations and is organized enough to soften the tourist distribution (Tureac & Turtureanu, 2008).

MEDITERRANEAN DIET

History of the Mediterranean Diet

The Mediterranean type of nutrition is a diet that originates from the Mediterranean region and dates back to 5 thousand years ago. The Mediterranean diet was influenced by the living conditions of different civilizations and showed some changes. In the historical process, it has taken the form of a part of the lifestyle of the people of the region it is in. Today, it is possible to list these regions as Turkey, Spain, France, Italy, Portugal, Greece and some of the Middle East regions. The fact that these regions are suitable for agriculture has been effective in the adoption of the Mediterranean Diet model. The efficiency of the production of various vegetables and fruits, especially olives, in these arable lands facilitated the adoption of the Mediterranean type diet model and made the applicability of this model increase. As a result of a comprehensive study conducted in 7 different Mediterranean countries in the 1950s, the Mediterranean Diet was defined and published in a research article, and the concept of "Mediterranean diet" was introduced to the literature. After this research study, studies on the Mediterranean diet gained momentum in the 1960s and studies were conducted on its effects on health. Studies on the effect of the Mediterranean diet on health have increased the importance of this diet model in terms of applicability and ensured that it remained out of date until today (Keys et al., 1984; Bach-Faig et al., 2011; Barbaros, 2015).

In the 1990s, the Mediterranean diet was started to be investigated as a sustainable diet model by Joan Dye Gussow (Gussow, 1995). This plant-based diet has also become an important subject of environmental sustainability studies as it is associated with a lower water footprint and greenhouse gas emissions compared to current Western diets (Tilman & Clark, 2014; Sáez-Almendros et al., 2013).

In 2009, the International Mediterranean Food Cultures Research Center (CIISCAM) organized an international conference titled "Mediterranean diet as a sustainable diet model" ("The Mediterranean diet as a sustainable diet model") in Parma, Italy was held. In this conference, the Mediterranean diet was evaluated in terms of nutrition, environmental, socio-cultural, economic and accepted as a sustainable nutrition model (Burlingame & Dernini, 2011). The conference was instrumental in the international recognition of the Mediterranean diet pyramid, which includes the foods identified with the Mediterranean diet, which has a less harmful effect on the environment (Bach-Faig et al., 2011).

In the symposium on 'biodiversity and sustainable diets' organized by the United Nations Food and Agriculture Organization (FAO) in 2010, sustainable diets were defined as follows: These are diets that have low environmental impacts that contribute to life. Sustainable diets are nutritionally adequate, safe and healthy; It is protective, respectful of biodiversity and ecosystems, culturally acceptable, accessible, economically fair and cost-effective while optimizing natural and human resources" (FAO/Bioversity, 2012).

Upon understanding the importance of the Mediterranean diet, it was accepted as an 'Intangible Cultural Heritage of Humanity by the United Nations Educational, Scientific and Cultural Organization (UNESCO) in 2010 (UNESCO, 2010a).

Definition and General Characteristics of the Mediterranean Diet

The Mediterranean diet is a diet model that reflects the common dietary styles of Mediterranean countries and is rich in macro and micronutrients (Dernini & Berry, 2015). The basis of the diet consists of fruits, vegetables, legumes, grains, oilseeds, fish, olive oil. Consumption of processed foods such as red meat, sausage and sausage is limited. The foods that form the basis of the Mediterranean diet are very rich in fiber and unsaturated fatty acids; wholemeal or whole-grain foods, moderately rich in nuts and legumes; saturated fat is low in trans fats. The nutrients that form the basis of the diet also show antioxidant and anti-inflammatory properties, thanks to the fact that they contain vitamins A, C, E, and selenium, which have positive effects on health (Dinu et al., 2018; Kürklü, 2021).

In this diet style, which represents the nutritional habits of Mediterranean societies, the use of high amounts of olive oil draws attention. The main feature that distinguishes the Mediterranean diet from many western diets developed today is its rich content of monounsaturated fatty acids (MUFA) (Kastorini, Milionis, Goudevenos, & Panagiotakos, 2010).

Research conducted in recent years; He states that the Mediterranean diet is a dietary model that reduces the risk of cardiovascular disease and obesity, has a protective effect against chronic diseases and some types of cancer, alleviates symptoms in case of illness and increases the quality of life (Trichopoulou et al., 2014; Rhazi et al., 2012). Mediterranean diet with local foods in season; is becoming a frequently preferred diet model in businesses in terms of its health benefits, socio-cultural, economic, environmental and sustainability.

Mediterranean Diet Nutrition Pyramid

The Mediterranean diet was updated in 2011 and a nutrition pyramid was created considering the daily needs of healthy adult people. Considering that the needs may change in various age and disease states, it is possible to increase the possible health benefits of the diet by creating differences in the foods in the nutrition pyramid and the portions of these foods. In this pyramid, food items are grouped and shown separately in meals, daily and weekly. Consumption of whole grains (1-2 servings), fruit (1-2 servings), vegetables (≥ 2 servings, raw or cooked) and olive oil at each meal is recommended. While it is recommended to consume daily low-fat milk and dairy products (2 servings), oilseeds (1-2 servings); The frequency of consumption of foods of animal origin such as at least 2 servings of legumes, fish and seafood (≥ 2 servings), white meat (2 servings), red meat (<2 servings) is included in the weekly form.

According to the pyramid, it is recommended to add onions, garlic, herbs and spices to meals to reduce daily salt consumption and increase flavor. Desserts at the top of the pyramid should be given as little space as possible (≤ 2 portions). In the Mediterranean diet nutrition pyramid, culinary activities such as sitting at the table with the family and preparing food together; the elements of the Mediterranean lifestyle such as regular physical activity, adequate rest, entertainment, moderate wine consumption according to local habits, and the use of seasonal, traditional, local and environmentally friendly foods are also included (Bach-Faig et al., 2011).

According to the Mediterranean diet model, the foods that should be consumed daily are given in Table 2, the foods that should be consumed weekly are given in Table 3 and the foods that should be consumed infrequently are given in Table 4 (Bach-Faig et al., 2011).

Mediterranean Diet; reducing low-density lipoprotein (LDL) represents influencing the microbiota. It stimulates autophagy (the process of digesting senescent and targeted cell structures by the cell's lysosome) thanks to its polyphenol sources such as olive oil, wine, grapes, and oilseeds. While this condition is well healed, it ensures that it is purified from accumulated stress and toxins. It has been a preferred nutrition model due to the effects of this nutrition model. Table 5 includes an example of the Mediterranean Diet menu (Baysal, 1996).

Figure 1. Mediterranean diet nutrition pyramid
Source: (Bach-Faig et al., 2011)

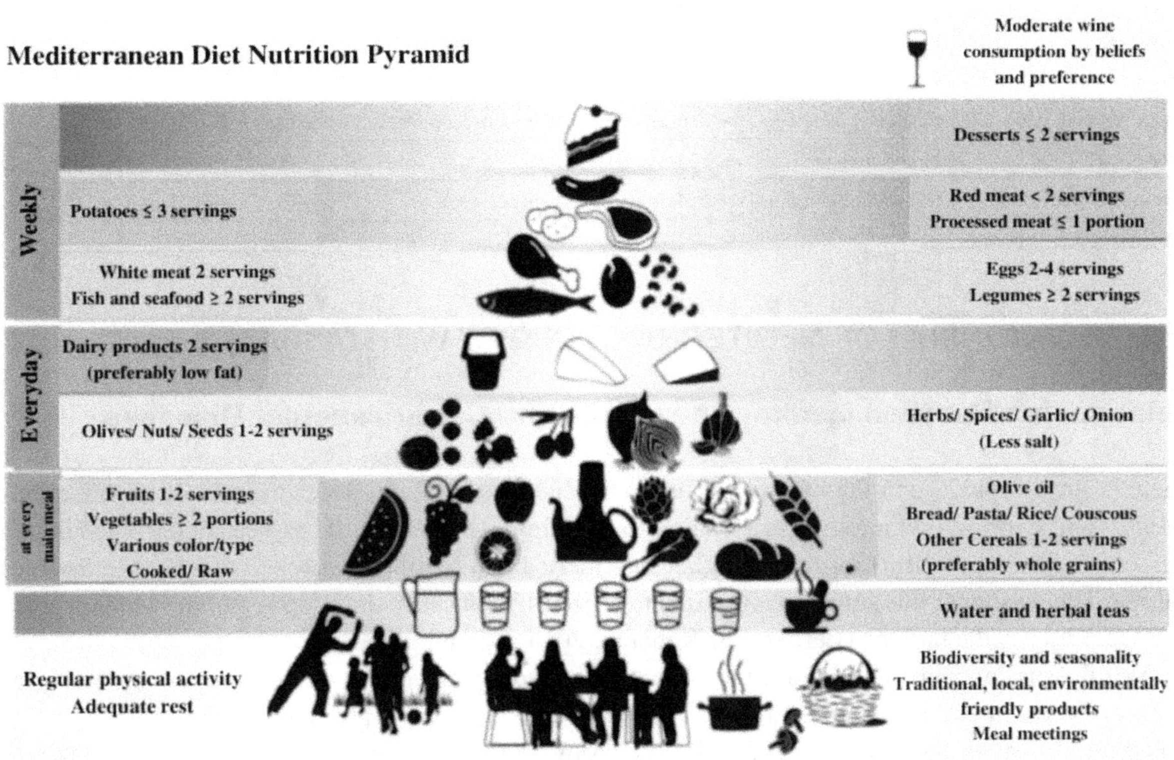

Table 2. Daily consumption according to the Mediterranean diet

Nutrients to be Consumed Daily	
Water	The water requirement is stated as 1.5-2 liters on average, depending on personal differences and seasonal conditions.
Grain	1-2 servings should be consumed at each main meal. It should preferably be wholemeal or whole grain.
Vegetable	Main meals should have 2 or more servings, with at least 1 serving raw. Vegetable consumption provides diversity in the intake of a wide range of vitamins, minerals and antioxidants.
Fruit	Each main meal should have 1-2 servings.
Dairy and dairy products	It should be consumed in 2 portions a day, preferably as half fat. It should be noted that excessive consumption will increase the intake of saturated fat.
Olive oil	It is the focus of the diet. It contains many beneficial nutritional components.
Nuts, seeds, olives	Approximately 1 handful can be consumed per day for snacks. It is a very rich group in terms of protein, fat, minerals, vitamins and fiber.
Salt	Spices can be preferred for flavoring to reduce salt consumption.

Source: (Bach-Faig, 2011)

Table 3. Foods to be consumed weekly according to the Mediterranean diet

Nutrients to be Consumed Weekly	
Legumes	Its intake of cereals increases the quality of plant-based protein.
Potato	Its consumption is 3 or fewer servings per week.
Eggs, seafood, poultry	Eggs are 2-4 servings, shellfish and fish are 2 or more servings, and white meat is 2 servings.
Red meat	Its consumption is less than 2 servings per week.
Processed meat	It should be consumed rarely, less than 1 serving per week.

Source: (Bach-Faig et al., 2011)

THE RELATIONSHIP OF MEDITERRANEAN DIET WITH DISEASES

Relationship Between Mediterranean Diet and Cardiovascular Diseases

There is strong evidence in the literature that the Mediterranean diet reduces the risk of cardiovascular diseases and improves symptoms. As a result of epidemiological studies, it has been determined that the incidence of coronary heart disease is higher in America and Northern Europe compared to Southern Europe. The reason for this situation is thought to be the differences in the adoption of the Mediterranean diet according to the regions (Barbaros & Kabaran, 2014; Strisciuglio et al.,, 2020).

A study was conducted to examine the effect of the Mediterranean diet on the course of carotid artery disease. According to the results of this study, it was observed that adopting the Mediterranean diet and making it a lifestyle significantly reduced the symptoms of carotid artery stenosis and prevented plaques that cause non-calcific atherosclerosis (Kuyumcu & Kuyumcu, 2019).

In a prospective study by Fung et al., approximately 76,000 women were included and followed for 20 years. As a result of this study, it was determined that women who adopted the Mediterranean diet had 22% less risk of developing cardiovascular diseases than those who did not (Fung et al., 2009).

The Relationship Between the Mediterranean Diet and Diabetes

The risk of obesity increases in people who do not adopt the Mediterranean diet model. Insulin resistance, which develops especially with abdominal fat and increasing beta-cell dysfunction, is the harbinger of Type 2 diabetes. Thanks to the high polyphenol content of the Mediterranean diet, its anti-inflammatory and antioxidant properties reduce the risk of Type 2 diabetes. In addition, the fact that this diet is rich

Table 4. Foods to be consumed rarely according to the Mediterranean diet

Foods that should be consumed infrequently
Bakery products, candy, sweets
Sugary juices and fizzy drinks
Foods high in unhealthy fats and fats (trans fat, etc.)

Source: (Bach-Faig et al., 2011)

in antioxidant vitamins, fiber and unsaturated fatty acids reduces oxidative stress and increases insulin sensitivity (Schröder, 2007; Wang et al., 2013).

Table 5. Mediterranean Diet Menu Examples

DAY	BREAKFAST	LUNCH	DINNER
1	Cheese (preferably low-fat) Olive Tomato 1 Bread (wholegrain/wholegrain) Herbal tea (Sage etc.)	White meat (chicken etc.) Bulgur pilaf seasonal salad Ayran	Lentil Soup Vegetable dish with olive oil Yogurt Bread
2	Tahini-molasses Bread Orange	Steamed/grilled fish seasonal salad Bread	Vegetable soup Olive oil wrap/stuffed Ashura
3	Egg Olive Tomatoes Bread Herbal tea	Stuffed Lentils seasonal salad Yogurt Bread	Yayla soup Kidney Beans with Olive Oil Fruit Bread
4	Tarhana soup Orange Bread	Chickpea stew Bulgur pilaf Salad/tzatziki	Forest Kebab (with garnish) Rice Pudding
5	Egg Olive Bread Tomatoes Orange	Lentil Meatballs Lettuce salad Ayran	Grilled/steamed fish vegetable salad Bread
6	Cheese Olive Bread Orange	Vegetable dish (with eggs) Rice Salad/tzatziki	Tandoori soup Vegetable dish with olive oil Pumpkin dessert
7	Omelet Bread Tomatoes	Veggie Pie Compote	Tarhana soup Olive oil wrap Fruit

Source: (Baysal, 1996).

A case-control study was conducted on the relationship between adherence to the Mediterranean diet and abdominal obesity and diabetes. According to the results of this study, it has been reported that people who adopt the Mediterranean diet model have a lower rate of fat around their waist. It has been determined that these people provide better weight control and have a lower risk of developing insulin resistance and type 2 diabetes (Romaguera, 2009). In a study evaluating the effects of adherence to the Mediterranean diet on biochemical parameters and anthropometric measurement results in people with type 2 diabetes, it was reported that increased adherence to the Mediterranean diet led to a significant decrease in body weight and thinning of the waist circumference. In this study, it was also seen that people who comply with the Mediterranean diet had positive changes in fasting blood glucose levels, insulin, total cholesterol, triglyceride, HDL and LDL cholesterol levels (Kündeş, 2019).

The Relationship Between the Mediterranean Diet and Obesity

In obesity, fiber is important in terms of affecting hormonal responses and providing satiety. The rich fiber content of the Mediterranean diet is effective in providing weight control, as it reduces the energy intake in the diet (Barbaros & Kabaran, 2014).

The Mediterranean diet is a diet model rich in monounsaturated fatty acids. As a result of the replacement of saturated fatty acids in the diet with the same amount of monounsaturated fatty acids, there is a significant decrease in body weight and fat mass (Piers et al., 2003).

In another study, the food consumption record of the participants was taken and their compatibility with the Mediterranean diet was evaluated and scoring was done. The 0,1,2,3,4,5 scoring system was matched with rare, frequent, very frequent, weekly and daily consumptions, respectively. This scoring system included unrefined grains, fruits, vegetables, legumes, olive oil, fish, red meat, poultry, dairy products and alcohol in the Mediterranean diet. The scoring system is designed for alcohol and 5 points are given for consumption less than 300 ml per day and 0 points for consumption more than 700 ml. At the end of the consumption record, the scores were evaluated between 0 and 55. It was determined that the body mass index (BMI) was 4 kg/m^2, the waist circumference 12 cm, and the hip circumference 17 cm lower on average, compared to those with a high dietary compliance score. It has been reported that the risk of weight gain decreases as adherence to the Mediterranean diet increases (Panagiotakos et al., 2006).

The Relationship Between the Mediterranean Diet and Alzheimer's Disease

The antioxidant content and anti-inflammatory effect of the Mediterranean diet, rich in vitamins A, C, and E, are effective in reducing oxidative stress and inflammatory response in Alzheimer's patients. B group vitamins also play an important role in preventing hyperhomocysteinemia. Since antioxidants increase cognitive functions in the frontal brain system, they have very beneficial effects in some disease states that trigger the uncontrolled proliferation of free radicals. According to some research results, increasing omega-3 fatty acid intake and decreasing total/saturated fatty acid intake were also associated with Alzheimer's (İplikçi & Ayhar, 2017).

The Relationship Between the Mediterranean Diet and Cancer

The fact that vegetables and fruits contain bioactive components such as antioxidants and polyphenols increases their usability in cancer patients due to their anticarcinogenic effects. A diet rich in vegetables such as broccoli, cauliflower and cabbage reduces the risk of pancreatic, stomach and colon cancer (Erman, 2007).

In the past years, Kim and Mason have conducted a retrospective study. The past food consumption records and frequencies of people diagnosed with cancer were examined, and it was determined that the frequency of consumption of fresh vegetables and fruits of these people was quite low (Kim & Mason, 1996).

Thanks to the antioxidant properties of squalene in olive oil, which is one of the main components of the Mediterranean diet, free radicals are neutralized in the tissue and reduce DNA damage. It also has a beneficial effect on preventing the formation of cancerous cells from normal cells. Oleic acid, the main fatty acid in olive oil, reduces the formation of molecules that can cause inflammation thanks to its antioxidant properties and prevents the tendency of cells to oxidation. Polyphenol reduces the risk of cancer thanks to its antimutagenic, antioxidant and antiproliferative properties (Bozbulut & Akbulut, 2016).

MEDITERRANEAN TYPE NUTRITION AND TOURISM ENTERPRISES

Food and tourism have always been in a relationship. However, the experiential potential of food has aroused increasing interest in food tourism among tourism service manufacturers and destination management organizations in recent years. In terms of scientific research, food tourism is also an expanding field, but critical studies of the conceptual and practical aspects of food tourism are still needed (Andersson et al., 2017).

The cuisine of a region is integrated with the culture of that region. The way the dishes are prepared and the flavors give an idea about the local culture and the people of the region (Şengül & Türkay, 2014). Mediterranean diet habits include cereals, olive oil, vegetables and fruits, aquatic products, spices, grape juice. Onion, garlic, leek, beet, okra are among the most consumed foods (Çevik, 1997). Mediterranean cuisine culture and eating habits also differ between the Eastern Mediterranean and Western Mediterranean regions. While foods such as tarhana and yogurt and foods made with these foods are in the foreground in the Western Mediterranean, meat dishes of the Southeastern Anatolian cuisine culture are common in the Eastern Mediterranean. In this region, there are mostly kebabs made with vegetables (Ertaş & Karadağ, 2013). Thanks to its favorable climate structure and fertile soils, orange, tangerine and grapefruit are important citrus fruits (Şengül, 2015).

Considering the health benefits of the Mediterranean type of nutrition model, it is suggested after the studies that the elements that will make the Mediterranean type nutrition model attractive in tourism establishments can be added (Şengül & Türkay, 2016). These suggestions can be listed as follows:

- In tourism establishments, the foods in the Mediterranean type nutrition model can be added by including phrases such as 'Mediterranean Cuisine Corner'.
- By creating a small shop concept where local foods such as tarhana, orange, cheese and olive varieties, which are suitable for the Mediterranean type of nutrition model and unique to the Mediterranean region, are sold, tourists can be provided to buy local foods from here.
- It can be ensured that qualified cooks who know the scope of the Mediterranean type of nutrition model and who can produce products unique to the Mediterranean region are trained in tourism establishments.
- By distributing informative brochures on the Mediterranean diet, tourists can be given the opportunity to be informed at tourism establishments, during their stay or their travels.
- Travel businesses can take tourists to markets where local products are sold.
- During the tours organized by travel companies, local foods of the Mediterranean region can be promoted and served.
- Geographical indication applications for the registration of gastronomic foods belonging to the Mediterranean region (Şengül & Türkay, 2016).

People's search for healthy food, wanting to make their lifestyles healthier, and not wanting to give up their habits in destinations with their families have started to create a condition for the places they will choose. Mediterranean diet is compatible with this healthy profile with its effects on diseases. It is of great importance for the businesses that provide services to health tourism that the foods are compatible with their profiles. There are also trips made just to change the diet. Such travels reveal alternative tourism branches. This shows that the quality of the food offered to the tourist is of undeniable importance for businesses. Health tourism breaks the taboo of mass tourism. Today, tourism has surpassed the trio of sea, sand and sun and has become a place where alternative tourism riches are evaluated and considered necessary (Sotiriadis, 2017).

For tourism businesses, Mediterranean Type Nutrition should be considered not only because of its health benefits, but also in terms of the role it plays in cultural fusion. Kitchens serve as a mirror for every culture. Considering that touristic movements are generally made for social purposes, it also stands out with its socially satisfying feature. Mediterranean diet is based on the system of keeping people together and eating together (Fragkiadakis, 2007).

The culture of spending time together is also common in the Mediterranean type of diet. Studies show that the gastronomic culture attracts the attention of tourists to a significant extent (Bucak & Aracı, 2013). According to the study conducted by Cohen and Avieli in 2004, it was measured to what extent the perception of food affects the attractiveness of the destination for tourists. It has been concluded that the food makes the destination attractive (Cohen & Avieli, 2004).

In 2011, Contento stated that tourists' food choice is affected by factors such as biological, psychological, sensory-affective factors, social and cultural norms, environmental stimuli, and food availability (Contento, 2011).

The study titled "The Importance of Local Cuisine in Tourism Marketing and the Example of Hatay Cuisine" by Cömert in 2014, it was aimed to contribute to the regional tourism related to the Mediterranean type nutrition model and food culture issues (Cömert, 2014).

The increasing number of tourists with food allergies and different dietary preferences draw attention to various problems in gastronomy and tourism service organizations. Therefore, the need to understand people's behavior helps tourism stakeholders adapt the changing nutritional paradigms to their activities (Donini et al., 2015).

CONCLUSION

Food and beverage service offers an important competitive opportunity for businesses that want to take a place in the tourism market. Mediterranean Type Nutrition has gained importance in tourism businesses due to food safety concerns, the search for a sustainable lifestyle, initiatives to highlight alternative tourism areas, and the increase in interest in the ecological environment. In 2010, the Mediterranean Diet was added to the list of intangible cultural heritage by UNESCO. It includes various vegetables, fruits, legumes and olive oil use. More beneficial effects were found than dietary interventions based on single dietary components. Studies are indicating that the Mediterranean Type Nutrition Model may have important duties in reducing the incidence of some important diseases and the risk of death, in terms of easy perception, psychological relaxation and longer life. Another feature that distinguishes this model from other diet models and makes it of particular importance for tourism businesses is that it has a socio-cultural structure. It is not just a dietary pattern, but a cultural model that includes the whole food chain.

The stages of selection, production, processing and distribution of nutrients are of great importance. It plays a major role in creating sustainable resources. It contains plenty of knowledge, experience, symbols, rituals and experiences. Local food has a symbolic value in the Mediterranean Diet and this value plays a role in its connection with cultural heritage. From an economic and environmental perspective, it plays a key role in development, biodiversity and sustainability. It is emphasized that reducing animal consumption and turning to plant food consumption will be beneficial both in terms of climate and the environment. The Mediterranean type of nutrition model plays a role not only in achieving individual and social health but also in the development of society as a whole.

REFERENCES

Alonso, A. D., & Liu, Y. (2011). The potential for marrying local gastronomy and wine: The case of the 'fortunate islands'. *International Journal of Hospitality Management*, 974–981. doi:10.1016/j.ijhm.2011.02.005

Andersson, T. D., Mossberg, L., & Therkelsen, A. (2017). Food and tourism synergies: Perspectives on consumption, production, and destination development. *Scandinavian Journal of Hospitality and Tourism*, *17*, 1–8. doi:10.1080/15022250.2016.1275290

Bach-Faig, A., Berry, E. M., Lairon, D., Reguant, J., Trichopoulou, A., & Dernini, S. (2011). Mediterranean diet pyramid today. Science and cultural updates. *Public Health Nutrition*, *14*(12), 2274–2284. PMID:22166184

Barbaros, B. (2015). *Yüksek, Orta ve Düşük Akdeniz Diyeti Uyumu Olan Bireylerde Oksidatif Stres ve Total Antioksidan Kapasitenin Belirlenmesi* [Determination of Oxidative Stress and Total Antioxidant Capacity in Individuals with High, Medium and Low Mediterranean Diet Compliance]. Yüksek Lisans Tezi, Doğu Akdeniz Üniversitesi, Beslenme ve Diyetetik Anabilim Dalı. Gazimağusa.

Barbaros, B., & Kabaran, S. (2014). Akdeniz Diyeti ve Sağlığı Koruyucu Etkileri [Mediterranean Diet and Health Protective Effects]. *Beslenme ve Diyet Dergisi*, *42*(2), 140–147.

Barış, Z. (2015). *Turizm İşletmelerinde Organik Gıda Kullanımı Algı ve Tutumlarının Araştırılması* [Investigation of Perceptions and Attitudes of Organic Food Use in Tourism Enterprises]. Yüksek Lisans Tezi. İzmir Katip Çelebi Üniversitesi Sosyal Bilimler Enstitüsü Turizm İşletmeciliği Anabilim Dalı.

Baysal, A. (1996). Sağlıklı Beslenme ve Akdeniz Diyeti [Healthy Eating and Mediterranean Diet]. *Beslenme ve Diyet Dergisi*, *25*(1), 21–29.

Bozbulut, R., & Akbulut, G. (2016). Kansere Karşı Koruyucu Yeni Bir Öge: Skualen [A New Anti-Cancer Element: Skualen]. *Beslenme ve Diyet Dergisi*, *44*(2), 160–168.

Bucak, T., & Aracı, Ü. E. (2013). Türkiye'de gastronomi turizmi üzerine genel bir değerlendirme [A general assessment of gastronomy tourism in Turkey]. *Balıkesir Üniversitesi Sosyal Bilimler Enstitüsü Dergisi*, *16*(30), 203–216.

Burlingame, B., & Dernini, S. (2011). Sustainable diets: The Mediterranean diet as an example. *Public Health Nutrition*, *14*(12A), 2285–2287. doi:10.1017/S1368980011002527 PMID:22166185

Çevik, N. K. (1997). *Türk Mutfağının Akdeniz Mutfak Kültürünün Genel Özellikleri Yönünden Değerlendirilmesinin Önemi* [Importance of Evaluating Turkish Cuisine in Terms of General Characteristics of Mediterranean Culinary Culture]. V. Milletlerarası Türk Halk Kültürü Kongresi.

Cohen, E., & Avieli, N. (2004). Food in tourism: Attraction and impediment. *Annals of Tourism Research, 31*, 755–778. doi:10.1016/j.annals.2004.02.003

Cömert, M. (2014). Turizm Pazarlamasında Yöresel Mutfakların Önemi ve Hatay Mutfağı Örneği [Importance of Regional Cuisines in Tourism Marketing and Hatay Cuisine Example]. *Journal of Tourism and Gastronomy Studies*, 64-70.

Contento, I. R. (2011). *Nutrition education linking research.* Jones and Bartlett Publishers.

Cosmescus, N. (2005). Formula 1: scena şi culisele marilor premii [Formula 1: the stage and the backstage of the big prizes]. Piteşti.

Dernini, S., & Berry, E. M. (2015). Mediterranean diet: From a healthy diet to a sustainable dietary pattern. *Frontiers in Nutrition, 2*, 15. doi:10.3389/fnut.2015.00015 PMID:26284249

Dinu, M., Pagliai, G., Casini, A., & Sof, F. (2018). Mediterranean diet and multiple health outcomes: An umbrella review of meta-analyses of observational studies and randomised trials. *European Journal of Clinical Nutrition, 72*, 30–43. doi:10.1038/ejcn.2017.58 PMID:28488692

Donini, L. M., Serra-Majem, L., Bulló, M., Gil, Á., & Salas-Salvadó, J. (2015). The Mediterranean diet: Culture, health, and science. *British Journal of Nutrition, 113*(2), 1–3. doi:10.1017/S0007114515001087 PMID:26148911

Dürdaroğlu, O., Barboroz, M., Mercan, Ş. O., & Avcıkurt, C. (2020). *Turistlerin Sağlıklı Beslenme ve Diyet Danışmanlığı Hizmeti Alma Motivasyonuyla Gerçekleştirdikleri Turizm Hareketlerinde Çekici ve İtici Faktörler: Sosyal Medya Destekli Bir Araştırma* [Attractive and Repulsive Factors in Tourism Movements carried out by tourists with the motivation to receive healthy nutrition and dietary consultancy services: a social media-supported research]. Seyahat ve Otel İşletmeciliği Dergisi.

Erman, Y. (2007). *Erkek ve Kadınların Diyet-Kanser İlişkisi Hakkında Bilgi ve İnanışları* [Knowledge and Beliefs of Men and Women about the Diet-Cancer Relationship]. Yüksek Lisans Tezi. Ankara Üniversitesi, Fen Bilimleri Enstitüsü, Ev Ekonomisi (Beslenme Bilimleri) Anabilim Dalı.

Ertaş, Y., & Karadağ, M. G. (2013). *Sağlıklı Beslenmede Türk Mutfak Kültürünün Yeri* [Place of Turkish Culinary Culture in Healthy Eating]. Gümüşhane Üniversitesi Sağlık Bilimleri Dergisi.

FAO/Bioversity. (2012). *Sustainable diets and biodiversity.* FAO: Directions and Solutions for Policy, Research and Action.

Fragkiadakis, G. A. (2007). *Combining tourism business and Mediterranean diet: An innovative approach.* Animal Hemagglutinins Technology View Project Archimedes Protect- Primary School Children Nutrition View Project.

Fung, T. T., Rexrode, K. M., Mantzoros, C. S., Manson, J. E., Willett, W. C., & Hu, F. B. (2009). Mediterranean diet and incidence of and mortality from coronary heart disease and stroke in women. *Circulation, 119*, 1093–1100. doi:10.1161/CIRCULATIONAHA.108.816736 PMID:19221219

Green, G. P., & Dougherty, M. L. (2009). Localizing linkages for food and tourism: Culinary tourism as a community development strategy. *Community Development (Columbus, Ohio)*, 148–158. doi:10.1080/15575330809489674

Gussow, J. (1995). Mediterranean diets: Are they environmentally responsible? *The American Journal of Clinical Nutrition*, 1383S–1389S. doi:10.1093/ajcn/61.6.1383S PMID:7754992

İçöz, O. (2014). *Seyahat Acentaları ve Tur Operatörlüğü Yönetimi* [Agencies and Tour Operator Management]. Turhan Kitabevi.

İplikçi, G., & Ayhar, N. Y. (2017). Beslenme Alzheimerdan Koruyucu Bir Faktör Müdür? [Is Nutrition a Protective Factor For Alzheimer's?] *Ankara Sağlık Bilimleri Dergisi, 1-2-3*, 1-11.

İrek, A. (1991). *Turizm İşletmelerinde Örgütlenme* [Organizing in Tourism Enterprises]. Yüksek Lisans Tezi. İstanbul Üniversitesi Sosyal Bilimler Enstitüsü.

Jang, S., & Cai, L. A. (2009). Travel motivations and destination choice: A study of British outbound market. *Journal of Travel & Tourism Marketing, 13*(3), 111–133. doi:10.1080/10548400209511570

Kastorini, C. M., Milionis, H. J., Goudevenos, J. A., & Panagiotakos, D. B. (2010). Mediterranean diet and coronary heart disease: Is obesity a link? - A systematic review. *Nutrition, Metabolism, and Cardiovascular Diseases, 20*, 536–551. doi:10.1016/j.numecd.2010.04.006 PMID:20708148

Kemer, A. K. (2011). *Otellerde Çalışan Mutfak Personelinin ve Aşçılık Alanında Yüksek Öğrenim* [Kitchen Staff working in Hotels and Higher Education in Cooking]. Basılmamış Yüksek Lisans Tezi. Gazi Üniversitesi Eğitim Bilimleri Enstitüsü.

Kernel, P. (2005). Creating and implementing a model for sustainable development in tourism and enterprises. *Journal of Cleaner Production, 13*(2), 151–164. doi:10.1016/j.jclepro.2003.12.023

Keys, A., Menotti, A., Aravanis, C., Blackburn, H., Djordevic, B., & Buzina, R. (1984). The seven countries study: 2,289 deaths in 15 years. *Preventive Medicine, 13*(2), 141–154. doi:10.1016/0091-7435(84)90047-1 PMID:6739443

Kim, Y. G., Eves, A., & Scarles, C. (2009). Building a model of local food consumption on trips and holidays: A grounded theory approach. *International Journal of Hospitality Management*, 423–431. doi:10.1016/j.ijhm.2008.11.005

Kim, Y.-I., & Mason, J. B. (1996). Nutrition chemoprevention of gastrointestinal cancers: A critical review. *Nutrition Reviews, 54*(9), 259–279. doi:10.1111/j.1753-4887.1996.tb03947.x PMID:9009668

Klenosky, D. B. (2003). The pull of tourism destinations: A means-end investigation. *Journal of Travel Research*, 385–395. doi:10.1177/004728750204000405

Kocaman, S. (2011). *Edirne İlinin Turizm Coğrafyası* [Tourism Geography of Edirne Province]. Doktora Tezi. Atatürk Üniversitesi Sosyal Bilimler Enstitüsü.

Korinth, A., Schiess, S., & Westenhoefer, J. (2010). Eating behaviour and eating disorders in students of nutrition sciences. *Public Health Nutrition*, 221–232. doi:10.1017/S1368980009005709 PMID:19433007

Kündeş, H. (2019). *Tip 2 Diyabetli Hastalarda Akdeniz Diyeti ve Dash Diyetine Uyumun Glisemik Kontrol Üzerine Etkisi* [Effect of Mediterranean Diet and Dash Diet Compliance on Glycemic Control in Patients with Type 2 Diabetes]. Yüksek Lisans Tezi. Hacettepe Üniversitesi Sağlık Bilimleri Enstitüsü, Toplu Beslenme Sistemleri Programı.

Kürklü, B. B. (2021). Akdeniz Diyeti ve Bileşenlerinin Osteoartrit ile İlişkisi [Relationship of Mediterranean Diet and Ingredients with Osteoarthritis]. *Beslenme ve Diyetetik Güncel Konular, 10*, 211–221.

Kuyumcu, A., & Kuyumcu, M. S. (2019). Akdeniz Diyeti ve Karotis Arter Hastalığı Arasındaki İlişki [The Relationship Between Mediterranean Diet and Carotis Artery Disease]. *Süleyman Demirel Üniversitesi Sağlık Bilimleri Dergisi, 10*(2), 99–104.

Oh, H.-C., Uysal, M., & Heung, C. (1995). Product bundles and market segments based on travel motivations: A canonical correlation approach. *International Journal of Hospitality Management, 14*(2), 123–137. doi:10.1016/0278-4319(95)00010-A

Öney, H. (2013). *Gastronomi turizmi. H. Öney içinde, Alternatif Turizm* [Gatronomy tourism. H. In Öney, Alternative Tourism]. Anadolu Üniversitesi Açıköğretim Fakültesi Yayınları.

Özgüç, N. (1996). *Turizm Coğrafyası* [Tourism Geography]. İstanbul Üniversitesi Edebiyat Fakültesi Coğrafya Dergisi.

Panagiotakos, D. B., Chrysohoou, C., Pitsavos, C., & Stefanadis, C. (2006). Association between the prevalence of obesity and adherence to the Mediterranean diet: The ATTICA study. *Nutrition (Burbank, Los Angeles County, Calif.), 22*, 449–456. doi:10.1016/j.nut.2005.11.004 PMID:16457990

Piers, L. S., Walker, K. Z., Stoney, R. M., Soares, M. J., & O'Dea, K. (2003). Substitution of saturated with monounsaturated fat in a 4-week diet affects body weight and composition of overweight and obese men. *British Journal of Nutrition, 90*(3), 717–727. doi:10.1079/BJN2003948 PMID:13129479

Rhazi, K. E., Nejjari, C., Romaguera, D., Feart, C., Obtel, M., Zidouh, A., & Gateau, R. B. (2012). Adherence to a Mediterranean diet in Morocco and its correlates: Cross-sectional analysis of a sample of the adult Moroccan population. *BMC Public Health, 12*, 345. doi:10.1186/1471-2458-12-345 PMID:22578133

Romaguera, D., Norat, T., Mouw, T., May, A. M., Bamia, C., Slimani, N., ... Cottet, V. (2009). Adherence to the Mediterranean diet is associated with lower abdominal adiposity in European men and women. *The Journal of Nutrition, 139*, 1728–1737. doi:10.3945/jn.109.108902 PMID:19571036

Sáez-Almendros, S., Obrador, B., Bach-Faig, A., & Serra-Majem, L. (2013). Environmental footprints of Mediterranean versus Western dietary patterns: Beyond the health benefits of the Mediterranean diet. *Environmental Health, 12*(1), 118. doi:10.3945/jn.109.108902 PMID:24378069

Sangpikul, A. (2008). Travel motivations of Japanese senior travelers to Thailand. *International Journal of Tourism Research, 10*(1), 81–94. doi:10.1002/jtr.643

Schröder, H. (2007). Protective mechanisms of the Mediterranean diet in obesity and type 2 diabetes. *The Journal of Nutritional Biochemistry, 18*, 149–160. doi:10.1016/j.jnutbio.2006.05.006 PMID:16963247

Şengül, S. (2015). *Akdeniz Bölgesi Mutfak Kültürü ve Yöresel Yemekleri* [Mediterranean Region Culinary Culture and Regional Dishes]. Beta Basım Yayım.

Şengül, S., & Türkay, O. (2014). *"Yöresel mutfaklar" eğitimi üzerinden kimlik, mutfak kültürü ve turizm gelişimine dair çıkarımlar: Mudurnu örneği* [Inferences on identity, culinary culture and tourism development through "regional cuisines" education: Mudurnu example]. Citta Slow Eko-Gastronomi Kongresi. Gökçeada.

Şengül, S., & Türkay, O. (2016). The evaluation of Mediterranean cuisine as a gastronomic tourism product. *Journal of Tourism and Gastronomy Studies*, *4*(Special issue), 86–99.

Sotiriadis, M. (2017). Pairing intangible cultural heritage with tourism: The case of Mediterranean diet. *Marios Sotiriadis*, *12*(3), 269–284. doi:10.1108/EMJB-07-2016-0019

Strisciuglio, C., Cenni, S., Serra, M. R., Dolce, P., Martinelli, M. A. S., & Miele, E. (2020). Effectiveness of Mediterranean diet's adherence in children with inflammatory bowel diseases. *Nutrients*, *12*(10), 3206. doi:10.3390/nu12103206 PMID:33092159

Taş, M., & Zengin, B. (2019). A research on university students' perception of destination image: The case of Kemaliye. *Research in International Business and Finance*, *1*(3), 205–216.

Taşdelen, A. H., Kara, A., Bozkurt, O., & Şaşmaz, Ö. (2020). Konaklama ve Seyahat Hizmetleri [Accommodation and Travel Services]. In Genel Turizm. Milli Eğitim Bakanlığı Yayınları.

Tilman, D., & Clark, M. (2014). Global diets link environmental sustainability and human health. *Nature*, *515*, 518–522. doi:10.1038/nature13959 PMID:25383533

TKD. (2021). https://sozluk.gov.tr/

Trichopoulou, A., Martínez-González, M. A., Tong, T. Y., Forouhi, N. G., Khandelwal, S., Prabhakaran, D., & Lorgeril, D. M. (2014). Definitions and potential health benefits of the Mediterranean diet: Views from experts around the world. *BMC Medicine*, *12*(1), 112. doi:10.1186/1741-7015-12-112 PMID:25055810

Tureac, C. E., & Turtureanu, A. (2008). Types and forms of tourism. *Economica*, *4*, 1.

UNESCO. (2010a). Representative list of the intangible cultural heritage of humanity. UNESCO.

Uysal, M., & Li, X. E. S.-T. (2008). Push–pull dynamics in travel decisions. In Handbook of Hospitality Marketing Management. Elsevier.

Wang, H., Livingston, K. A., Fox, C. S., Meigs, J. B., & Jacques, F. P. (2013). Yogurt consumption is associated with better diet quality and metabolic profile in American men and women. *Nutrition Research (New York, N.Y.)*, *33*(1), 18–26. doi:10.1016/j.nutres.2012.11.009 PMID:23351406

Yoon, Y., & Uysal, M. (2005). An examination of the effects of motivation and satisfaction on destination loyalty: A structural model. *Tourism Management*, 45–56. doi:10.1016/j.tourman.2003.08.016

Yousefi, M., & Marzuki, A. (2012). Travel motivations and the influential factors: The case of Penang, Malaysia. *Anatolia: An International Journal of Tourism and Hospitality Research*, *23*(2), 169–176. doi:10.1080/13032917.2012.662906

Chapter 5
Portuguese Mediterranean Diet as a Future Trend for Healthy and Sustainable Restaurant Consumption:
A Post-COVID-19 Perception

Helena Real
TOXRUN - Toxicology Research Unit, University Institute of Health Sciences, IUCS-CESPU, Portugal & Portuguese Association of Nutrition, Portugal

Rui Rosa Dias
https://orcid.org/0000-0002-5934-8429
ISAG - European Business School, Portugal & Research Center in Business Sciences and Tourism (CICET-FCVC), Portugal

ABSTRACT

The COVID-19 pandemic brought new challenges to the way the world thinks about the sustainability of the planet and also brought new opportunities to emphasize conviviality, familiar links, cultural heritage, and exchange of knowledge on intergenerational gastronomy. This chapter will explore the opportunity to promote the Mediterranean diet as a solution to the improvement of food habits in the post-COVID-19 era. It will explore the future trends of Mediterranean diet concept use, specially focused on food industry/distribution and tourism as an inspiration to impulse more healthy and sustainable consumption in Portuguese restaurants. Gastronomy and tourism also will be presented as a sustainability safeguard possibility for the Mediterranean diet.

INTRODUCTION

The COVID-19 pandemic brought new challenges to the understanding of food sustainable concept,

DOI: 10.4018/978-1-7998-9148-2.ch005

but also new opportunities to emphasize conviviality, familiar links, cultural heritage, and exchange of knowledge on intergenerational gastronomy. At the same time the agri-food sector (from production to consumption) was forced to introduce new ways to interact with the costumers. The very fast grow of all FDA -_food delivery applications (Uber Eats, Glovo, take ways, GrubHug and many others) are one example of it. New production and consumption patterns were assumed by all agri-food stakeholders, to face this new an uncertain global reality. In the same way, a radical change in the system in which we live is necessary (Petrini, 2017). In this context, restaurants and tourism in general were also under pressure by the costumers, which is why it could be an opportunity for these sectors to redesign new, healthier and more sustainable proposals, all closer to the cultural roots. To reestablish the Humanity "well-being", per capita consumption should be reduced. The climate, the food and the table will design an equation of balance which the "new" agri-food age, will forcing to the consumption adjustment (Rosa, 2017).

The main approach of this chapter was based on the historical exploration of the Mediterranean Diet and the various advantages of its adherence, as well a presentation of the main trends of the Mediterranean Diet concept uses and a graphic model of its presentation.

This chapter is supported in four dimensions: a) The importance of the Portuguese Mediterranean Diet; b) Food Trends; c) Healthy and Sustainable food consumption and d) Post Covid perception.

The specific main objectives of this book chapter are:

- Define and understand the Portuguese Mediterranean Diet as food consumption trend;
- Identify the main Mediterranean Diet factors that could be pointed as healthy and sustainable options;
- Make a theoretical reflection about how will be the new behavior consumption at the restaurant level in post covid era;
- Theorize concerning the impact of the Mediterranean Diet at the hospitality sector.

As methodology, this chapter is based on a review of the literature on the topics under analysis, using, in parallel, some specific scientific publications by the authors, following research projects carried out in the area. It was also used some application data from a Delphi study previously developed by the authors.

BACKGROUND

The COVID-19 pandemic brought a lot of challenges to the world. In all countries around the world, people were deprived of being with their families or friends, going to restaurants, traveling, among many other ways of society living. Some recent studies appointed the impact about COVID-19 effect at the food consumption in the isolation and lockdown period (Ammar et al., 2020; Coulthard et al., 2021; DGS, 2020b) and the future population perspective about tourism and food restaurant consumption in particular (Bhalla et al., 2021). Some authors perspective the higher search for nature travelling, as a kind of psychotherapeutic healing to those who suffered from mental health issues during COVID-19 (Bhalla et al., 2021) and others discuss about the COVID-19 transformational opportunity in the tourism sector (Sigala, 2020).

A recent publication from 2019 of the EAT-Lancet Commission on healthy diets from sustainable food systems highlighted the importance of the relation between food, environment, and health. They concluded that it is essential to have a transformation of food systems to preserve the planet and to feed an exponentially growing population in a balanced way (Willett et al., 2019).

The Mediterranean Diet can be part of the strategy to promote the food systems transformation, because it represents a very well-studied cultural model, with a positive impact on health, in the environment preservation, and cohesion between people (Dernini et al., 2017; Estruch & Bach-Faig, 2018; Martinez-Lacoba, et al., 2018; Real, et al., 2021a). The distinction awarded to the millenary Mediterranean Diet as an Intangible Cultural Heritage of Humanity by the United Nations Educational, Scientific and Cultural Organization brings new opportunities to promote this model and contribute to its preservation (UNESCO, 2013a).

The term "Mediterranean Diet" was firstly suggested by Ancel Keys in the 50s/60s of the 20th century, based on the results obtained in the Seven Countries Study (Keys, 1970, 1980; Trichopoulou & Lagiou, 1997). However, this reference would only be published in 1975, in the book "How to eat well and stay well. The Mediterranean way" from Ancel Keys and his wife Margaret. In this book they presented the main aspects of the diet population from Greece, Italy and the Mediterranean coast of France and Spain. This publication was preceded by another with a similar name "How to eat well and stay well" (1959), where food was not suggested as a different element of Mediterranean population (Moro, 2016). The Keys's research founds that the way of eating and living in the Mediterranean were different and impulses a better state of health. Nevertheless, the Mediterranean Diet had been prepared for thousands of years, since the stabilization of the climate in this region, about 8500 years ago (Berry, et al., 2011; Carvalho, 2015; Graça, 2015). The Neolithic brought several changes, mainly due to the ability to domesticate animals and cultivate plants, with the oldest records of domestication being precisely in the Mediterranean and Near East (Carvalho, 2015). Its typical climate and geography have also enabled the implantation of certain species of plants that are still characteristic of the Mediterranean, as the olive tree and the vine (Graça, 2014 ; Graça, 2015). Later, it was even possible to grow other plants because of the irrigation mechanisms brought by the Arabs (Graça, 2014; Graça, 2015). In addition, this region had an adaptive capacity to make grow species brought from other parts of the world, such as rice, corn, tomato, beans, pumpkin, pepper or potatoes (Carvalho, 2015; Moreno, 2015). On the other hand, along the years, the Mediterranean Sea allowed the exchange of knowledge, food, people, animals and utensils between Phoenicians, Greeks, Romans, Arabs and other peoples, which contributed to a unique and specific diversity and richness of this area (Graça, 2014; Graça, 2015; Trichopoulou & Lagiou, 1997).

Vegetable products were one of the predominant components of the diet of Mediterranean people and the smaller animals, with the ability to look for food in the inhospitable environments, were more common than bigger animals and represented a way to obtain another food resources such as cheese (Graça, 2015). In fact, there were in this region a multiplicity of resources that complemented each other, from fishing, hunting or livestock (Moreno, 2015). These people also used various food preservation techniques (salting, drying, smoking) to ensure greater food durability throughout the year (Graça, 2015; Moreno, 2015). They also included in their habits some seasonal foods or food of opportunity, such as small fruits (wild fruits and oleaginous fruits), aromatic herbs, snails and bivalves (Graça, 2015).

Through the Study of the Seven Countries of Ancel Keys, which intended to assess the prevalence of Cardiovascular Diseases and their determinants, it was possible to verify that the Mediterranean countries included in the study, Greece, Italy and Yugoslavia, had a different diet type and a different lifestyle from countries not included in this geographical area, which sparked the interest to better understand the advantages of the Mediterranean food pattern. Food in these countries had many characteristics in common, especially the strong presence of olive oil, commonly associated with the consumption of a good amount of vegetables. In Greece, values of fat consumption were reported in the order of 40% and in Italy about 30% of the total energy consumed. Besides these foods, it was possible to find cereals, especially wheat, as well as olives and fruits. The ratio between monounsaturated and saturated fat in these countries was much higher than in the other countries considered in the Seven Country Study (Keys, 1980). In addition, there was also a moderate consumption of alcohol – principally from wine -, high consumption of vegetables and fruits, cereals - including bread - low consumption of meat and meat derivatives, with preference given to fish in the zones coastlines, and moderate consumption of dairy products (Keys, 1980; Trichopoulou & Lagiou, 1997).

Since the 60s of the last century, scientific studies have an exponentially arise and demonstrate mostly the various benefits of the Mediterranean Diet, especially in terms of health. In addition to the recognized effect on the prevention of Cardiovascular Diseases (Keys, 1980; Keys & Keys, 1975; Trichopoulou et al., 2005, 2009; Zazpe et al., 2011), the Mediterranean Diet is now associated with the prevention of other chronic diseases such as obesity (Garcia-Fernandez et al., 2014; Gotsis et al., 2015), the Diabetes *Mellitus* (Koloverou et al., 2014; Martinez-Gonzalez et al., 2008), the Metabolic Syndrome (Babio et al., 2009; Esposito et al., 2007; Kastorini et al., 2011; Kesse-Guyot et al., 2013), some kind of cancers (Ginter & Simko, 2015; Tyrovolas & Panagiotakos, 2010; Verberne et al., 2010) or the neurodegenerative diseases (Huhn et al., 2015; Schulz & Deuschl, 2015; Solfrizzi et al., 2011).

In addition to food, there are several factors that may contribute to the multiple health and longevity benefits pointed out to the Mediterranean Diet in the literature.

The presence of snacks throughout the day may be one of the examples recently inserted in the habits of the populations of the Mediterranean, in result of the society evolution (Bellisle, 2009; Radd-Vagenas et al., 2017). On the other hand, the time and number of meals and their composition influence the state of satiety and also can influence the health (Bellisle, 2009). The siesta, which is disused in the Mediterranean, can also contribute to these benefits (Bellisle, 2009; Naska et al., 2007). Another important aspect is the circumstance that accompanies the meals, whether in terms of their culinary preparation, with fresh and seasonal products, or in terms of the conviviality around the table at the meal time. The pleasure of eating, associated with conviviality, eating at the table, with friends and/or family has always been a factor present in the initial speech associated with the Mediterranean Diet, as pointed out by Ancel Keys (Keys & Keys, 1975; Nestle, 1995). Nowadays, there are several studies that report that children's food consumption in front of the television is associated with a higher Body Mass Index (Bellisle, 2009; Moreira et al., 2010; Vik et al., 2013). The Mediterranean Diet would thus be assumed as a food model inserted in a complex lifestyle, imbued with characteristics such as the sea, the sun and the siesta (Bellisle, 2009).

It is necessary to reinforce that the success of the Mediterranean Diet is not only due to the way of eating. It may even represent a philosophy of what to choose for a meal and what to drink from it, along with an adequate practice of physical activity (Bonaccio et al., 2012). A reflection by Joan Reguant-Aleix and collaborators also mentions that the richness of the Mediterranean Diet consists in its simplicity, naturalness and imagination in the combinations between foods and flavors, which create different dishes with the same basic ingredients, with few resources and a lot of diversity (Reguant-Aleix et al., 2009). There are other important factors that contribute to the success of this model, related with some Mediterranean region specifications, such as the weather conditions that induce a higher content of antioxidants in plant foods, the socioeconomic conditions that induced a context of frugality in consumption (Mackenbach, 2007) or the specific culture of this area that influences gastronomy, cooking techniques or traditions (Altomare et al., 2013). It is even theorized that if one of the factors fails, the entire mechanism may collapse (Bonaccio et al., 2012).

Other areas beyond health, began to emerge in the Mediterranean Diet panorama, such as the sustainability, the gastronomy, the sustainable production of food, the conviviality, the tourism, among others (Real et al., 2021). For instance, Mediterranean Diet is frequently presented as a case study for the development of tools to assess and improve the sustainability of diets and dietary patterns (Dernini et al., 2013; Burlingame et al., 2015; Willett et al., 2019). Mediterranean Diet also suited well with the most recent international recommendations, as the sustainable development Goals.

In 2017, the following authors wrote a paper supported in a study with some experts from the Algarve region in Portugal, about the concept perceptions, with the aim of building an adequate and easily transmitted Mediterranean Diet definition (Real & Graça, 2019; Real & Graça, 2020). This study was expanded to an identification of the major future trends in Mediterranean Diet use by 2028, in Portugal (Real et al., 2021). Culture, education, environment, health, food industry/distribution and tourism were identified as the future trends of Mediterranean Diet use areas. This projection of trends can allow us to identify opportunities to implement the Mediterranean Diet concept, adjusted to the end consumer needs.

The authors also validated, with a Delphi panel, a graphic model about the Mediterranean Diet concept, published in 2021 (Real et al., 2021), based on four dimensions: food, conviviality, food sustainable production and lifestyle. "Food" lists elements as the central role of water on food and cooking; the chemical composition of food; the commensality; essentially vegetarian model (spotlight on cereals, legumes, vegetables, fruits, herbs, nuts and seeds); the frugality; the local production; the preservation models (salting, drying, pickles); the seasonality; the simplicity of cooking; the food transformation; or the trilogy bread-wine-olive oil. The "sustainable food production" combines factors related to agriculture (biodiversity, local production, production of species adapted to local ecosystems, sustainable production, traditional production); related to animal husbandry; or fishing. The "conviviality" support elements as the community; eating together; the family; the festivals and celebrations; sharing knowledge; the flavors and traditions; or the social and communication exchanges. The fourth dimension, the "lifestyle" translates into a concern with fresh and local products, seasonality and sustainable production; the Mediterranean way of living; the physical activity; the protection and enjoyment of the Mediterranean landscape; the respect and communion with nature; the nap; or the values (creativity, hospitality, intercultural dialogue, neighborliness, respect for diversity, sharing). The conjugation of these attributes represents a cultural model, with impact on areas such as the environment (climate; flora; food and sustainability), economy (circular economy; exchange; agreement and mutual respect; markets), health (health promotion food pattern; nutrition), and tourism (agro-tourism; hospitality; restaurants) (Real et al., 2021).

The graphic model proposed by Real et al can be used as a pedagogical tool, which can improve initiatives to safeguard and to promotion of Mediterranean Diet concept. It represents a contribute to the broadening of the discussion on the Mediterranean Diet, through the clarification of its concept and applicability, especially from the point of view of the various stakeholders in the area, as producers or concept users. Obtaining a high level of agreement for the future use of a conceptual model on the definition of the Mediterranean Diet will facilitate the creation of tools that can contribute to this same greater applicability. However, it is important to highlight the difficulties of use this model because of the coexistence of very different areas in the same conceptual model. In this way, it will be important to raise awareness for a more conscious use of the concept of the Mediterranean Diet by those involved on its promotion. During more than 40 years the Mediterranean Diet investigation and promotion were essentially centered in the food pattern and its influence in health (Real & Graça, 2019), so the concept may have become more reductive over the years. The Mediterranean Diet distinction from United Nations Educational, Scientific and Cultural Organization as intangible cultural heritage of humanity opened up new perspectives and highlighted the importance of other pillars besides food. However, it brought an enormous challenge, which was to combine all the elements in the same concept, as a way to preserve it.

MAIN FOCUS OF THE CHAPTER

This chapter will explore the COVID-19 effects on food consumption during the lockdown period and the opportunity to promote the Mediterranean Diet as a solution to improve new food habits. It will be explored the future trends of Mediterranean Diet concept use, specially focused on food industry/distribution and tourism as an inspiration to impulse more healthy and sustainable consumption in Portuguese restaurants. Gastronomy and tourism also will be presented as a sustainability safeguard possibility for Mediterranean Diet.

Issues, Controversies, Problems

In 2019, EAT-Lancet Commission published a reflection about the healthy diets from sustainable food systems, where this Commission highlighted the importance of the relation between food, environment and health (Willett et al., 2019). Nowadays, the humans are the dominating driver of change on our planet and the population is growing (Willett et al., 2019), so this report can improve the discussion about the better strategies to enhance the food systems and the population diets. The report authors suggested the Mediterranean Diet as a possible solution to this situation, because it represents a very well-studied cultural model, which allows a positive impact on health, in the environment preservation, and cohesion between people. For the same reasons Mediterranean Diet also fits well with some other current world agenda, that is the Sustainable Development Goals, adopted by all United Nations Member States in 2015 (United Nations, 2015). It represents 17 Sustainable Development Goals mainly about people, the planet, peace, prosperity, and partnership (United Nations, 2015). Mediterranean Diet is also been used to the adaptation of national food guides, as the Portuguese example – the Mediterranean Food Wheel (DGS, PNPAS, DGC, & FCNAUP, 2016).

Although it is currently common to see that the Mediterranean Diet is often used as a recommendation for use or as an example of good practices, in fact, we have been observing a decrease in its adherence over time. A worldwide study about the trends of Mediterranean Diet adherence between the periods 1961–1965, 2000–2003 and 2004–2011 shows that the Mediterranean Diet significantly decreased between 1961–65 and 2000–03, and between 2004–2011 there was a stabilization of the values and an increase among 16 countries (Vilarnau et al., 2018). However, it must be specially considered the decrease in the Mediterranean countries, which can improve the reflection about the safeguard action in these countries. In Portugal the values of adherence to the Mediterranean food pattern were reported to be about 18%, according to data from the last national food and physical activity survey (Lopes et al., 2017). However, a recent study from Directorate-General for Health verified that the Portuguese population had an increase on the Mediterranean Diet adherence to 24% (DGS, 2020a).

It is important to promote and preserve this millenary model, so is imperious to find strategies that achieve those goals. The fact that the Mediterranean Diet concept definition is complex and not always passed in the best way, can contribute to this loss of adherence. This also complicates the information transmission to the population and the safeguard of the correct model. Nowadays, UNESCO promotes the consensual Mediterranean Diet definition (UNESCO, 2013a), although it is quite extensive and complex.

2020 was a year marked by the emergence of the pandemic by COVID-19. This pandemic fostered reflection on many issues, including environmental ones, beyond all the habits changes inherent to the lockdown and social isolation necessary at this stage. In Portugal, a study from Directorate-General from Health verified that 45.1% of Portuguese population changed their food habits and 41.8% had the perception that the changes were for the worse, during the first lockdown (DGS, 2020b). The reasons given for these changes include modifications in working hours (17.6%) and in the food purchase model (34.3%). On the other hand, the changes in habits also seemed to be due to the stress experienced (18.6%) and changes in appetite (19.3%). A third group of factors included the relationship with the economic situation (10.3%) (DGS, 2020b). In general, there was an increase in the home cooking meals and a decrease of pre-prepared meals (40.7%) and takeaway (43.8%). There was also an increase in the consumption of fruit (29.7%) and vegetables (21%). In addition, most respondents revealed that they started to cook more (56.9%), but also to snack more (31.4%) (DGS, 2020b). The results obtained allowed to identify two types of patterns in terms of behavior in the area of eating and physical activity. On the one hand, a protective health pattern, where higher levels of physical activity were verified, along with a non-increase in the consumption of snacks and healthy eating in general. This pattern was associated with a higher level of education and better socioeconomic conditions. On the other hand, it was verified the existence of a risk pattern, where low levels of physical activity and simultaneously increased consumption of snacks and soft drinks were verified. This pattern was also associated with a lower level of education and poor financial situation (DGS, 2020b).

The literature also pointed some food habits changes in other countries worldwide, associated to the lockdown situation and social isolation, as examples of studies from Italy (Di Renzo et al., 2020; Scarmozzino, 2020), Spain (Rodríguez-Pérez et al., 2020), França (Deschasaux-Tanguy et al., 2021).

The effect of these changes on children takes on a special highlight, because may aggravate childhood obesity situations, as suggested in the study by Nicodemo et al (Nicodemo et al., 2021). Still, it is possible to show that in many cases, in different countries, there was an increase in the consumption of food made at home and more importance was given to cooking at home (Deschasaux-Tanguy et al., 2021; DGS, 2020b; Di Renzo et al., 2020; Rodríguez-Pérez et al., 2020; Scarmozzino, 2020). This kind of habits change could improve some Mediterranean values, as the family meeting at food table, the traditions related to food and food recipes, the recognition of local production and the national gastronomy. The commensality or eating together with other people it is an important manifestation of human society, including the cultural aspects and the people health (Fischler, 2021).

The COVID-19 pandemic may have brought about some kind of transformation in the way people value their cuisine, their cultural roots and their identity. It also important to investigate the extent of changes in the commensality habits, because it was changed by the government lockdown imposition and restrictions in the number of people in the restaurant tables (Yngve et al., 2021).

Certainly these aspects can also be used by tourism and restaurants in order to improve and enhance the gastronomic offer and cultural expressions that convey the Mediterranean Diet concept. Alongside a greater recognition of identity, there was also a greater concern with health, as the Mediterranean Diet is also a health vehicle, it will be an important combination to be explored by Tourism and certainly, specifically by restaurants.

Mediterranean Diet: Concept and Future Trends of Concept Use, in Our Perspective

The perception of the Mediterranean Diet associated with culture and other contexts, that are difficult to measure, brings new opportunities but also some difficulties for those who want to use the concept pedagogically. The connection of the concept to health, allowed an easier transmission of those opportunities, nevertheless, it represents a small part of the equation. In this sense, it is expected that various areas of knowledge interconnect and create synergies in this context of passing on the message to the population, namely to children, as it is one of the most recommended ways of promoting and preserving the Mediterranean Diet.

This chapter subtopic presents some results of an investigation of these authors about the identification of the major future trends in Mediterranean Diet, in Portugal and its contribution to the inspiration of other countries. Culture, education, environment, health, food industry/distribution and tourism were identified as the future trends of Mediterranean Diet use areas and it was published 19 trends.

The trends published by Real et al, 2021 (Real et al., 2021), in the area of Food industry/Distribution and Tourism, listed the following tendencies:

a) In 2028, there will be a fiscal incentive that will allow the food products typical of the Mediterranean Diet to be cheaper than those considered outside this food standard.
    i. In Portugal, currently there are rates applied to foods with lower nutritional value, such as soft drinks, but there isn´t a policy about lower prices for Mediterranean food. Around the world there are some good examples about this kind of policy. For instance, in Australia there is an exemption from tax for basic food (including fresh fruit and vegetables) (Veerman JL, 2013).
b) In 2028, one of the communication trends of i n the agri-food industry will be the inclusion of the concept of Mediterranean Diet.
    i. In 2019 the main trends in product communication were based on sustainability, with focus at the level of environmentally friendly packaging, more natural products and from more sustainable production, with a deviate from the artificial (data provided by PortugalFoods). In 2021, according to the international specialized consultancy Mintel, Covid-19, exponentially accelerated the innovation trends that were foreseen for the next decade, due to the profound impact it has on society. The major trends to 2021 identified were: Wellness - feed the mind; Value, that redefine the product quality (the Ethical expectations accelerate 'value with values' and the experience in retail and restaurants evolves, because people wants more hygiene and safety); and organizing into communities for socialization (PortugalFoods, 2021).
c) In 2028, 50% of the large distribution chains will use the concept of Mediterranean Diet for the communication of their products and services.
    i. Currently only one distribution company uses expressively the Mediterranean Diet concept in its communication about food products and services. Currently there are five large distribution companies in Portugal that represent at least 70% of the national market (data provided by Portuguese Association of Distribution Companies).
d) In 2028, 5% of Portuguese tourist hotel accommodations will seek to incorporate the concepts of Mediterranean Diet into the provision of space and services, such as meals.
    i. There is no data to report the current number of Units that incorporate these principles, but we estimated that it is less than 1%. In 2017, Pordata reported the existence of 5,840 tourist

accommodations in Portugal. Since that date, there has been an increase until 2019, with a subsequent reversal of the growth trend, possibly due to the COVID-19 pandemic (Pordata, n.d.).

e) In 2028, there will be a quality seal that certifies the use of the concept of Mediterranean Diet for Portuguese restaurants.
   i. In Spain it is used a seal in restaurants with the Mediterranean Diet concept and join a network recognized by the Mediterranean Diet Foundation (Mediterranea). There is currently no recognized seal attesting to the use of the Mediterranean Diet concept in Portugal.

f) In 2028, 10% of the new tourist offerings will position themselves and communicate their insertion in the concept of Mediterranean Diet, together with a model of sustainable tourism and/or a vehicle of conviviality through the promotion of festivities and traditions.
   i. Nowadays there is a varied tourist offer within the scope of gastronomic routes, like the thematic visits to the harvest, the fruit picking, the olive picking, the festivities associated with collective celebrations, among others. However, there is no reference to the Mediterranean experience.

g) In 2028, there will be more than 500 agro-tourism units in Portugal.
   i. In 2017, there were 230 agro-tourism units in Portugal, representing an increase of 94 units since 2007, according to Pordata. Since that date, there has been an increase until 277 units in 2019, but in 2020 there was a decrease to 237 units, possibly due to the COVID-19 pandemic (Pordata, n.d.).

The trend "In 2028, 5% of Portuguese tourist hotel accommodations will seek to incorporate the concepts of Mediterranean Diet into the provision of space and services, such as meals" had the biggest agreement percentage from the experts of the Delphi panel developed. The trend about agro-tourism units and the trend about the tourism offerings had also a high agreement percentage and it is in the list of the top ten positions. In this way, it was possible to verify that almost all the panel experts consider that tourism can be an area of potential insertion and recognition of the Mediterranean Diet.

Renna et al recently hypothesized the possibility of a "next generation" of food products creation that promote the Mediterranean Diet, from the fusion of agrobiodiversity and innovations in both gastronomy and agri-food sector, where the sustainability, high-quality production and traditions were linked (Renna M, 2021).

Kim and Eves developed a motivational scale to understand the motivational dimensions of local food consumption, which has five dimensions: cultural experience, excitement, interpersonal relation, sensory appeal and health concern (Kim & Eves, 2012). These motivations can help the tourism sector to create new communication and innovative activities to enhance the tourist adherence to local food options.

Mediterranean Diet principles is considered as a global health tool but is also a cultural model where the conviviality is a central idea, reinforcing the importance to have pleasure of shared meals. Probably the description of a "palatable" diet is the key characteristic that promote the difference of this diet from others (Phull, 2015). This importance of pleasure and conviviality has a perfect match with sector of hospitality, so the Mediterranean Diet can have a good margin of progression in this sector.

Mediterranean Diet: Gastronomy and Tourism, a Sustainability Safeguard Possibility

Nowadays, "sustainability" is currently a hot topic, being part of the political agendas in general and of several intervention programs in the community. The Mediterranean Diet is presented by several investigators as a model that promotes more sustainable eating habits and behaviors. This topic will address the main indicators of the Mediterranean Diet that make it an example of food sustainability.

Gastronomy contains a strong cultural component of a region or a country. The Mediterranean Diet has a gastronomy characteristic, which preserves the nutrients of the food, are intended to be frugal, produced in the proximity and in season. Therefore, it will be something to be promoted by the restaurants, as a vehicle for the local identity and traditions of each region or country. At a time when people give greater value to this type of offer, due to the restrictions it had in the current pandemic, it could be a proposal to increase demand and the economy. In this context, there are several examples that reveal good practices and can be presented in this subtopic.

Through gastronomy, traditions, festivities, hospitality, crafts and culture in general linked to the Mediterranean Diet, the tourism sector itself can be reinvented and leveraged with the Mediterranean Diet and may be an axis still expanding in this regard.

Actually, the sustainability is one of the main drives to the younger people and some new restaurants are aware of this concern following this trend. In addition, other benefits that restaurants can achieve with these changes should be highlighted: comply with legislation in terms of single-use plastics or selective collection of bio-waste; savings resulting from the elimination/reduction of edible food waste (FoodSave, n.d.); or saving on energy costs by implementing energy efficiency measures (Trust, 2008). These kind of changes also can be responsible for tangible economic benefits for restaurants, in the short to medium term, which can act as an incentive for change and strategic planning when developing the restaurant concept. Sustainable or thematic restaurants can catch the attention of more conscientious consumers, but they can also inspire other restaurants to change. Nowadays, there are some sustainable restaurants examples in Portugal or in other countries that are recognized by their specifications: "A Cozinha" Restaurant (Guimarães, Portugal) (Restaurante A Cozinha, n.d.); "Kitchen Dates" Restaurant (Lisboa, Portugal) (Restaurante Kitchen Dates, n.d.); "Nolla" Restaurant (Helsinki, Finland) (Nolla Restaurant, n.d,), "Silo" Restaurant (London, United Kingdon) (Silo Restaurant, n.d.) or "Azurmendi" Restaurant (Larrabetzu, Spain) (Azurmendi Restaurant, n.d.).

On the other hand, there are many distinctions or recognitions, national or international, that can be attributed to restaurants that have more sustainable characteristics. Exemples of these seals in Portugal are: "Embrulha" Project (Lipor, n.d. a); "Dose Certa" Seal ("Lipor, n.d."); "Restaurante FOOD" Seal (Food Programme, n.d.); "100% local" Seal (Loco, n.d.); "Alimentação Saudável no Ensino Superior" Seal of Excellence (Nutrimento, 2019); "Selo Saudável" (Almeida et al., 2016). Examples of these seals around the world are: Food that doesn't cost the Earth (Advance London, 2020); Green Globe (Green Globe, n.d.); Green Key (Green Key, n.d.); Checklist of good environmental practices for food services (Colares et al., 2018); Healthy Acting Seal - Food Services (Agir Sustentável, n.d.); Sustainability indicators in foodservice (Maynard, 2020).

Regarding the distinction applied to Mediterranean restaurants, there is a distinction from Spain, currently comprising more than 740 establishments, present in more than 160 Spanish municipalities. This kind of restaurants represents the oldest chain of Mediterranean restaurants, and its main objective is to promote Mediterranean food as a healthy food model to be promoted in restaurants. Restaurants in this chain - *Establecimientos Promotores de la Alimentación Mediterránea* - must follow a set of criteria associated with Mediterranean food, appearing in an interactive list, easily accessed by users who want to access this type of offer (Amed, n.d.).

In Portugal, there is also an approach to restaurants that follow a Mediterranean style, being present at Algarve Mediterranean Diet Route, an initiative framed in the Safeguarding Plan for the Mediterranean Diet, promoted by the In Loco Association and other members of the Regional Committee of the Mediterranean Diet (Rota da Dieta Mediterrânica, n.d.).

A more complete approach for restaurants could include the criteria linked to Mediterranean restaurants and the sustainability criteria presented in the examples above. It could thus function as a more complete approach and more truly focused with the broader concept of the Mediterranean Diet, which also considers sustainability issues. Recently, was published a checklist with the main sustainability criteria for application in Portuguese or European restaurants. It intends to represent a tool that can help the restaurants that wants to have or improve a sustainable concept (Real, et al., 2021).

The use of the Mediterranean Diet concept can has an important impact at the hospitality sector. There are multiples examples about the use of the concept in the restaurants or applied to tourism. The authors list in this chapter 15 suggestions of this application:

- Use more seasonal, fresh and local products in restaurants and hotels, in their food offer.

Sustainable production represents one of the pillars of the Mediterranean Diet (Real et al., 2021). In this context, promoting the consumption of seasonal, fresh, and local products allows for a reduction in the environmental impact of food, as well as boosting the local economy and a commitment to the surrounding society. There are currently numerous examples around the world that seek to enhance this type of offer, and an example is presented in the hospitality from Greece. The "Greek Breakfast" was promoted by the Hellenic Chamber of Hotels and represents a way to support the local producers and boost the local economy. The local culinary and local products were incorporated in the breakfast offer, enabling and enhancing an important relationship between local producers and hotels, promoting, further, an enrichment of the cultural and gastronomic experience of the users (Kyriakaki, et al., 2016).

- Increase the use of national products and products from a more sustainable production.

In line with the previous topic, the consumption of more products from sustainable production is one of the principles from Mediterranean Diet, so the restaurants and the tourism have an important role in this matter.

- Create unique and authentic experiences for tourists in the restaurants, hotels or in the production fields that supply the restaurant/hotel.

The supply of quality food products whose production is known is an asset for any restaurant that wishes to highlight these aspects. Knowing how to tell the story of each food served demonstrates an involvement appreciated by many customers. On the other hand, restaurants and hotels are also starting to suggest the involvement of customers in the food production process with the producer or simply visiting the production units. The grape and olive picking are older examples, but it is beginning to be a practice that can be extended to other productions, like bread, cheese, vegetable production, among others.

The "Greek Breakfast" project also brings some examples of these practices and the synergies between producers and tourism agents (Kyriakaki et al., 2016).

- Communicate the most sustainable food choices on the breakfast, lunch and dinner menus.

Changes made in a restaurant can always be communicated to the consumers, as this way they are more aware of its efforts, but also because they can feel more committed to the advances that the restaurant can achieve.

- Use the menu as a vehicle to communicate the Mediterranean Diet to people. The menu represents how the restaurant presents its food offer to consumers, but it can also represent a vehicle of communication, where the origin of the products, their nutritional value, elements about the history of the products, exchange with the culture can be inserted of the region, etc. The menu can therefore act as a good way to improve the food literacy of consumers and tourists in particular.
- Increase the presence of plant-based products and reduce animal-based foods in hotels and restaurants.

The population is increasingly seeking to have a greener and more flexible diet were the meat products lose strength. However, a recent study in Australia shown that people prefer increase their whole grain consume, over meat, instead of increasing the consumption of foods that try to mimic the meat consumption experience (Estell et al., 2021). This kind of information can be useful for the food industry and also to restaurants that want to develop the offer that best suits consumer preferences.

Try to use seals that distinguish the concept of Mediterranean Diet or the concept of sustainability.

Local gastronomy can contribute to a region or country brand. However, each restaurant can look for its own distinction, taking in account its particular characteristics and its stage of development of a certain concept or commitment. Sustainability has been increasingly promoted, and there are currently numerous seals in this area, as described above.

Create spaces for food production in hotels or restaurants (next to, on the roof, etc.) to minimize food travel.

The rooftop production is not new. It represents a smart way to embrace urban agriculture and promote a food miles decrease, inspiring people to eat local and seasonal. Rooftop to Tabletop can be a trend and also the production of their own food, in proximity. The existence of a vegetable garden associated with the restaurant is even one of the parameters valued in the Green Key seal ("Green Key,"). In addition, this culture can be a place where the restaurant or hotel deposits the result of the composting that can be carried out promoting a circular economy.

Promote the circular economy through the elimination of waste, waste recovery, use of by-products, among others.

The Circular Economy is materialized in the grouping by flows (of materials, energy and water), which are closed, promoting continuous reuse and repair (Advance London, 2020; Stahel, 2016). This approach, in addition to avoiding the creation of waste, can also lead to a greater appreciation of food products and their surroundings. The theme has been given great prominence and is currently the target of an intervention strategy by the European Union (European Commission, 2020).

Currently there are circularity analysis tools ("Circularity Check"), within the scope of the City Loops project, and it is expected that a version applied to the restaurant sector will be developed (Cityloops, n.d.).

Reduce food waste through the reduction of portions served, appealing to frugality.

The Food and Agriculture Organization (FAO) estimates that annually 1/3 of food for human consumption produced worldwide is wasted (FAO, 2011). In the European Union, it is estimated that the equivalent of 20% of all food produced is wasted (United Nations, 2017). At the level of restaurants, it is estimated that food waste arises from situations such as food preparation, food deterioration and leftovers from customers' dishes (Wrap, 2013). The reduction of food waste represents one of the main challenges in all sectors, so the catering sector must also continue to work towards the best possible management of this matter.

Favor the gastronomic experience of tourists based on the local food supply, culture and traditions of the region.

Tourist agents are beginning to realize that tourists value knowledge of the region and seek to discover local flavors and aromas. The Mediterranean area, well known for its gastronomic aspect, has been exploring this approach very well. An article from Sánchez-Torres et al demonstrates how the Mediterranean Diet culture affects chefs in Barcelona restaurants, revealing their involvement and commitment to promoting this cultural and food model (Sánchez-Torres et al., 2017).

Increase the number of restaurants, hotels and agro-tourism units that demonstrate greater concern with sustainability.

With the growing interest in sustainability issues, the tourism offer that incorporates its principles has also been growing. The pandemic brought greater awareness to issues such as the environment, proximity to nature, interest in traveling and getting to know different cultures, which can help boost new concepts and hotel units.

A recent study in Italy shows that the environmental sustainability, among other motivations, can influence Italian tourists' frequency to experience local food and beverage in agri-tourism destinations (Testa et al., 2019).

Develop tourist guides and routes linked to the concept of the Mediterranean Diet and/or Mediterranean gastronomic experiences.

This kind of offer can be include or be included in other route or guide type developed by the tourism regions. However, two examples could be the AMED - Establecimientos Promotores de la Alimentación Mediterránea – in Spain (Amed, n.d.) and the Mediterranean Diet Route in the Algarve region, from Portugal (Tavira).

Create passports for the experimentation of restaurants and hotels that follow the principles of the Mediterranean Diet.

The idea of the passport has been increasingly promoted as a tourist tool. In fact, it is a way to draw attention to a set of places, cultures, traditions, and other aspects that create curiosity to know the various points indicated in the passport in order to stamp it. Also noteworthy is the interest that this type of tool can have for the whole family, as it can thus involve and motivate family members of different ages. In Portugal, one of the passports that has attracted the most interest is the National Road 2 passport (a road that crosses the country from north to south, covering practically the entire longitudinal axis of the country) (RotaN2, 2021). In addition to the passport being stamped in 35 municipalities, there is a tourist guide that identifies the main places around the National Road 2 across the country, presenting aspects related to culture, traditions, gastronomy, hotels, the landscape, among others (Foge Comigo, 2021). This guide can represent a good outline of what can be found in terms of the Mediterranean Diet on this route through the country. Although no such identification is made, many of the aspects of the concept are described in this Guide, showing that it is possible to find the manifestation of the Mediterranean Diet throughout the Portuguese territory.

Increase and/or maintain the existence of cultural and food festivals and fairs.

Many festivals or fairs held throughout the year often have a character associated with the gastronomy and/or culture. In addition to representing an ode to the culture and traditions of each region, they can also act as good drivers for local restaurants and hotels. The pandemic brought about the cancellation of many of these festivities in the last two years, so people around the world probably gave more value to these celebrations, to the conviviality associated with them, to sharing and to highlight what best has each region or country. People started to give more value to the places where they lived and the tourist offer of their country. In this sense, the organization of these festivals or fairs is increasingly becoming a tourist tool to be promoted in other countries, but also for tourists residing in their own country. In Portugal, linked to the Mediterranean Diet, the Mediterranean Diet Fair, promoted by the Municipality of Tavira, represents a way of highlighting the cultures, traditions, gastronomy, research and projects of the Algarve area within the scope of promotion and safeguarding the Mediterranean Diet (Mediterrânica, n.d.).

Figure 1presents the systematization of the suggestions above, about the use of the Mediterranean Diet concept in the restaurants or applied to tourism, based in some examples from Portugal and other places in the world.

SOLUTIONS AND RECOMMENDATIONS

The choice of the tourist destination can be motivated by the gastronomy. This so-called "Gastronomic Tourism" is estimated to be responsible for 15% of the chosen places (Berbel-Pineda, 2019). The gastronomy can influence the choice of the place and the tourist gastronomy experience can influence the satisfaction of their destination (Berbel-Pineda, 2019).

Eating local foods can be a secondary experience to tourists in destinations that are not famous or promoted as a good local gastronomy. However, there is a potential food role in directly providing or indirectly supporting the tourists experiences, so it can be interesting to promote the gastronomic experience when this is not the primary motivation to visit (Almeida & Garrot, 2017).

Gastronomy can help the local economy, the products of each region, the social and cultural development of a region, creating new opportunities for future generations. In this sense, it will be essential to promote a "brand" associated with the cuisine of each region or country, which can translate the culture and the history of those regions. On the other hand, it will be possible to promote a greater connection between tourists and food producers and the places where they are sold or cooked (Trihas, et al., 2015). In the other hand, if we could associate the Mediterranean Diet to this concept "brand" of, we can promote more healthy and sustainable choices.

Roberto Burdese, from the Slow Food movement in Italy, wrote in 2012 that the Mediterranean Diet is not just a set of foods associated with a well-defined geographic space. It is also an important educational tool for more conscious consumption (CIHEAM, 2012). So, it will be important for health professionals, from the agri-food sector and from the catering/tourism sector to understand the concept of the Mediterranean Diet in order to be able to guide their customers, products, services or initiatives along this axis.

Figure 1. Fifteen suggestions about the use of the Mediterranean Diet concept in the restaurants or applied to tourism
Source: Real & Rosa (2021).

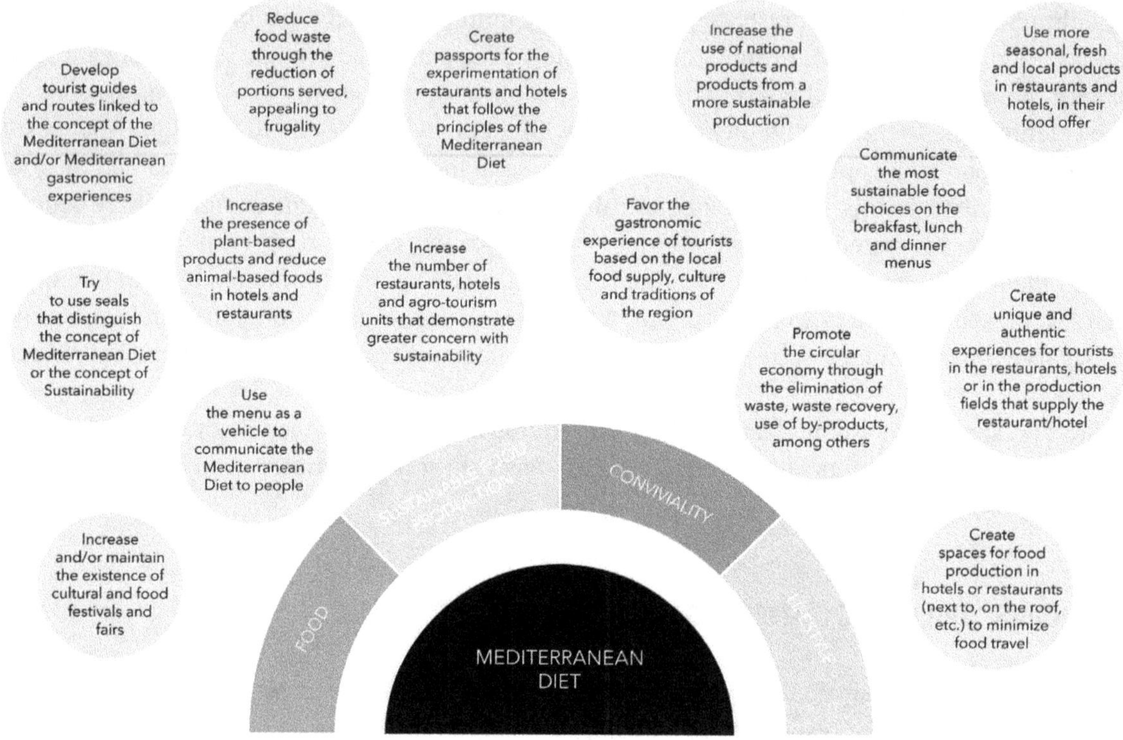

FUTURE RESEARCH DIRECTIONS

Considering the accelerated growth of communication around the Mediterranean Diet since the distinction awarded by UNESCO and the increase in its visibility and importance by the population, one of the problems to be highlighted leads us to potential misuses of the concept of the Mediterranean Diet. A study carried out by Real et al (Real & Graça, 2019), with professionals from various areas of the Algarve, Portugal, assessed their perceptions about the misuse of the Mediterranean Diet concept. It was found, in the first instance, that it was not a subject on which they had given much thought in the scope of discussions on the Mediterranean Diet. However, the identification of areas such as food, gastronomy, the food industry and food advertising was transversal, motivated above all by the commercial use of the concept, as the main areas where there would be a more inappropriate or distorted use of the Mediterranean Diet concept. This identification resides mainly in the areas involved with the food component of the concept, which can be seen, as it is the most explored and recognized area.

In this way, if we want to promote a growth in the use of the concept in the area of gastronomy and tourism, it will also be important to bet on an adequate training plan for professionals in the sector for a more adequate use of the concept and work on a plan to monitor the use of the concept in order to prevent it from becoming distorted. Besides that, it can be also important to invest in the basis of Nutritionists and Tourism formation, about nutrition and health tourism, because a better work will be made by properly formed and conscious professionals.

Future research could focus in the impact of Mediterranean Diet concept use in the tourism sector, with exploration of activities with greater impact, the perceptions of tourists and their appreciation of the tourist offer, in order to improve the approach.

It would also be interesting to study the relationship of the Mediterranean Diet with Slow Food's universal principles, since both concepts respect and promote seasonal and local production and consumption. In this context, we would widen the fields of intervention, ensuring new thoughts.

CONCLUSION

This chapter could represent an incentive to other researchers and professionals to explore some areas from Mediterranean Diet that are not yet very well explored and also inspire other academics and practitioners to embrace sustainable food studies approaches. In addition, in terms of gastronomy and tourism, it can represent a reinforcement of its value propositions and a way of enhancing the products and services of each region, which can not only impulse the local economy and culture, but also improve the safeguard of the Mediterranean diet essence in each region.

Travelling can influence positively the physical, emotional, and spiritual well-being of people, so this sector could be improved in the future and the demand tourism type could be different. Perhaps, also, something like "the Mediterranean tourism" offer could arise in this context. Eating, as explained before, is part of the culture of each country. Additionally, there are a real opportunity to explore the Mediterranean Diet not only as a contribute to the culture, but mainly, as a gastronomy tourism offer with his own idiosyncrasies.

The distinction of UNESCO as Intangible Cultural Heritage of Humanity has consolidated the Mediterranean Diet as a cultural and dynamic model that must be preserved in the context of a society that evolves over time, in order to guarantee the safeguarding of the basic premises of the concept for this do not become distorted in the course of the evolution of society. And this in itself is an immense challenge for which professionals from different sectors must be aware and prepared.

It will be essential to understand the future impacts of the concept of the Mediterranean Diet on tourism and gastronomy, ensuring that this concept can be used properly and not misrepresented. It will be essential to use it in order to promote the sector and highlight aspects that promote the concept, but also contribute to its safeguarding and correct sharing with future generations.

On the other hand, there is a need to apply a model that the tourism sector understands, in order to use it properly, and that makes use of a historical, ancestral legacy, currently validated by UNESCO, using it for the common good and the balances of the agri-food value chain. In this sense, it might be useful to emphasize the need to promote a collaborative rather than competitive sense with regard to the agri-food and gastronomy sector.

It will be very important to reinforce the axis of gastronomic sustainability as a way to develop the future.

Finally, with this Mediterranean diet holistic approach, there are a real opportunity to encourage the policy makers to change the way that we are seeing the agri-food sector and the gastronomy.

ACKNOWLEDGMENT

This research received no specific grant from any funding agency in the public, commercial, or not-for-profit sectors.

REFERENCES

Advance London. (2020, June). *Food that doesn't cost the earth.* https://www.foodmadegood.org/wp-content/uploads/2020/06/Food_that_doesnt_cost_the_earth_Full_guide-1.pdf

Agir Sustentável. (n.d.). https://agirsustentavel.com.br/

Almeida, A., & Garrod, B. (2017). Experiences with local food in a mature tourist destination: The importance of consumers' motivations. *Journal of Gastronomy and Tourism*, *2*(3), 173–187. doi:10.3727/216929717X14870140201116

Almeida, A. M., Gouveia, A. T., Prazeres, C., Ramos, S., & Graça, P. (2016). *Manual "Selo Saudável"* [Manual "Healthy Seal"]. Academic Press.

Altomare, R., Cacciabaudo, F., Damiano, G., Palumbo, V. D., Gioviale, M. C., & Bellavia, M. (2013). The mediterranean diet: A history of health. *Iranian Journal of Public Health*, *42*(5), 449–457. https://www.researchgate.net/publication/242018034 PMID:23802101

Amed. (n.d.). *Amed de establecimientos promotores de la Alimentación Mediterránea* [Amed of establishments promoting Mediterranean Food]. http://www.amed.cat/home.php

Ammar, A., Brach, M., Trabelsi, K., Chtourou, H., Boukhris, O., & Masmoudi, L. (2020). Effects of COVID-19 Home Confinement on Eating Behaviour and Physical Activity: Results of the ECLB-COVID19 International Online Survey. *Nutrients*, *12*(6). Advance online publication. doi:10.3390/nu12061583 PMID:32481594

Azurmendi Restaurant. (n.d.). https://azurmendi.restaurant

Babio, N., Bullo, M., & Salas-Salvado, J. (2009). Mediterranean diet and metabolic syndrome: The evidence. *Public Health Nutrition*, *12*(9a), 1607–1617. doi:10.1017/S1368980009990449 PMID:19689829

Bellisle, F. (2009). Infrequently asked questions about the Mediterranean diet. *Public Health Nutrition*, *12*(9a), 1644–1647. doi:10.1017/S1368980009990498 PMID:19689834

Berbel-Pineda, J., Palacios-Florencio, B., Ramírez-Hurtado, J. M., & Santos-Roldán, L. (2019). Gastronomic experience as a factor of motivation in the tourist movements. *International Journal of Gastronomy and Food Science*, *18*, 100171. doi:10.1016/j.ijgfs.2019.100171

Berry, E. M., Arnoni, Y., & Aviram, M. (2011). The Middle Eastern and biblical origins of the Mediterranean diet. *Public Health Nutrition*, *14*(12a), 2288–2295. doi:10.1017136898001 1002539 PMID:22166186

Bhalla, R., Chowdhary, N., & Ranjan, A. (2021). Spiritual tourism for psychotherapeutic healing post COVID-19. *Journal of Travel & Tourism Marketing*. Advance online publication. doi:10.1080/10548408.2021.1930630

Bonaccio, M., Iacoviello, L., de Gaetano, G., & Moli-Sani, I. (2012). The Mediterranean diet: The reasons for a success. *Thrombosis Research*, *129*(3), 401–404. doi:10.1016/j.thromres.2011.10.018 PMID:22100317

Burlingame, B., Gitz, V., & Meybeck, A. (2015). *Mediterranean food consumption patterns: diet, environment, society, economy and health* [White Paper]. FAO.

Carvalho, A. F. (2015). Domesticação e difusão: as origens da Dieta Mediterrânica [Domestication and dissemination: the origins of the Mediterranean Diet]. In Dimensões da Dieta Mediterrânica: Património Cultural Imaterial da Humanidade. Universidade do Algarve.

CIHEAM. (2012). *Mediterra 2012. The Mediterranean Diet for Sustainable Regional Development*. Presses de Sciences Po.

Cityloops. (n.d.). https://cityloops.eu/cities/porto

Colares, L. G. T., Figueiredo, V. O., Ferreira, A. A., & Oliveira, A. G. M. (2018). Good environmental practices check list for food services: Elaboration, content validation and inter-rater reliability. *Brazilian Journal of Food Technology*, *21*(e2017066).

Coulthard, H., Sharps, M., Cunliffe, L., & van den Tol, A. (2021). Eating in the lockdown during the Covid 19 pandemic; self-reported changes in eating behaviour, and associations with BMI, eating style, coping and health anxiety. *Appetite*, *161*, 105082. doi:10.1016/j.appet.2020.105082 PMID:33476651

Dernini, S., Berry, E. M., Serra-Majem, L., La Vecchia, C., Capone, R., & Medina, F. X. (2017). Med Diet 4.0: The Mediterranean diet with four sustainable benefits. *Public Health Nutrition*, *20*(7), 1322–1330. doi:10.10171368980016003177 PMID:28003037

Dernini, S., Meybeck, A., Burlingame, B., Gitz, V., Lacirignola, C., & Debs, P. (2013). Developing a methodological approach for assessing the sustainability of diets: The Mediterranean diet as a case study. *New Medit*, *12*(3), 28–36.

Deschasaux-Tanguy, M., Druesne-Pecollo, N., Esseddik, Y., de Edelenyi, F. S., Allès, B., & Andreeva, V. A. (2021). Diet and physical activity during the coronavirus disease 2019 (COVID-19) lockdown (March-May 2020): Results from the French NutriNet-Santé cohort study. *The American Journal of Clinical Nutrition*, *113*(4), 924–938. doi:10.1093/ajcn/nqaa336 PMID:33675635

DGS,, PNPAS, DGC, & FCNAUP. (2016). *Roda da alimentação Mediterrânica: cartaz da roda da alimentação Mediterrânica* [Mediterranean food wheel: Mediterranean food wheel poster] [Infographic]. https://nutrimento.pt/activeapp/wp-content/uploads/2016/06/Cartaz-A3-13_med.pdf

DGS. (2020a). *Estudo de adesão ao padrão alimentar mediterrânico* [Study of mediterranean food pattern adhering]. https://nutrimento.pt/activeapp/wp-content/uploads/2020/10/Estudo-de-adesa%CC%83o-ao-padra%CC%83o-alimentar-mediterra%CC%82nico.pdf

DGS. (2020b). *Inquérito sobre alimentação e atividade física em contexto de contenção social* [Survey on food and physical activity in a context of social restraint]. https://alimentacaosaudavel.dgs.pt/active-app2020/wp content/uploads/2021/01/REACTCOVID_Survey_Retificado.pdf

Di Renzo, L., Gualtieri, P., Pivari, F., Soldati, L., Attinà, A., & Cinelli, G. (2020). Eating habits and lifestyle changes during COVID-19 lockdown: An Italian survey. *Journal of Translational Medicine*, *18*(1), 229. doi:10.118612967-020-02399-5 PMID:32513197

Esposito, K., Ciotola, M., & Giugliano, D. (2007). Mediterranean diet and the metabolic syndrome. *Molecular Nutrition & Food Research*, *51*(10), 1268–1274. doi:10.1002/mnfr.200600297 PMID:17879992

Estell, M., Hughes, J., & Grafenauer, S. (2021). Plant Protein and Plant-Based Meat Alternatives: Consumer and Nutrition Professional Attitudes and Perceptions. *Sustainability*, *13*(3), 1478.

Estruch, R., & Bach-Faig, A. (2018). Mediterranean diet as a lifestyle and dynamic food pattern. *European Journal of Clinical Nutrition*. Advance online publication. doi:10.103841430-018-0302-z PMID:30487557

European Commission. (2020). *A new Circular Economy Action PlanFor a cleaner and more competitive Europe*. https://eur-lex.europa.eu/resource.html?uri=cellar:9903b325-6388-11ea-b735-01aa75ed71a1.0017.02/DOC_1&format=PDF

FAO. (2011). *Global Food Losses and Waste. Extent, Causes and Prevention*. https://www.fao.org/docrep/014/mb060e/mb060e00.pdf

FAO. (n.d.a). *What is meant by the term "sustainability"?* https://www.fao.org/3/ai388e/AI388E05.htm

Fischler, C. (2021). Commensality, society and culture. *Social Sciences Information. Information Sur les Sciences Sociales*, *50*, 528–548. doi:10.1177/0539018411413963

Lipor. (n.d.). *Dose Certa* [Certain Dose]. https://www.lipor.pt/pt/sensibilizar/100-desperdicio/desperdicio-alimentar/

Lipor. (n.d.a). *Embrulha. e os seus restaurantes à distância de um clique!* [Wraps. and your restaurants just a click away!]. https://www.lipor.pt/pt/noticias/embrulha-e-os-seus-restaurantes-a-distancia-de-um-clique/

Comigo, F. (2021). *Portugal de norte a sul pela mítica estrada nacional 2* [Portugal from north to south by the mythical national road 2]. Foge Comigo.

FoodSave. (n.d.). http://www.foodsave.org/wp-content/uploads/2014/07/FoodSaveFacts.pdf

Food Programe. (n.d.). *Recomendações* [Recommendations]. https://www.food-programme.eu/pt/ferramentas/article/recomendacoes

Garcia-Fernandez, E., Rico-Cabanas, L., Rosgaard, N., Estruch, R., & Bach-Faig, A. (2014). Mediterranean diet and cardiodiabesity: A review. *Nutrients*, *6*(9), 3474–3500. doi:10.3390/nu6093474 PMID:25192027

Ginter, E., & Simko, V. (2015). Recent data on Mediterranean diet, cardiovascular disease, cancer, diabetes and life expectancy. *Bratislavské Lekárske Listy*, *116*(6), 346–348. doi:10.4149/BLL_2015_065 PMID:26084734

Gotsis, E., Anagnostis, P., Mariolis, A., Vlachou, A., Katsiki, N., & Karagiannis, A. (2015). Health benefits of the Mediterranean diet: An update of research over the last 5 years. *Angiology*, *66*(4), 304–318. doi:10.1177/0003319714532169 PMID:24778424

Graça, P. (2014). Dieta Mediterrânica: um desafio para a cooperação entre a cultura e a saúde nos próximos anos [Mediterranean Diet: a challenge for cooperation between culture and health in the coming years]. In A. Romano (Ed.), *A Dieta Mediterrânica em Portugal: Cultura, Alimentação e Saúde* (pp. 332–335). Universidade do Algarve.

Graça, P. (2015). Dieta Mediterrânica: uma realidade multifacetada [Mediterranean Diet: a multifaceted reality]. In *Dimensões da Dieta Mediterrânica: Património Cultural Imaterial da Humanidade*. Universidade do Algarve. http://hdl.handle.net/10400.1/6728

Green Globe. (n.d.). https://greenglobe.com/standard/

Green Key. (n.d.). https://greenkey.abae.pt/

Huhn, S., Kharabian Masouleh, S., Stumvoll, M., Villringer, A., & Witte, A. V. (2015). Components of a Mediterranean diet and their impact on cognitive functions in aging. *Frontiers in Aging Neuroscience*, *7*, 132. doi:10.3389/fnagi.2015.00132 PMID:26217224

Kastorini, C. M., Milionis, H. J., Esposito, K., Giugliano, D., Goudevenos, J. A., & Panagiotakos, D. B. (2011). The effect of Mediterranean diet on metabolic syndrome and its components: A meta-analysis of 50 studies and 534,906 individuals. *Journal of the American College of Cardiology*, *57*(11), 1299–1313. doi:10.1016/j.jacc.2010.09.073 PMID:21392646

Kesse-Guyot, E., Ahluwalia, N., Lassale, C., Hercberg, S., Fezeu, L., & Lairon, D. (2013). Adherence to Mediterranean diet reduces the risk of metabolic syndrome: A 6-year prospective study. *Nutrition, Metabolism, and Cardiovascular Diseases*, *23*(7), 677–683. doi:10.1016/j.numecd.2012.02.005 PMID:22633793

Keys, A. (1970). Coronary heart disease in seven countries. Summary. *Circulation*, *41*(4, Suppl), I186–I195. PMID:5442782

Keys, A. (1980). *Seven countries: a multivariate analysis of death and coronary heart disease*. Harvard University Press.

Keys, A., & Keys, M. (1975). *How to eat well and stay well. The Mediterranean way*. Doubleday Publishing.

Kim, Y. G., & Eves, A. (2012). Construction and validation of a scale to measure tourist motivation to consume local food. *Tourism Management*, *33*(6), 1458–1467. doi:10.1016/j.tourman.2012.01.015

Koloverou, E., Esposito, K., Giugliano, D., & Panagiotakos, D. (2014). The effect of Mediterranean diet on the development of type 2 diabetes mellitus: A meta-analysis of 10 prospective studies and 136,846 participants. *Metabolism: Clinical and Experimental*, *63*(7), 903–911. doi:10.1016/j.metabol.2014.04.010 PMID:24931280

Kyriakaki, A., Zagkotsi, S., & Trihas, N. (2016). Gastronomy, tourist experience and location. The case of the greek breakfast. Tourismos. *An International Multidisciplinary Journal of Tourism*, *11*(3), 227–261. doi:10.26215/tourismos.v11i3.489

Loco. (n.d.). *Projeto*. http://www.in-loco.pt/pt/projectos/20190301/100-local/

Long, M. (2004). *Culinary Tourism*. University Press of Kentucky.

Lopes, C. T. D., Oliveira, A., Severo, M., Alarcão, V., Guiomar, S., & Mota, J. (2017). *Inquérito Alimentar Nacional e de Atividade Física* [National Food and Physical Activity Survey]. IAN-AF 2015-2016: Relatório de Resultados. www.ian-af.up.pt

Mackenbach, J. P. (2007). The Mediterranean diet story illustrates that "why" questions are as important as "how" questions in disease explanation. *Journal of Clinical Epidemiology*, *60*(2), 105–109. doi:10.1016/j.jclinepi.2006.05.001 PMID:17208115

Martinez-Gonzalez, M. A., de la Fuente-Arrillaga, C., Nunez-Cordoba, J. M., Basterra-Gortari, F. J., Beunza, J. J., & Vazquez, Z. (2008). Adherence to Mediterranean diet and risk of developing diabetes: Prospective cohort study. *BMJ (Clinical Research Ed.)*, *336*(7657), 1348–1351. doi:10.1136/bmj.39561.501007.BE PMID:18511765

Martinez-Lacoba, R., Pardo-Garcia, I., Amo-Saus, E., & Escribano-Sotos, F. (2018). Mediterranean diet and health outcomes: A systematic meta-review. *European Journal of Public Health*, *28*(5), 955–961. doi:10.1093/eurpub/cky113 PMID:29992229

Maynard, D. C., Zandonadi, R. P., Nakano, E. Y., & Botelho, R. B. A. (2020). Sustainability indicators in restaurants: The development of a checklist. *Sustainability*, *12*, 4076. doi:10.3390u12104076

Mediterrânica. (n.d.). *D. Feira da Dieta Mediterrânica*. https://www.dietamediterranica.pt/?q=pt/node/759

Moreira, P., Santos, S., Padrao, P., Cordeiro, T., Bessa, M., & Valente, H. (2010). Food patterns according to sociodemographics, physical activity, sleeping and obesity in Portuguese children. *International Journal of Environmental Research and Public Health*, *7*(3), 1121–1138. doi:10.3390/ijerph7031121 PMID:20617022

Moreno, I. (2015). Culturas Mediterrânicas e sistemas alimentares: continuidades, imaginários e novos desafios [Mediterranean cultures and food systems: continuities, imaginaries and new challenges]. In *Dimensões da Dieta Mediterrânica: Património cultural imaterial da humanidade*. Universidade do Algarve. http://hdl.handle.net/10400.1/6728

Moro, E. (2016). The Mediterranean Diet from Ancel keys to the UNESCO cultural heritage. A pattern of sustainable development between myth and reality. *Procedia: Social and Behavioral Sciences*, *223*(Supplement C), 655–661. doi:10.1016/j.sbspro.2016.05.380

Naska, A., Oikonomou, E., Trichopoulou, A., Psaltopoulou, T., & Trichopoulos, D. (2007). Siesta in healthy adults and coronary mortality in the general population. *Archives of Internal Medicine*, *167*(3), 296–301. doi:10.1001/archinte.167.3.296 PMID:17296887

Nestle, M. (1995). Mediterranean diets: Historical and research overview. *The American Journal of Clinical Nutrition*, *61*(6, Suppl), 1313s–1320s. doi:10.1093/ajcn/61.6.1313S PMID:7754981

Nicodemo, M., Spreghini, M. R., Manco, M., Wietrzykowska Sforza, R., & Morino, G. (2021). Childhood obesity and COVID-19 lockdown: Remarks on eating habits of patients enrolled in a food-education program. *Nutrients*, *13*(2). Advance online publication. doi:10.3390/nu13020383 PMID:33530570

Nolla Restaurant. (n.d.). https://www.restaurantnolla.com/

Nutrimento. (2019, July 15). *Alimentação saudável no ensino superior -selo excelência* [Healthy eating in higher education-seal excellence]. Nutrimento. https://nutrimento.pt/noticias/alimentacao-saudavel-no-ensino-superior-selo-excelencia/

Petrini, C. (2017). *Food Trends – Agri-Food Trends in Portugal to 2027 according to the Principles of Slow Food: Scientific Handbook*. Slow food.

Phull, S. (2015). The Mediterranean diet: Socio-cultural relevance for contemporary health promotion. *The Open Public Health Journal, 8*, 35–40. doi:10.2174/1874944501508010035

Pordata. (n.d.). https://www.pordata.pt/

PortugalFoods. (2021). *Trends 2021 – Setor Agroalimentar* [Trends 2021 - Agri-Food Sector]. https://qualifica.portugalfoods.org/wp-content/uploads/2021/02/trends2021-portugalfoods.pdf

Portuguese Association of Distribution Companies. (n.d.). http://aped.pt/

Radd-Vagenas, S., Kouris-Blazos, A., Singh, M. F., & Flood, V. M. (2017). Evolution of Mediterranean diets and cuisine: Concepts and definitions. *Asia Pacific Journal of Clinical Nutrition, 26*(5), 749–763. doi:10.6133/apjcn.082016.06 PMID:28802282

Real, H., Dias, R. R., & Graça, P. (2021a). Mediterranean Diet conceptual model and future trends of its use in Portugal. *Health Promotion International, 36*(2), 548–560. doi:10.1093/heapro/daaa056 PMID:32696052

Real, H., & Graça, P. (2019). Marcos da História da Dieta Mediterrânica, desde Ancel Keys [Landmarks of the History of the Mediterranean Diet, from Ancel Keys]. *Acta Portuguesa de Nutrição, 17*, 6–14.

Real, H., & Graça, P. (2020). Perceções sobre o conceito de Dieta Mediterrânica e a sua evolução - perspetivas de profissionais do Algarve-Portugal [Perceptions about the concept of Mediterranean Diet and its evolution - perspectives of professionals from the Algarve-Portugal]. *Revista Española de Comunicación en Salud, 11*(2), 203–216. doi:10.20318/recs.2020/5306

Real, H., Soares, S., Ferreira, C., Marques, L., & Gonçalves, T. F. (2021). Referencial de critérios para checklist de avaliação da sustentabilidade em restaurantes [Benchmark criteria for checklist of sustainability assessment in restaurants]. *Acta Portuguesa de Nutrição, 26*, 18–30. doi:10.21011/apn.2021.2604

Reguant-Aleix, J., Arbore, M. R., Bach-Faig, A., & Serra-Majem, L. (2009). Mediterranean heritage: An intangible cultural heritage. *Public Health Nutrition, 12*(9a), 1591–1594. doi:10.1017/S1368980009990413 PMID:19911466

Renna, M. M. F., Serio, F., & Gonnella, M. (2021). The Mediterranean diet between traditional foods and human health through culinary examples. In G. Cm (Ed.), *Gastronomy and food science* (pp. 75–100). Elsevier.

Restaurante A Cozinha. (n.d.). https://restauranteacozinha.pt/

Restaurante Kitchen Dates. (n.d.). https://kitchendates.pt/

Rodríguez-Pérez, C., Molina-Montes, E., Verardo, V., Artacho, R., García-Villanova, B., Guerra-Hernández, E. J., & Ruíz-López, M. D. (2020). Changes in dietary behaviours during the COVID-19 outbreak confinement in the Spanish COVIDiet study. *Nutrients, 12*(6). Advance online publication. doi:10.3390/nu12061730 PMID:32531892

Rosa, R. (2017). *Food Trends – Agri-Food Trends in Portugal to 2027 according to the Principles of Slow Food: Scientific Handbook*. Academic Press.

Rota da Dieta Mediterrânica. (n.d.). http://www.dietamediterranica.pt/?q=pt/node/604

RotaN2. (2021, May). *Passaporte Rota da Estrada Nacional 2* [National Road Route Passport 2]. https://www.rotan2.pt/merchandise/4

Sánchez-Torres, J., Arroyo-Cañada, F., Solé-Moro, M., & Argila-Irurita, A. (2017). The Mediterranean Diet: Culture and its impact on consumer habits in Barcelona restaurant. *Revista CEA, 3*(6), 47–54. doi:10.22430/24223182.657

Scarmozzino, F. V. F. (2020). Covid-19 and the subsequent lockdown modified dietary habits of almost half the population in an Italian sample. *Foods, 9*, 675. doi:10.3390/foods9050675 PMID:32466106

Scarpato, R. (2002). Gastronomy as a tourist product: the perspective of gastronomy studies. In A. Hjalager & G. Richards (Eds.), *Tourism and Gastronomy*. Routledge.

Schulz, J. B., & Deuschl, G. (2015). Influence of lifestyle on neurodegenerative diseases. *Der Nervenarzt, 86*(8), 954–959. doi:10.100700115-014-4252-y PMID:26105162

Sigala, M. (2020). Tourism and COVID-19: Impacts and implications for advancing and resetting industry and research. *Journal of Business Research, 117*, 312–321. doi:10.1016/j.jbusres.2020.06.015 PMID:32546875

Silo Restaurant. (n.d.). https://silolondon.com

Solfrizzi, V., Panza, F., Frisardi, V., Seripa, D., Logroscino, G., Imbimbo, B. P., & Pilotto, A. (2011). Diet and Alzheimer's disease risk factors or prevention: The current evidence. *Expert Review of Neurotherapeutics, 11*(5), 677–708. doi:10.1586/ern.11.56 PMID:21539488

Stahel, W. R. (2016). The circular economy. *Nature, 531*, 435–438. doi:10.1038/531435a PMID:27008952

Testa, R., Galati, A., Schifani, G., Di Trapani, A. M., & Migliore, G. (2019). Culinary tourism experiences in agri-tourism destinations and sustainable consumption—Understanding Italian Tourists' Motivations. *Sustainability, 11*(17), 4588. doi:10.3390u11174588

Trichopoulou, A., Bamia, C., & Trichopoulos, D. (2005). Mediterranean diet and survival among patients with coronary heart disease in Greece. *Archives of Internal Medicine, 165*(8), 929–935. doi:10.1001/archinte.165.8.929 PMID:15851646

Trichopoulou, A., Bamia, C., & Trichopoulos, D. (2009). Anatomy of health effects of Mediterranean diet: Greek EPIC prospective cohort study. *BMJ (Clinical Research Ed.), 338*, b2337. doi:10.1136/bmj.b2337 PMID:19549997

Trichopoulou, A., & Lagiou, P. (1997). Healthy traditional Mediterranean diet: An expression of culture, history, and lifestyle. *Nutrition Reviews*, *55*(11), 383–389. doi:10.1111/j.1753-4887.1997.tb01578.x PMID:9420448

Trihas, N., Kyriakaki, A., & Zagkotsi, S. (2015). *Local Cuisine and agricultural products as a means of enhancing tourist's gastronomic experiences in Greece* [Conference session]. 1st International Conference on Experimential Tourism, Santorini. https://imic2015.conferences.gr/wp-content/uploads/TRIHAS.pdf

Trust, C. (2008). *Food preparation and catering Increase carbon savings without compromising on quality*. https://www.dcproducts.co.uk/documents/carbon_trust_food_preparation_and_catering-2.pdf

Tyrovolas, S., & Panagiotakos, D. B. (2010). The role of Mediterranean type of diet on the development of cancer and cardiovascular disease, in the elderly: A systematic review. *Maturitas*, *65*(2), 122–130. doi:10.1016/j.maturitas.2009.07.003 PMID:19656644

UNESCO. (2013a). ITH/13/8.COM/Decisions.

UNESCO. (2013b). *Mediterranean diet*. http://www.unesco.org/culture/ich/en/RL/mediterranean-diet-00884

United Nations. (2015). https://sdgs.un.org/goals

United Nations. (2017). *World Population Prospects 2017*. https://www.un.org/development/desa/pd/sites/www.un.org.development.desa.pd/files/files/documents/2020/Jan/un_2017_world_population_prospects-2017_revision_databooklet.pdf

Veerman, C. L. (2013). Removing the GST exemption for fresh fruits and vegetables could cost lives. *The Medical Journal of Australia*, *199*(8), 534–535. PMID:24138373

Verberne, L., Bach-Faig, A., Buckland, G., & Serra-Majem, L. (2010). Association between the Mediterranean diet and cancer risk: A review of observational studies. *Nutrition and Cancer*, *62*(7), 860–870. doi:10.1080/01635581.2010.509834 PMID:20924961

Vik, F. N., Bjornara, H. B., Overby, N. C., Lien, N., Androutsos, O., & Maes, L. (2013). Associations between eating meals, watching TV while eating meals and weight status among children, ages 10-12 years in eight European countries: The ENERGY cross-sectional study. *The International Journal of Behavioral Nutrition and Physical Activity*, *10*, 58. doi:10.1186/1479-5868-10-58 PMID:23675988

Vilarnau, C., Stracker, D. M., Funtikov, A., da Silva, R., Estruch, R., & Bach-Faig, A. (2018). Worldwide adherence to Mediterranean Diet between 1960 and 2011. *European Journal of Clinical Nutrition*. Advance online publication. doi:10.103841430-018-0313-9 PMID:30487566

Willett, W., Rockstrom, J., Loken, B., Springmann, M., Lang, T., & Vermeulen, S. (2019). Food in the anthropocene: The EAT-lancet commission on healthy diets from sustainable food systems. *Lancet*, *393*(10170), 447–492. doi:10.10160140-6736(18)31788-4 PMID:30660336

Wrap. (2013). *An overview of waste in the UK hospitality and food service sector*. https://www.wrap.org.uk/sites/files/wrap/Overview%20of%20Waste%20in%20the%20UK%20Hospitality%20and%20Food%20Service%20Sector%20FINAL.pdf

Yngve, A., Neuman, N., Haapala, I., & Scander, H. (2021). The Project Collection Food, Nutrition and Health, with a Focus on Eating Together. *International Journal of Environmental Research and Public Health, 18*(4), 1572. doi:10.3390/ijerph18041572 PMID:33562352

Zazpe, I., Bes-Rastrollo, M., Ruiz-Canela, M., Sanchez-Villegas, A., Serrano-Martinez, M., & Martinez-Gonzalez, M. A. (2011). A brief assessment of eating habits and weight gain in a Mediterranean cohort. *British Journal of Nutrition, 105*(5), 765–775. doi:10.10170007114510004149 PMID:21138604

KEY TERMS AND DEFINITIONS

Gastronomic Tourism: This concept was firstly presented by Long, in 1998 and intended to indicate that a tourist can experience new cultures through food. In this case, the gastronomic activities can the first motivation to travel and must be unique and memorable (Long, 2004; Trihas et al., 2015).

Gastronomy: The term refers to the study of "good eating" and was published in this way on a French dictionary in 1835 (Scarpato, 2002).

Chapter 6
COVID-19, Food Safety, Risk Assessment, and Future Approaches in the Food Industry

Ayşen Çoban Dinçsoy
Istanbul Gedik University, Turkey

ABSTRACT

COVID-19, which is thought to be transmitted from infected animals to humans as a zoonotic disease, plays an important role in food safety and public health. Inappropriate, defective, and unsuitable food safety systems ignoring it play a key role in the spread of the disease. Controlling food safety from farm to fork provides access to healthy and high quality foods in the protection of health and the prevention of diseases. Making new effective and applicable standards and principles in the food safety system worldwide may be preventing possible diseases, risks, especially at epidemic level. In this context, official authorities, food businesses, and supply chain officials should take an active role to reform the process. Consequently, both food safety and free trade principles may be more efficient under common global food safety systems for public health.

INTRODUCTION

COVID-19 caused the 2020 pandemic that a disease thought to be transmitted from animals to humans, is still insufficient of control all over the world. There is drastic proof that the infected human and animal is in a food harvest, preparation and serving processes lacking hygiene standards and proper controls, to be sold as food and beverage. On the other hand, there is important evidence that an improper, inadequate, incomplete and excessively tolerant food safety system was the key to the original transmission of the disease. According to reported studies that coronavirus could maintain viable in the air (up to 3 hours), on copper (up to 4 hours), on cardboard (up to 24 hours), plastic and stainless steel (up to 72

DOI: 10.4018/978-1-7998-9148-2.ch006

hours). Additionally, importance of cooking or processing foods to reduce the hazard of coronavirus, as a proactive approach to prevent foodborne illness.

First of all, food safety plays a vital role in protecting public health: a good system of food safety regulation permits us to work on disease prevention, both by protecting and encouraging the production of healthy and quality foods and by ensuring the protection of health in all steps of the supply chain, from the beginning steps of production to the latest consumption. Secondly, the protection of food safety is a worldwide problem. In order to carry on enjoying the comparative advantages offered by the international market, the present model of trade regulation must be equipped. The food industry has been showed prospering in the detraction of new infections in many cases, and particularly when implemented perpetually: keep physical distancing, using mask and gloves, active risk based regulation systems. In order to implement related and applicable reformations may be performed. Possible points are in below.

1. To revise the Codex Alimentarius into an organisation that objectives global food safety standards with the unique and specific aim of protecting health and without charge trade principles
2. To implement the cautiousness principle globally and in global industrial areas
3. To perform the obligatory labelling tool, definitive traceability during all the production chain and also give an information about both guiding consumers and manufacturers for more process details
4. To constitute a valid and common system of controls under the government administration, protect and sustain principles, rules, procedures, coordination and supervision at the maximum level.

In this study, it is aimed to examine the precautions and new generation technologies related to the risks and dangers in the food industry in food safety during the coronavirus pandemic period.

BACKGROUND

While the restaurant service industry plays a vital role in country economy, it requires a successful economic management plan to be a long-term business (Gnonlonfoun, 2017). The last health outbreak led to by COVID-19 immediately made into an economic collapse by virtue of the restraint affected by most countries to manage successfully the distribution of pandemic (IMF, 2020). The unforeseen outbreak has also notably imposed the gastronomy, and ultimately, the restaurant service sector has been mainly one of the damaged (Muller, 2020).

Restaurants were constrained to shut down totally or service with many restrictions, and merely takeaway and delivery options were permitted. Even though restaurant owners have experienced and survived against to different crises, the COVID-19 crisis is thought that will be destructive impacts on the global economy recently (Alonso-Almeida et al., 2018; Kim et al., 2020).

According to other research, the economic impact on the restaurant service industry can be minimized if correct strategies are applied to understand the characteristics of the COVID-19 pandemic. Consequently, it is necessary to identify the effects of the sector in order to enhance economic difficulties and sustainability through and then by the crisis (Kim et al., 2020).

In December 2019, an outbreak of pneumonia of unknown causes was reported in the Wuhan region of China. Epidemiologically, it was stated that the first cases were related to the Huanan seafood wholesale, where aquatic products and live animals were sold (Liu et al., 2020).

It was named 2019–novel Coronavirus (2019–nCoV) following the discovery of an unknown betacoronavirus from lower respiratory tract samples of patients by next-generation sequencing technique (Zhu et al., 2020). When the virus was examined by electron microscopy, it was determined to have a diameter of 60 to 140 nm with characteristic spikes of 9 to 12 nm, similar to the Coronoviridae family (Zhu et al., 2020). Phylogenetically, it was stated that the novel coronavirus showed ~79% and ~50% similarity for SARS and MERS, respectively, as well as ~88% similarity to the coronavirus strain (Lu et al., 2020).

On February 11, 2020, the WHO declared a novel 2B betacoronavirus that led to viral pneumonia and proclaimed COVID-19 for the disease standard name (WHO, 2020c). It was named SARS-CoV-2 by the International Committee on Virus Taxonomy (ICTV) (Gorbalenya et al., 2020; Hui et al., 2020).

In spite of widespread worldwide attention, COVID-19 pandemic that started in China's Wuhan territory and promptly spread all through the nation and after that over 50 other nations around the world, was pronounced a widespread by the World Health Organization (WHO) on March 11, 2020 (WHO, 2020c).

The COVID-19 pandemic has come approximately 226 million affirmed cases globally, with over 4.5 million deaths and one-third the world's population quarantined (WHO, 2021). By the WHO, zoonotic transmission (diseases that can be transmitted from animal to human) is responsible for 75% of lastly emerging infectious diseases (WHO, 2014a). Direct contact with an infected animal or ingestion of undercooked animal food products are among the main reasons for the spread of the disease (Goodwin et al., 2012). In addition, for instance, it was stated that 90% of the patients with sporadic hepatitis E viral infection in Hokkaido, Japan resulted from the consumption of undercooked pork liver (Yazaki et al., 2003).

There is no data yet on the transmission of SARS-CoV-2 due to contamination of water and food by the Centers for Disease Control and Prevention (CDC) and WHO (Centers for Disease Control and Prevention, 2020a). In addition to the preparation of food on contaminated surfaces, the preparation of food by an infected person and packaging affects the risk of transmission of the virus (Galanakis, 2020).

Physical contact and food sharing were led to COVID-19 at a conference that carried on in Singapore in 2020 (Pung et al., 2020), referring that food could be a vital role related to SARS-CoV-2 infection. As a consequence, it's vital to figure out the precise transmission path including food path, and to use scientific data to come up with new strategies to stop the virus from spreading quickly and to revise global COVID-19 standards in the face of a pandemic.

Insufficient hygiene, sanitation, and residence and food facility infrastructure have been a major source of concern among vulnerable groups during COVID-19 pandemic. In addition, although it is known that the usable water source is vital in providing basic hygiene WASH (water, sanitation and hygiene) to protect from COVID-19, it is stated that approximately 40% of the world does not have access to basic hand washing at home (Armah et al., 2018).

Many epidemics have been seen in the world in previous periods. The COVID-19 pandemic is one of today's epidemics. This worldwide epidemic has brought some social and economic problems affecting human life (Gössling et al., 2020). With the spread of COVID-19 around the world, some measures have begun to be taken. Travel restrictions, accommodation and food and beverage sector restrictions were introduced (Dunford et al., 2020). Therefore, the accommodation, food and beverage sector has experienced an economic lost. For example, the gross domestic product (GDP) of the United States (USA) decreased by 4.8% annually in the first quarter of 2020 (Bureau of Economic Analysis, 2020). In another example, China's GDP decreased by 6.8% in the first quarter (Cheng, 2020) and hotel occupancy rates in China have decreased by 90% (Nicola et al., 2020). According to the National Agency Restaurant Association (the industry's main trade association), sales fell 47% between 1-22 March and

7 of 10 enterprises had to lay off personnel. On the other hand, 3% of restaurants have been completely closed due to COVID-19 since March 2020 and restaurant industry's sales have decreased approximately $120 billion during the first quarter of 2020 (National Restaurant Association, 2020).

Additionally, actions to reduce social distance are prohibited in the COVID-19 epidemic. Restaurants have been closed and the takeaway method has been adopted. The USA aimed to support restaurant businesses with a $150 billion bailout (Ozili & Arun, 2020). Consumers have started to prefer immune-boosting products during the epidemic period. Businesses that produce according to this need are making profits (Galanakis et al., 2020).

Hotels and restaurants have reduced staff numbers. In developed countries, solutions such as tax reductions, incentive aid packages and flexible loans were offered to business owners (Williams & Price, 2020).

THE MAIN FOCUS OF THE CHAPTER

COVID-19 Transmission

At the start of the pandemic, two-thirds of the 41 recorded cases said they visited the Huanan seafood wholesale market, where live animals are sold in addition to seafood and meat products, thus spreading the virus from animals to humans (Guan et al., 2020; Harapan et al., 2020; Naserghandi et al., 2020). Additionally, several patients infected with SARS-CoV-2 who visited a seafood market in China and caught pneumonia stated that they may have consumed contaminated animals or birds where live animals were sold (Wang et al., 2020). On the other hand, the role of the seafood market in the transmission of COVID-19 is not totally known (Harapan et al., 2020).

Although the ways of known transmission are respiratory droplets spread by speaking, sneezing and coughing, direct contact with infected people is also a strong possibility. Moreover, fecal-oral transmission, contaminated items, and vertical transmission between mother and fetus are some of the other known possibilities (Naserghandi et al., 2020; Wang et al., 2020). Recently, other putative SARS-CoV-2 transmission routes, such as fecal-oral transmission, have been identified in young children as gastrointestinal symptoms and occult infections (Chan et al., 2020).

Transmission by Food and Food Contact Material

Although the knowledge about the role of surfaces and food in transmission in COVID-19 infection is limited, recent studies have indicated that the SARS-CoV-2 virus can be transmitted by food, water, and frequently used fomites and environmental surfaces (Garraturo et al., 2020; Kampf et al., 2020). On the other hand, Wu et al. (2020) stated that the virus can be transmitted through food and/or water as a result of detecting SARS-CoV-2 in urine and stool samples taken from laboratory-confirmed COVID-19 patients.

Consequently, if the respiratory tract excretions of infected COVID-19 patient contact with food, the food may become an infected fomites, and contact with contaminated fomites by other persons is more likely to enter the respiratory epithelium of the virus as a result of contact with other sensory organs, especially with unhygienic hands (Bundesinstitut für Risikobewertung, 2020; Centers for Disease Control and Prevention, 2020).

Transmission by Food Contact Surfaces

The survival time of the virus varies greatly according to the virus and surface type, temperature and relative humidity (Kampf et al., 2020). Although SARS-CoV-2 is also known to survive for 2 days on stainless steel and 3 days on plastic at 21–23°C and 40% relative humidity, no virus was determined on copper or cardboard after 4 and 24 hours, respectively (van Doremalen et al., 2020). Consumer standards for food shopping, processing, and preparation have been recommended by the Food and Drug Administration (2020). To prevent transmission of SARS-CoV-2, food handlers, not excepting food service industry personnel and consumers, should directly abide by appropriate sanitation and hygiene practice standards and meet at least minimum requirements for food safety. Due to the fact that the COVID-19 virus can survive as a virion on inanimate substances for up to 72 hours, it is necessary to properly implement Food Hygiene Practices from farm to fork, after the virus's life cycle was completed in an infected individual (van Doremalen et al., 2020).

Container surfaces, packaging materials, countertops, conveyor belts, interiors of transport vehicles and all other food workstations that human interaction with food may occur should be among the priority points food handlers need to start to stop the spread of COVID-19. As a result, adherence to public health norms is critical, including adequate usage of personnel protective equipment and regular hand washing when shopping for goods, additionally the use of hand sanitizers, wearing masks and gloves, and paying attention a minimum distance of 6 feet between workers. There are various disinfectants and cleaning materials available for sale open to consumer access. In addition to claiming effectiveness against coronaviruses or noroviruses, disinfectant labels should also be effective for SARS-CoV-2. Additionally, United States Environmental Protection Agency (EPA) disinfectant labels provide detailed recommendations on contact time, concentration, and acceptable surfaces for application (Environmental Protection Agency, 2020).

Survival and Inactivation of SARS-CoV-2 on Inanimate Surfaces

The importance of SARS-CoV-2 transmission via contaminated surfaces has recently been postulated. Numerous viruses, included SARS-CoV and MERS-CoV, have been shown to live for hours or even months on different surfaces as plastic, metal, wood, and stainless steel. Clearly, if exposed to unsanitary conditions, the surfaces of a number of foods are possible vital role for viral transmission (Mullis et al., 2012). SARS-CoV-2 survival on various surfaces according to current literatures is below (Table 1).

It is known that viruses, mainly SARS-CoV-2, can be transmitted to the environment through indirect contact (fomites) (Centers for Disease Control and Prevention, 2020a; WHO 2020b), notably if people contact contaminated surfaces and following that touch the sense organs without washing their hands. Warnes et al. (2015) proved that human coronavirus 229E (HuCoV-229E), which is of the same family as SARS-CoV-2, can survive minimum 5 days on the surfaces of various food contact materials (e.g. Teflon, PVC).

SARS-CoV-1, MERS coronavirus or human coronavirus (HuCoV-229E) were determined that coronaviruses survived up to 9 days on inanimate surfaces (Kampf et al., 2020). SARS-CoV-2 can also be persistence during 72 hours on plastic and stainless steel, during 4 hours on copper, and during 24 hours on cardboard in experimental conditions under 21 to 23°C and 40% humidity (Van Doremalen et al., 2020).

Table 1. SARS-CoV-2 survival on various surfaces

Virus	Surface	Inoculum	Temperature (°C)	Stability	Log reduction	Reference
SARS-CoV-2	Steel	$10^{3.7}$	NA*	3-4 days	3.2	Van Doremalen et al., 2020
	Copper	$10^{3.2}$	NA*	4-8 h	1.7	
	Tissue paper	$10^{7.8}$	22	30 min	5.5	
	Banknote	$10^{7.8}$	22	2 d	6	
	Cardboard	$10^{2.5}$	NA*	24 h	2	
	Plastique	$10^{3.7}$	NA*	3-4 days	3.2	
	Glass	$10^{5.8}$	22	4 days	5.8	Alex Chin et al., (2020)
	Plastique	$10^{5.8}$	22	7 days	5.8	
	Wood	$10^{5.6}$	22	1-2 days	5.6	
	Paper	$10^{4.8}$	22	3 h	4.8	

*NA: Not applicable

Pezzotti et al. (2020) determined that after 1 minute of application, silicon nitride (Si_3N_4) inactivates 99% of the SARS-CoV-2 virus and is as efficient as copper. In conclusion, the use of Si_3N_4 for personnel protective equipment and other materials may be a successful strategy to reduce viral spreading.

In addition, RatnesarShumate et al. (2020) found that it was sensitive to artificial sunlight as a result of mixing SARS-CoV-2 in saliva in the laboratory environment and that 90% of the virus was inactivated in less than 7 minutes. However, in the experimental setting, Chin et al. (2020) stated that no infectious virus was obtained from different surfaces at different times at 22°C. It has also been proven that the virus is more stable on flat surfaces. It proved the absence of contagious virus on glass and banknotes after 4 days and on plastic and stainless steel after 7 days. In addition, it is not clear that the results obtained are in laboratory conditions and to what extent they will reflect the risk of transmission of the virus in the real environment.

By van Doremalen et al. (2020) that SARS-CoV-2 has a lower survivability on copper surfaces and is deactivated for 2 hours. The absence of experiments on components of foods, as well as about food regulatory requirements, continue to be obstacles of new food packaging. The development of antiviral biopolymers and related application in the food industry is still a work in progress. For example, using nanomaterial coatings or films comprising copper, silver, and zinc nanoparticles to avoid contamination of food packaging surfaces and hence limit SARS-CoV-2 transmission has recently been reported (Sportelli et al., 2020).

Additionally, more research is needed to determine the relationship between the consumption of SARS-CoV-2 contaminated food and the transmission possibility of infection and the effectiveness of nano-based biopolymer materials in the production of antiviral active packaging. About 90% of SARS-CoV-2 infections originated in symptomatic, non-symptomatic, and asymptomatic patients, while another 10% come from inanimate areas. As a result, person-to-person transmission is the highest risk in the food industry, such as manufacturing, retail and food service. Therefore, it demonstrates significance of using protective equipment special for personnel, hand hygiene and social distance (Bode et al., 2020).

SURVIVING AND DEACTIVATING OF SARS-CoV-2 IN FOOD AND FOOD SUPPLY SYSTEM FOR RESTAURANTS

Pathogens, including viruses, can be inactivated or reduced in the environment and on food using thermal and non-thermal inactivation methods (Dvorak, 2008; Skara & Rosnes, 2016). Foodborne viruses have been inactivated using various thermal treatments on food products (Bosch et al., 2018). Inactivation of viruses and bacteria can be accomplished using both dry (e.g. oven) and humid heat (e.g. steam) (Bosch et al., 2018; Dvorak, 2008; Kampf et al., 2020a).

According to Kampf et al. (2020a) some different types of coronavirus, including SARS-CoV-1 and MERS-CoV in liquid media, were decreased by at least 4 logs by using heat treatments such as 60°C 30 minutes, 65°C 15 minutes or 80°C 1 minute. After 5 minutes heat treatment at 70°C, Chin et al. discovered that SARS-CoV-2 decreased by an average of 7 logs (Chin et al., 2020).

In addition, ANSES (2020) studied food heat treatment and determined that exposing the food to 63°C for 4 minutes would be sufficient to destroy the virus.

Chin et al. (2020) discovered a similar resilience to refrigeration when they reported no decline in infectious SARS-CoV-2 in carrying medium kept at 4°C for 14 days, which was the length of the experiment. As a result, the virus is more likely to persist in certain media at refrigerated temperatures for longer periods of time. Given that freezing is often not thought of as a technique of virus destruction in food, SARS-CoV-2 is likely to withstand freezing.

Fisher et al. (2020) discovered that the infectious SARS-CoV-2 can remain in the refrigerator at 4, 20 and 80°C for 3 weeks in infected chicken, pig and salmon. Mullis et al. (2012) also found that bovine coronavirus in lettuce chilled at 4°C remained infectivity of virus for at least 14 days. Additionally, coronavirus 229E persisted during 2 days on lettuce refrigerated at 4°C before their numbers decreasing rapidly, but was not found on the strawberries, possibly due to acid (Yépiz-Gómez et al., 2013).

Despite the fact that heat, sunlight and UV light can be inactivated for virus (Chin et al., 2020; Pressman et al., 2020; WHO, 2020b), approximately 20°C, different pH levels also can be inactivated SARS-CoV-2 (Oberholzer & Febbo, 2020; Peeri et al., 2020). Non-thermal physical disinfection options include ultraviolet (UV) light, pulsed light, ionizing radiation, high pressure, cold plasma, and high-intensity ultrasound (Deng et al., 2019). The cold plasma technique is a new disinfection technique that uses different inert gases, does not contain chemicals, and is highly preferred because it is not harmful to the environment (Filipic et al., 2020).

Food safety, nutrition/quality, and manufacturing allocation are all addressed by the food system, which is broad, multifaceted, and highly linked (Abbaspourrad et al., 2017; Clancy, 2017).

COVID-19 affects food supply and demand in direct relation to food security (Chen et al., 2020). Throughout the COVID-19 pandemic, all stages of the food chain have been severely damaged, including fresh vegetables, fruits, baked goods, perishable foods and food grains (Ivanov & Dolgui, 2020).

The COVID-19 pandemic has damaged one of the four main columns of the food system: food security (Galanakis, 2020).

Recommended food safety procedures from farm to fork during a possible pandemic for the food industry are summarized in Table 2 for each stage of the food chain.

Employees are categorized according to treatment conditions, personal hygiene, surface disinfection, workplace cleaning, food preparation and distribution, and social distancing. Although all parts of the food chain have established safety requirements, the last stage (consumption) is the most risky source of infection, but contains the highest safety concern at the consumer level. Similarly, it is vital for the food

Table 2. Safety guidelines for COVID-19 Pandemic at each stage of the food industry

1. Medical condition (FDA, 202a)	Stay home if feeling unwell or sick	Agricultural food product Handling during post-harvesting Processing & Packaging Storage & Marketing Distribution & Retail Consumption
	Check for symptoms as upper respiratory tract, body temperature, fatigue	
	Close mouth and nose included coughing or sneezing, throw away of tissues and properly wash hands	
2. Personnel hygiene (FDA, 202a)	Properly washing with soap and water minimum 20 seconds	Agricultural food product Handling during post-harvesting Processing & Packaging Storage & Marketing Distribution & Retail Consumption
	Frequent use of alcohol-based hand sanitizers	
	Avoiding touch the eyes, nose and mouth with unwashed hands	
3. Disinfection of surfaces (FDA, 2020a)	Disinfected high touch surfaces with proper products (70% ethanol)	Handling during post-harvesting Processing & Packaging Storage & Marketing Distribution & Retail Consumption
	Frequent use of sanitizers according to label instruction	
	Follow the protective measure	
4. Working on food-processing environments (Liu, 2020)	Provide PPE	Processing & Packaging Storage & Marketing Consumption
	Proper disinfection of toilet areas	
	Develop open plan workspaces	
	Limit the number of personnel in a food preparation area	
	Apply the windows ventilation	
5. Preparation of food (FDA, 2020a)	Properly wash the fruits and vegetables before consumption	Consumption
	Cooking the food over 70°C	
	Separate raw and cooked foods	
6. Delivery of food (FDA, 2020a)	No touch deliveries is good	Storage & Marketing Distribution & Retail Consumption
	Controls & maintain the time and temperature	
	Assure that transport containers are cleaned and sanitized	
7. Social distancing (Kissler, 2020)	Maintain social distance at least 2 m between people	Consumption
	Do not participate to group events	
	Stay away from crowded areas and gathering	

*Adapted from Han et al. (2021)

industry that the food on consumers' plates is safe and does not pose a health problem at any stage of the process. In addition, there are preventive measures that are widely used before food consumption. For example, at the beginning of the crisis, restaurants, cafeterias and health authorities which many Central European stopped selling uncooked meat products to take precaution against viruses and pathogens, despite scientific evidence of SARS-CoV-2 being transmitted through food (Euractiv, 2020).

Because of the COVID-19 pandemic, there is a growing demand for food and beverages on the internet (FAO, 2020). Under such stringent lockdown conditions, where most logistics activities have ceased, a food scarcity is unavoidable. Online food service companies such as Zomato and Swiggy have been restricted due to the risk of infection (Narayanan et al., 2020). Mullis et al. (2012) investigated the stability of bovine coronavirus in chilled lettuce leaves and discovered that the virus could be detected for at least 14 days and became more stable at lower temperatures and relative humidity. This suggested that contaminated vegetables might be a possible route for zoonotic transmission. Human coronavirus (HuCoV) 229E had a similar effect on lettuce leaves stored at 4°C, with the virus decreasing by 0.2 $\log_{10}$ after 2 days and no effect after 4 days. It is seen that this situation can survive on fresh products even storage temperature, and possible zoonotic contamination may occur, especially through fresh fruit that cannot be subjected to heat treatment (Yépiz-Gómez et al., 2013).

According to Dai et al. (2020) reported that SARS-CoV-2 determined viable 8 days at 4°C and survived 2 days at 25°C in salmon, confirmed that it depends on temperature, raising the potential risk of infection that SARS-CoV-2 infectivity is mostly found in fish or seafood stored and transported at low temperatures.

While no evidence determined that the virus can be transmitted through food to prevent human-to-human transmission of SARS-CoV-2, maintaining adequate personnel hygiene, including hand hygiene and safe waste management, may be a good approach. In addition, food products of animal origin stored at low temperatures should be examined for SARS-CoV-2 to ensure food safety.

Control Measures for the SARS-CoV-2 in the Food System

The existence of SARS-CoV-2 in food manufacturing and distribution has created significant uncertainty in the food service sector (Oliveira et al., 2020). Because no particular antiviral medications, treatments, or vaccines for SARS-CoV-2 have been approved, precaution focuses on personnel hygiene, including sufficient cleaning of surroundings and food-contact surfaces (Yang, 2020), as well as social distancing (Makroo et al., 2020).

Heat Inactivation

Several studies have indicated that numerous environmental factors, particularly temperature and relative humidity, influence coronavirus persistence, and that these factors might be employed as public intervention strategies. If the kinetic of inactivation and minimum virus infectivity are determined, heat inactivation may be addressed and accomplishedly implemented for food safety (Steardo et al., 2020).

SARS-CoV-2 can be inactivated by heating it to 56–67°C for 60–90 minutes, according to Henwood (2020). These findings suggest that food prepared under standard cooking temperature (70°C) is presumably safe from SARS-CoV-2.

Despite the fact that SARS-CoV-2 is heat-labile, some restrictions could be possible to apply heat treatment to inactivate SARS-CoV-2. Furthermore, many fresh food products cannot be inactivated by heat. Freezing is another common method of preserving many foods, but the virus can still be transmitted through frozen foods. Consequently, discovering a viable alternative to minimize the hazard of viral transmission by uncooked foods is a challenging research topic for the future.

Ultraviolet (UV) Treatment

In addition to infection precaution and control strategies to prevent COVID-19 transmission, WHO has outlined strategies for treating COVID-19, prolonging the life of medical equipment, and disinfecting items (WHO, 2020a). It is predicted that developments related to UV light will provide robot-controlled

Table 3. Ethanol Usage with Different Percentage and Time for SARS-CoV-2

Analysis type	Concentration	Application time	Loss of viral infectivity
Suspension test	20%	30 s	1.08
	20%	1 min	1.33
	20%	3 min	1.75
	20%	5 min	1.92
	30%	30 s	4.42
	30%	1 min, 3 min, 5 min	≥ 4.75
	40, 50, 60, 75%	30 s, 1 min, 3 min, 5 min	≥ 4.75

microbial decontamination in surface disinfection and food safety applications in health care facilities. Because the maximum absorption wavelength of DNA molecules is about 260 nm, viruses are determined to be particularly sensitive to UV at wavelengths close to 253.7 nm (UVC range) (Quevedo-Leon et al., 2020).

These data is important extremely for improving effective UV-based approaches to minimize the spread of COVID-19. More UV-based inactivation experiments are required to determine crucial aspects such as application time and dose for correct validation as well as quantification before germicidal UV-based approaches can be used on a large scale.

Inactivation via Chemical Disinfectants

Although there is insufficient data on the eradication of the virus from food surfaces, disinfection is recognized as the best strategy to minimize the transmission of COVID-19 infection in public and domestic settings. There are disinfectants available to the consumer, which are generally inexpensive, simple to apply, and can be applied on a large scale on frequently contacted surfaces. Precise biocidal chemicals as chlorine and its derivatives and ethanol are known to act rapidly against coronaviruses, including SARS-CoV-2 (Quevedo-Leon et al., 2020).

Although human and different animal coronaviruses can be performed in 1 minute with 62-71% ethanol, 0.5% hydrogen peroxide or 0.1% sodium hypochlorite as surface disinfectant (Zuber & Brüssow, 2020). According to Table 3, it has been stated that SARS-CoV-2 disinfection can only be carried out with ethanol at different percentages and times specified (Yin et al., 2020)

However, Alex et al. (2020) tested the viruscide activities of different disinfectants for SARS-CoV-2 under various conditions; following 5 minutes of application with household bleach, ethanol (70%), povidone-iodine, chloroxylenol (0.05%), chlorhexidine (0.05%), or benzalkonium chloride, no transmitted virus particles were detected (0.1%). On the other hand, hand soap needed 15 minutes to totally deactivate SARS-CoV-2.

Yoshimoto et al. (2020) used the deactivation effects of acetic acid and vinegar on SARS-CoV-2 at the rates of 4% and 6%, respectively, in terms of food safety. They found that it effectively reduced the viral amount by more than 4 $\log_{10}$ after 5 minutes following both uses.

Although 70% ethanol concentration was previously recommended for surface cleaning to inactivate viruses (WHO, 2014b); currently, medical disinfectants are recommended for cleaning and disinfection of environmental surfaces to prevent COVID-19 infection. In addition, it has been stated by WHO (2020d) that the use of 5% NaClO or 0.5% H_2O_2 at a ratio of 1:100 works well. In addition, it is critical to determine optimal dosages and procedures to effectively decrease the risk of SARS-CoV-2 infection.

On the other hand, air disinfection is possible to apply in all food areas. Two known functional solutions are the application of filters or UV light including the storage area (Nardell et al., 2020).

RISK ASSESSMENT OF SARS-CoV-2 BY FOOD AND FOOD PACKAGING TRANSMISSION

One of the known knowledges related the emergence of COVID-19 is the characterization of the disease as zoonotic. In addition, there is strong evidence that it is shaped by the consumption of infected animals sold as food. With the pandemic, the world is more actively interacting with food safety. Some inferences can be made on the subject. First of all, although the issue of food safety is global, if regulations are ineffective, its effects can be felt on a global scale. Actual or anticipated risks under regulations can have negative consequences for food systems and supplier trade (Bevilacqua, 2020).

The ability of SARS-CoV-2 to survive through the digestive tract following food, contact, or airborne transmission is not fully known. SARS-CoV-2 is understood to utilize angiotensin-converting enzyme 2 (ACE) receptors produced in lung epithelial and mucosal cells and intestinal epithelial cells (Hamming et al., 2004). It is not known whether SARS-CoV-2 can enter the circulation infect people and infection dose through the gastrointestinal tract due to food consumption.

Food manufacturing companies use different levels of Food Safety Management Systems (FSMS) according to local standards. In any case, a firm foundation of Good Hygiene Practices (GHP) is required to improve and implement of successful FSMS (CAC, 2009).

Published FSMS based on HACCP principles should be closely followed and updated by the entire food industry. If the HACCP rules are not followed in food establishments, the expert who will communicate with the public health authorities in emergency crisis situations should work permanently at the workplace. Informative posters on designated soapy hand washing areas, warm running water, hand sanitizers, and proper hand washing and sanitation should be provided for workers. In addition, physical distance of 6 feet must be maintained between workers (Kimball et al., 2020; Pan et al., 2020; Tong et al., 2020; Wei et al., 2020; Yu et al., 2020).

As a result of the studies carried out during the pandemic period, some critical points have been determined in the current regulation of food safety. Four recommendations are made for more comprehensive risk prevention and health protection. The first consists of reforming the Codex Alimentarius Commission (CAC) without recourse to government instruments to increase global health protection. Second, it is related to its dissemination and implementation all over the world as it is used and applied in the EU and member countries. Third, mandatory labeling is possible for a more effective application. There should be more comprehensive information than the information stated today and should include more and more details. Finally, it is the control system that relates to the food safety aspect of the regulatory framework. It is a set of activities that includes shared rules and procedures from local governments to global scale (Bevilacqua, 2020).

A New Codex Alimentarius

In the main objectives set out in the CAC charter, the international regulations of food security show an unbalance among public health and without charge trade. CAC needs to continue one of its two main objectives is to set standards for agri-food products to ensure a strong protection of human health. The CAC should be allowed to set standards that determine when, how and why agri-food markets endanger public health while facilitating international trade, and whether they are closed or restricted. CAC needs to maintain one of either main objective is to set standards for agri-food products to ensure a high level of protection of human health. The explicit difference between the CAC and the World Trade Organisation (WTO) is that it must be able to act in balance on the same pitch for separate targets. Although the CAC is fundamentally included in the WTO system, WHO is hitherto improved itself as an autonomous regulatory method with different purposes and functions to those referred to the CAC (Bevilacqua, 2020).

Global Cautious Rules

The second objective of proposal concerns the comprehensive implement of the cautious principle on a world basis. This measure seems to be the desired legal and theoretically possible way. Moreover, this principle of precaution should be affirmed in terms of non-European law. There are three reasons for using it globally. The first is its active functioning in the EU legally. To date, taking precautions is widely used to protect health and the environment by many states. Mechanisms between Member States while the common market continues to function in an orderly manner, without the principle of protectionism, the court of justice may judge if the measures are used unlawfully or disproportionately. The global application of the cautious principle, legally enforceable to support decision-making criteria that can be utilised in spesific areas and the protection of certain interests. It aims to harmonize rules around the world by regulating risks. Thus, it provides national governments with a well-founded option in preventing epidemics, pandemics or other diseases. It should therefore be essential that the precautionary principle be central. This will provide the competent authorities with more precautionary and interest-based decision-making and stronger and more effective protection (Bevilacqua, 2020).

Obligatory Labelling and Traceability

The third proposition for the food industry sector is about training from the training and traceability section of trainings. First, food production techniques that are not clearly and appropriately stated today must be mandatory, understandable and clearly stated on the label (e.g. intensive aquaculture production of meat products, use of genetically modified organisms, use of chemicals, use of artificial preservatives and dyes). While this is costly to large food producers, it can cause economic damage by creating distrust of consumers towards food. On the other hand, mandatory enforcement may encourage eco-friendly agriculture and consumption of organic or more natural products. Second, labels should restructure the origin of products, where they were harvested, processing and packaging of applied chemicals so that they are truthful and easily understood throughout the supply chain. Labeling enables states and related quality control safety systems to compete with the effect of race to the top. Third, if the true indication of the source is mislabeled or suitably protected from contamination, the quality of products can be correlated with specific geographical features. At the same time, this situation encourages competition with manufacturers who aim not only to reduce costs but also to increase quality, by ensuring that qualified production techniques are preserved and preferred more than those that have become standard. Although some products have special geographic protection and are currently developing in Europe, quality products are not yet common worldwide due to counterfeiting, cheating or inadequate prevention and control. These proposals may cause few disadvantages that should be mentioned. First, *readability* is understood as redundant information on the label. The indications stated on the label should be investigated in accordance with the average consumer information. According to many views, global food safety authorities are reminded to include competent specialists to determine detailed, broad-based, understandable and effective labeling systems. Second is that the *information is inaccurate*. This hazard lead to a cost for government agencies liable for food safety: administrative regulations should assure that label information is truthful. Such an action requires important sources, a large scale and open control organization (Bevilacqua, 2020).

Moreover, the European legislator demands full traceability of food "from the field to the table" (European Parliament and Council (EC) No 178/2002, 2002). Solution-oriented, new methods are required to ensure food safety on a large scale and at a high level. On the subject, blockchain technology offers great potential. So far, this claim has been met with verification of documents that are easily manipulated, such as shipping documents and invoices. In the 2011 Enterohemorrhagic *Escherichia coli* (EHEC) epidemic in Germany, more than 3,800 people were infected with dangerous intestinal bacteria. As a precaution, supply chain restructuring and recall practices are complex and impossible on a global scale, but rather difficult and time-consuming. At first, Spanish cucumbers were cited as the causative agent, followed by Egyptian vegetable sprouts as the real culprits. In this process, more people became ill than they should have, due to the fact that the main source could not be identified. This is the case in many food-borne outbreaks (Federal Institute for Risk Assessment, 2012).

Blockchain technology is used for different purposes in the food industry (Galvez et al., 2018; Kim & Laskoski, 2018). It has a wide application area and is closely related to traceability. While traceability with traditional methods lasts for days; by blockchain technology, the entire food chain can be monitored in a few seconds (Yiannas, 2018).

In addition, it enables the full identification and elimination of the source of contamination. In this way, it is possible to reduce or eliminate food crises much faster. Additional smart sensors (temperature recorders whose data records are transferred directly to the blockchain) can be used in the automation and digitization of documents to ensure data security (Fuertes et al., 2016; Kuswandi et al., 2011).

On the other hand, it has been stated that 10% of the food products offered for consumption are fake and traceability can be achieved with blockchain technology (Johnson, 2014). At the same time, the emergence of life-threatening situations such as allergies can be prevented in food harvest, main and auxiliary materials used in food production and production methods. In addition, it provides digital recording and monitoring for sustainability for consumer confidence in terms of animal welfare and environmental conditions (Kim & Laskoski, 2018).

In addition, food waste can be reduced by early detection of contaminated products. On the other hand, shelf life of foods can be monitored during transportation and storage (Yiannas, 2018). At the same time, blockchain technology provides transparency to the consumer directly with barcode or QR codes on food-related data on the packaging (Yiannas, 2018). This approach provides both space-saving features and the opportunity to read quickly with smart applications (Esteki et al., 2019; Pigini & Conti, 2017).

Figure 1. Comparison of Traditional Model and Blockchain Technology in Food Chain
Source: Xu et al., 2020

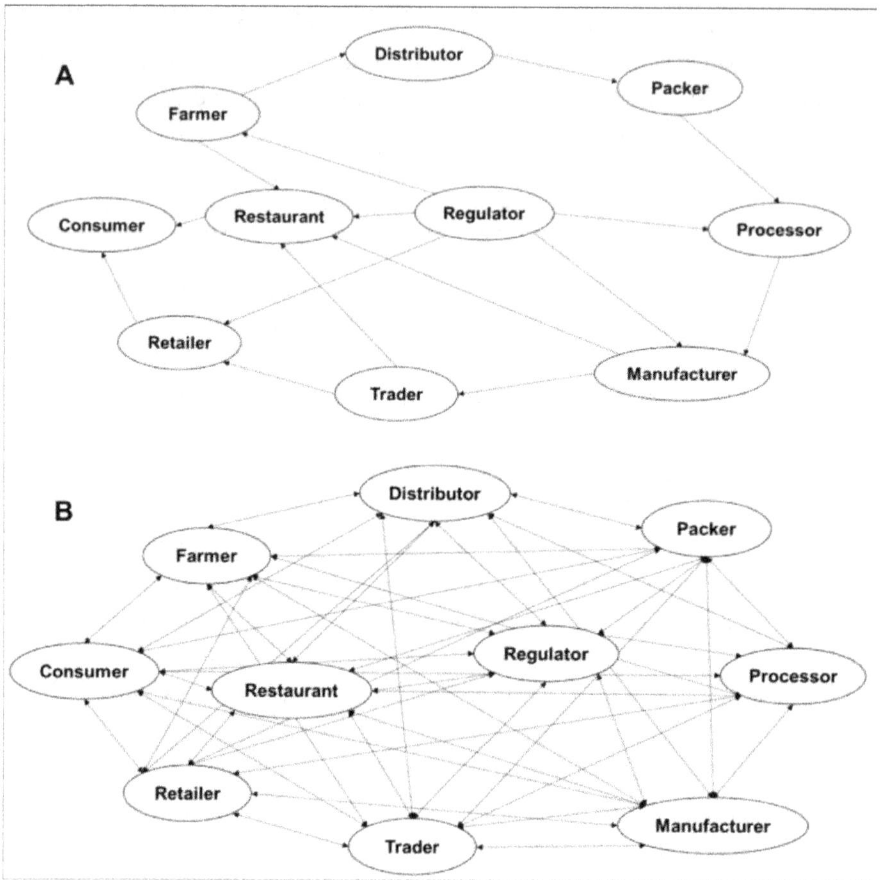

Finally, it provides additional time advantage with the automation system in logistics processes. It eliminates the need for paperwork and provides simultaneous control of all production and process flows (paper based invoice etc.). It also offers the opportunity to order when it falls below a certain amount with its smart refrigerators (Floarea & Sgârciu, 2016).

Global Approach of Food Safety Controls

In the last stage, regarding the control system, the project envisaged in the establishment of the said system on a global scale, the common framework should follow the European model, EU Regulation 625/2017. This has the dual purpose of rationalizing and simplifying. The regulatory framework for controls does not bring different agri-food issues into a single establishing an EU-wide coherent discipline for the text and the organization of these activities reference to the entire food supply chain. This paradigm is a significant model about international food regulation. On the other hand, In order to ensure the system in agricultural food, it should be ensured that it is applied at the local level for all common rules, together with its international application. This model is especially important in terms of trade so that the common sector can be organized and centralized on a global scale and the protection of domestic policies in the local sense and to ensure that the system develops as a single model. In this respect, the discipline is multi-level, dividing certain tasks between departments. The responsibility of the EU and the CAC should be provided by the WHO, with a determined unit in the WHO for organizing and auditing national checkouts, in addition to perform the worldwide model accepted all over the world. Finally, in case of compliance with the common model, safety patents are granted to the products in terms of commercial and economic aspects, especially to protect health (Bevilacqua, 2020).

SOLUTIONS AND RECOMMENDATIONS

For the future of the restaurant industry, 4 different groups were determined as "management", "fund", "hygiene" and "social commitment" after the epidemic. For "management", it is focused on redesigning the catering operation to increase their profitability and establish a long-term business by diversifying their activities with regular investments and short-term returns. "Fund" is directly related to the working capital situation. For "hygiene", the necessity of respecting health and increasing the safety perception of consumers came to the fore. Finally, the theme of "social engagement" focuses on the happiness of the business, employees and customers, and the necessity of retraining the restaurant industry with new habits and lifestyles. Entrepreneurs need government policies that allow them to pay operational costs and taxes, such as loans, which they can access more easily during the pandemic, which is a concern in restaurant businesses. With the pandemic, the new normal plan of the enterprises is the re-creation of the service with the application of package service and the application of hygiene standards. Restaurant owners consider these arrangements, combined with a smart communication strategy that leads to a new target audience, as great applications for the future. Gradual workstations are considered an effective solution to prevent physical distance difficulties in food processing facilities. Food deliveries are more frequently preferred over the internet to reduce human interaction during the pandemic. This also ensures that customers and salespeople are physically separated. Knowledge of appropriate food handling procedures should also be clearly stated at this stage. Although distribution companies have started to offer contactless home delivery services, new measures are required as there is contact of packaging and paper money between the consumer and the retail chain in order to reduce the risk of contamination. In addition, the correct use of gloves, disinfectants and disinfectants can also be effective in reducing the risk of disease transmission and virus spread (FAO & WHO, 2020; Food and Drug Administration, 2020a).

Retail food businesses must ensure that responsible people contribute and take care to maintain safe farm-to-table food mobility. Consumer confidence and confidence in the reliability and accessibility of food safety must be maintained. Retailers can help in providing hygienic conditions by sharing products with visual content and hygienic applications such as wipes, disinfectants and disinfectants. The minimum distance between persons can be achieved with reference marked points. Food samples for demonstration purposes should not be used in this process and plexiglass can be used to avoid interaction with authorities at checkout points (FAO & WHO, 2020; Food and Drug Administration, 2020a).

In addition to hygiene measures, shopping trolleys can be wiped before touching the baskets, shopping bags should be disposable, disinfected if reused, and it should be emphasized that respiratory rules should be observed especially in public environments. In order to reduce market interaction for culinary shopping, the use of existing foodstuffs should be prioritized according to their expiration dates. Careful attention should be paid to the safe handling of food. This point is among the most important issues at every stage of food safety. Examining food processing procedures, it is noted that while coronaviruses cannot withstand heat treatment, if frozen without washing, the virus can survive for up to two years (Bundesinstitut für Risikobewertung, 2020).

In addition, if the fruits and vegetables are not sufficiently heat treated or consumed raw, it is necessary to wash them by rubbing. Before using canned foods, the lids should be cleaned. Disinfection of the kitchen equipment, surfaces and refrigerator used, and cross-contamination during storage should be considered. In addition, adequate cooking time, temperature and thawing methods should be applied. In terms of food safety and quality, packaging processes must be active and smart. Concerns about the survival of SARS-CoV-2s on package surfaces have led to products using antiviral polymers and biopolymers in packaging. In addition, the efficacy of polymers and biopolymers against hepatitis A virus (HAV) and human norovirus (HuNoV) has been scientifically proven (Randazzo et al., 2018).

FUTURE RESEARCH DIRECTIONS

Food safety management systems must be strictly implemented at every stage from farm to fork to prevent food contamination during and after the COVID-19 pandemic. Although coronaviruses cannot reproduce in food (viruses need living host cells to gain viability), additional research is needed to reduce the hazards associated with handling raw meat, fish, shrimp and eggs. Additionally, more investigations should be carried out in order to develop safe food packaging materials and methods of preventing cross-contamination. In summary, after the pandemic, comprehensive food system studies focusing on resilience can be done. In this area, the main challenges are sustainability, circular economy, energy and water efficiency, environmentally friendly products and process techniques. Developing food systems that are durable enough to keep functioning is one of the most difficult tasks in crisis planning.

The ten factors identified by Hecht et al. (2019) to contribute to organizational flexibility for food systems include emergency crisis planning, personnel training, personnel engagement, food oversupply, infrastructure and location of food suppliers, service providers, insurance and post-factum learning. Possible reform for a stronger future in food security consists of the active participation of key stakeholders such as industry, politicians, governments and consumers. Agri-food systems can be diverted to the consumption of local food to reduce reliance on long distance shipping and delivery by third parties to avoid carbon footprints. Safety habits formed during the epidemic and safe food practices should be encouraged to support changes. To strengthen the food supply chain's ability to be prepared potential outbreaks in the future, the role of food scientists and technologists in management and decision-making strategies should be considered. Because the danger of COVID-19 may continue, a truth and rapid determination procedure for SARS-CoV-2 must be established. The COVID-19 pandemic has identified SARS-CoV-2 in food production, packaging and delivery systems, posing new challenges for maintaining food safety. In the post-COVID-19, the food chain and its workers will rely more for norms and regulations such as good hygiene practices, cleaning, sanitation and providing social distancing between food, along with food supply, relevant bioanalytical tools.

CONCLUSION

As a result, state governments should focus on maintaining social food programs and take necessary measures to prevent the spread of the virus. Restaurants should adopt more proactive techniques than passive techniques in emergency crises. During the pandemic period, it is stated that consumers are primarily concerned about cleaning practices about ensuring safety. Although the COVID-19 pandemic is the beginning of a new era in the food industry, it seems difficult to determine the epidemiology and exact transmission route of the virus. As a result, examining the current and potential impact of the virus in terms of food safety is a vital challenge for governments, the food industry and public health worldwide. The effects of some preferred applications such as heat treatment, chemical disinfectants and UV on the inactivation of the virus continue to be investigated. More research, including international standards, is needed to protect food safety and prevent and control emergency crisis situations. In addition, demographic, sociological and economic changes caused by the increase in the human population and the deterioration of the natural habitat of living things affect the possible emergence of new viruses, but there is no clear information about the evolution of old viruses and the transmission route from different hosts or reservoirs. All human, animal and environmental health should work under the One Health concept to communicate and collaborate effectively in infectious disease management.

REFERENCES

Abbaspourrad, A., Padilla-Zakour, O., Wiedmann, M., Moraru, C. I., & Goddard, J. M. (2017). Taking a systems approach to clean label challenges. *Food Technology*, *71*(11), 32–43.

Alex Chin, J. C., Perera, M., Hui, K., Yen, H. L., Chan, M., & Poon, L. (2020). Stability of SARS-CoV-2 in different environmental conditions. *The Lancet Microbe*, *1*(1). Advance online publication. doi:10.1101/2020.03.15.20036673

Alonso-Almeida, M. D. M., Bagur-Femenias, L., Llach, J., & Perramon, J. (2018). Sustainability in small tourist businesses: The link between initiatives and performance. *Current Issues in Tourism*, *21*, 1–20. doi:10.1080/13683500.2015.1066764

ANSES. (2020, March 9). *Opinion on an Urgent Request to Assess Certain Risks Associated with COVID-19*. https://www.anses.fr/fr/system/files/SABA2020SA0037-1EN.pdf

Armah, F. A., Ekumah, B., Yawson, D. O., Odoi, J. O., Afitiri, A. R., & Nyieku, F. E. (2018). Access to improved water and sanitation in sub-Saharan Africa in a quarter century. *Heliyon (London)*, *4*, e00931. doi:10.1016/j.heliyon.2018.e00931

Bevilacqua, D. (2020). Food Safety, Coronavirus and Risk Prevention: Future Perspectives in Four Proposals. *European Journal of Risk Regulation*, *13*(1), 56–77. doi:10.1017/err.2020.81

Bode, M., Craven, M., Leopoldseder, M., Rutten, P., & Wilson, M. (2020). *Contact Tracing for COVID-19: New Considerations for Its Practical Application*. https://www.mckinsey.com/industries/public-and-social-sector/our-insights/contact-tracing-for-covid-19-new-considerations-for-its-practical-application

Bosch, A., Gkogka, E., Le Guyader, F. S., Loisy-Hamon, F., Lee, A., & van Lieshout, L. (2018). Foodborne viruses: Detection, risk assessment, and control options in food processing. *International Journal of Food Microbiology*, *285*, 110–128. doi:10.1016/j.ijfoodmicro.2018.06.001 PMID:30075465

Bundesinstitut für Risikobewertung. (2020). *Can The New Type Of Coronavirus Be Transmitted Via Food And Objects?* https://www.bfr.bund.de/en/can_the_new_type_of_coronavirus_be_ transmitted_via_food_and_objects_-244090.html

Bureau of Economic Analysis. (2020, April 20). *Gross Domestic Product, 1st Quarter 2020 (advance Estimate).* https://www.bea.gov/news/2020/gross-domestic-product-1st-quarter-2020-advance-estimate

CAC. (2020). *General Principles of Food Hygiene CAC/RCP 1-1969.* https://www.fao.org/fao-who-codexalimentarius/sh-proxy/en/?lnk=1&url=https%253A%252F%252Fworkspace.fao.org%252Fsites%252Fcodex%252FStandards%252FCXC%2B1-1969%252FCXC_001e.pdf

Centers for Disease Control and Prevention. (2020). *How Coronavirus Spreads.* CDC.

Centers for Disease Control and Prevention. (2020a). *Updates COVID19 Transmission Webpage to Clarify Information About Types of Spread.* https://www.cdc.gov/media/releases/2020/s0522-cdc-updates-covid-transmission.html

Chan, J. F. W., Yuan, S., Kok, K. H., To, K. K. W., Chu, H., & Yang, J. (2020). A familial cluster of pneumonia associated with the 2019 novel coronavirus indicating person-to-person transmission: A study of a family cluster. The Lancet, 395, 514–523. , 10223. doi:10.1016/S0140-6736(20)30154-9

Chen, S., Brahma, S., Mackay, J., Cao, C., & Aliakbarian, B. (2020). The role of smart packaging system in food supply chain. *Journal of Food Science*, *85*(3), 517–525. doi:10.1111/1750-3841.15046 PMID:32056210

Cheng, J. (2020). *China Records First Ever Contraction in Quarterly GDP on Coronavirus.* https://www.wsj.com/articles/china-set-to-report-plunge-in-first-quarter-gdp-11587086697

Chin, A. W. H., Chu, J. T. S., Perera, M. R. A., Hui, K. P. Y., Yen, H. L., & Chan, M. C. W. (2020). Stability of SARS-CoV-2 in different environmental conditions. *The Lancet Microbe*, *1*, e10. doi:10.1016/S1473-3099(20)30113-4 PMID:32835322

Clancy, K. (2017). Digging deeper: Transdisciplinary and systems approaches to food security. *Journal of Agriculture, Food Systems, and Community Development*, *7*(4), 13–16. doi:10.5304/jafscd.2017.074.012

Dai, M., Li, H., Yan, N., Huang, J., & Zhao, L. (2020). Long-term survival of salmon-attached SARS-CoV-2 at 4oC as a potential source of transmission in seafood markets. *bioRxiv*. Advance online publication. doi:10.1101/2020.09.06.284695

Deng, L. Z., Mujumdar, A. S., Pan, Z., Vidyarthi, S. K., Xu, J., & Zielinska, M. (2019). Emerging chemical and physical disinfection technologies of fruits and vegetables: A comprehensive review. *Critical Reviews in Food Science and Nutrition*, *60*, 2481–2508. doi:10.1080/10408398.2019.1649633 PMID:31389257

Dunford, D., Dale, B., Stylianou, N., Lowther, E., Ahmed, M., & De la Torres Arenas, I. (2020). *Coronavirus: The World in Lockdown in Maps and Charts.* https://www.bbc.com/news/world-52103747

Dvorak, G. (2008). *Disinfection 101.* Center for Food Security and Public Health. http://www.cfsph.iastate.edu/Disinfection/Assets/ Disinfection101.pdf

Environmental Protection Agency. (2020). *List N: Disinfectants for Use Against SARS-CoV-2.* https://www.epa.gov/coronavirus/about-list-n-disinfectants-coronavirus-covid-19-0

Esteki, M., Regueiro, J., & Simal-Gándara, J. (2019). Tackling fraudsters with global strategies to expose fraud in the. *Food Chain*, *18*(2), 425–440. doi:10. 1111/1541-4337.12419 PMID:33336950

Euractiv. (2020). *No evidence of COVID-19 transmission through food, says EFSA*. https ://www.euractiv.com/section/coronavirus/news/no-evidence-of-COVID-19-transmission-through-food-says-efsa/

European Parliament and Council Regulation (EC) No 178/2002 (2002). https:// publications.europa.eu/en/publication-detail/-/publication/b18e2753-3f89-4e5e8b03-6c6a9d340ac1/language-en

Federal Institute for Risk Assessment. (2012). *EHEC outbreak 2011 Investigation of the outbreak along the food chain*. https://mobil.bfr.bund.de/cm/350/ehec-outbreak-2011-investigation-of-the-outbreak-along-the-food-chain.pdf

Filipic, A., Gutierrez-Aguirre, I., Primc, G., Mozetic, M., & Dobnik, D. (2020). Cold plasma, a new hope in the field of virus inactivation. *Trends in Biotechnology*, *38*, 1278–1291. doi:10.1016/j.tibtech.2020.04.003 PMID:32418663

Fisher, D., Reilly, A., Kang Eng Zheng, A., Cook, A. R., & Anderson, D. E. (2020). Seeding of outbreaks of COVID-19 by contaminated fresh and frozen food. *bioRxiv*. Advance online publication. doi:10.1101/2020.08.17.255166

Floarea, A., & Sgârciu, V. (2016). Smart refrigerator: A next generation refrigerator connected to the IoT [Conference session]. 8th international conference on electronics, computers and artificial intelligence (ECAI), Ploiesti. 10.1109/ECAI.2016.7861170

Food and Agriculture Organization of the United Nations & World Health Organization. (2020). *COVID-19, And Food Safety. Guidance For Food Businesses*. https://www.who.int/publications/i/item/covid-19-and-food-safety-guidance-for-food-businesses

Food and Agriculture Organization. (2020). *Q&A: COVID-19 pandemic–impact on food and agriculture*. https://www.fao.org/2019-ncov/q-and-a/impact-on-food-and-agriculture/en/

Food and Drug Administration. (2020). *Shopping for food during the COVID-19 pandemic -information for consumers*. Author.

Fuertes, G., Soto, I., Carrasco, R., Vargas, M., Sabattin, J., & Lagos, C. (2016). Intelligent packaging systems: Sensors and nanosensors to monitor food quality and safety. *Journal of Sensors*, *8*. Advance online publication. doi:10.1155/2016/4046061

Galanakis, C. M. (2020). The food systems in the era of the coronavirus (COVID-19) pandemic crisis. *Foods*, *9*(4), 523. doi:10.3390/foods9040523 PMID:32331259

Galvez, J. F., Mejuto, J. C., & Simal-Gandara, J. (2018). Future challenges on the use of blockchain for food traceability analysis. *Trends in Analytical Chemistry*, *107*, 222–232. doi:10.1016/j.trac.2018.08.011

Garraturo, F., del Guidice, C., Morelli, M., Cerullo, V., Libralato, G., & Galdiero, E. (2020). Persistence of SARs-CoV-2 in the environment and COVID-19 transmission risk from environmental matrices and surfaces. *Environmental Pollution*, *265*, 115010. doi:10.1016/j.envpol.2020.115010 PMID:32570023

Gnonlonfoun, R. (2017). *Restaurants Owner Strategies for Financial Sustainability Beyond 5 Years* [Doctoral dissertation]. Walden University.

Goodwin, R., Schley, D., Lai, K. M., Ceddia, G. M., Barnett, J., & Cook, N. (2012). Interdisciplinary approaches to zoonotic disease. *Infectious Disease Reports*, *4*(2), e37. doi:10.4081/idr.2012.e37 PMID:24470951

Gorbalenya, A. E., Baker, S. C., Baric, R. S., Groot, R. J. D., Drosten, C., & Gulyaeva, A. A. (2020). Severe acute respiratory syndrome-related coronavirus: The species and its viruses-a statement of the coronavirus study group. *bioRxiv*. Advance online publication. doi:10.1101/2020.02.07.937862

Gössling, S., Scott, D., & Hall, C. M. (2020). Pandemics, tourism and global change: A rapid assessment of COVID-19. *Journal of Sustainable Tourism*, *29*(1), 1–20. doi:10.1080/09669582.2020.1758708

Guan, W., Ni, Z., Hu, Y., Liang, W., Ou, C., & He, J. (2020). Clinical characteristics of coronavirus disease 2019 in China. *The New England Journal of Medicine*, *382*, 1708–1720. doi:10.1056/NEJMoa2002032 PMID:32109013

Hamming, I., Timens, W., Bulthuis, M. L., Lely, A. T., Navis, G., & van Goor, H. (2004). Tissue distribution of ACE2 protein, the functional receptor for SARS coronavirus. A first step in understanding SARS pathogenesis. *The Journal of Pathology*, *203*, 631–637. doi:10.1002/path.1570 PMID:15141377

Han, S., Kumar Roy, P., Iqbal Hossain, M., Bun, K. H., Choi, C., & Ha, S. D. (2021). COVID-19 pandemic crisis and food safety: Implications and inactivation strategies. *Trends in Food Science & Technology*, *109*, 25–36. doi:10.1016/j.tifs.2021.01.004 PMID:33456205

Harapan, H., Itoh, N., Yufika, A., Winardi, W., Keam, S., & Te, H. (2020). Coronavirus disease 2019 (COVID-19): A literature review. *Journal of Infection and Public Health*, *13*, 667–673. doi:10.1016/j.jiph.2020.03.019 PMID:32340833

Hecht, A. A., Biehl, E., Barnett, D. J., & Neff, R. A. (2019). Urban food supply chain resilience for crises threatening food security: A qualitative study. *Journal of the Academy of Nutrition and Dietetics*, *119*(2), 211–224. doi:10.1016/j.jand.2018.09.001

Henwood, A. F. (2020). Coronavirus disinfection in histopathology. *Journal of Histotechnology*, *43*(2), 102–104. doi:10.1080/01478885.2020.1734718 PMID:32116147

Hui, D. S., Azhar, E. I., Madani, T. A., Nitoumi, F., Kock, R., & Dar, O. (2020). The continuing 2019-nCoV epidemic threat of novel coronaviruses to global health – the latest 2019 novel coronavirus outbreak in Wuhan, China. *International Journal of Infectious Diseases*, *91*, 264–266. doi:10.1016/j.ijid.2020.01.009 PMID:31953166

International Monetary Fund. (2020). *World Economic Outlook, April 2020: The Great Lockdown*. https://www.imf.org/en/Publications/WEO/Issues/2020/04/14/weo-april-2020

Ivanov, D., & Dolgui, A. (2020). Viability of intertwined supply networks: Extending the supply chain resilience angles towards survivability. A position paper motivated by COVID-19 outbreak. *International Journal of Production Research*, *58*(10), 2904–2915. doi:10.1080/00207543.2020.1750727

Johnson, R. (2014). *Food fraud and "economically motivated adulteration" of food and food ingredients*. Congressional Research Service. https://fas.org/sgp/crs/misc/R43358.pdf

Kampf, G., Todt, D., Pfaender, S., & Steinmann, E. (2020). Persistence of coronaviruses on inanimate surfaces and their inactivation with biocidal agents. *The Journal of Hospital Infection*, *104*(3), 246–251. doi:10.1016/j.jhin.2020.01.022 PMID:32035997

Kampf, G., Voss, A., & Scheithauer, S. (2020a). Inactivation of coronaviruses by heat. *The Journal of Hospital Infection*, *105*, 348–349. doi:10.1016/j.jhin.2020.03.025 PMID:32243951

Kim, H. M., & Laskowski, M. (2018). Agriculture on the Blockchain: Sustainable Solutions for Food, Farmers, and Financing. In D. Tapscott (Ed.), *Supply Chain Revolution, Barrow Books 2018*. doi:10.2139srn.3028164

Kim, J., Kim, J., Lee, S. K., & Tang, L. (2020). Effects of epidemic disease outbreaks on financial performance of restaurants: Event study method approach. *Journal of Hospitality and Tourism Management*, *43*, 32–41. doi:10.1016/j.jhtm.2020.01.015

Kimball, A., Hatfield, K. M., Arons, M., James, A., Taylor, J., Spicer, K., & ... . (2020). Asymptomatic and presymptomatic SARS-COV-2 infections in residents of a long-term care skilled nursing facility. *Morbidity and Mortality Weekly Report*, *69*, 377–381. doi:10.15585/mmwr.mm6913e1 PMID:32240128

Kuswandi, B., Wicaksono, Y., Jayus, A. A., Heng, L. Y., & Ahmad, M. (2011). Smart packaging: Sensors for monitoring of food quality and safety. *Sensing and Instrumentation for Food Quality and Safety*, *5*(3), 137–146. doi:10.1007/ s11694-011-9120-x

Liu, Y., Ning, Z., Chen, Y., Guo, M., Liu, Y., Gali, N. K., Sun, L., Duan, Y., Cai, J., Westerdahl, D., Liu, X., Xu, K., Ho, K. F., Kan, H., Fu, Q., & Lan, K. (2020). Aerodynamic analysis of SARS-CoV-2 in two Wuhan hospitals. *Nature*, *1–4*. Advance online publication. doi:10.103841586-020-2271-3 PMID:32340022

Lu, R., Zhao, X., Li, J., Niu, P., Yang, B., Wu, H., & Wang, W. (2020). Genomic characterisation and epidemiology of 2019 novel coronavirus: Implications for virus origins and receptor binding. *Lancet*, *395*(10224), 565–574. doi:10.1016/S0140-6736(20)30251-8 PMID:32007145

Makroo, H., Majid, D., Siddiqi, M. A., Greiner, R., & Dar, B. N. (2020). *COVID-19 pandemic and its implication on food system*. Preprints., doi:10.20944/preprints202008.0321.v1

Muller, C. (2020, May 19). Will Dine-In Restaurants Survive the Pandemic? *Futurity*. https://www.futurity.org/ restaurants-after-COVID-19-pandemic-2369632/

Mullis, L., Saif, L. J., Zhang, Y., Zhang, X., & Azevedo, M. S. (2012). Stability of bovine coronavirus on lettuce surfaces under household refrigeration conditions. *Food Microbiology*, *30*(1), 180–186. doi:10.1016/j.fm.2011.12.009 PMID:22265299

Narayanan, L., Pandit, M., Basu, S., Karmakar, A., Bidhan, V., & Kumar, H. (2020). *Impact of lockdown due to COVID-19 outbreak: Lifestyle changes and public health concerns in India*. Preprints., doi:10.20944/preprints202006.0129.v1

Nardell, E. A., & Nathavitharana, R. R. (2020). Airborne spread of SARS-CoV2 and a potential role for air disinfection. *Journal of the American Medical Association*, *324*(2), 141–142. doi:10.1001/jama.2020.7603 PMID:32478797

Naserghandi, A., Allameh, S. F., & Saffarpour, R. (2020). All about COVID-19 in brief. *New Microbes and New Infections*, *35*, 100678. doi:10.1016/j.nmni.2020.100678 PMID:32292590

National Restaurant Association. (2020). *New Research Details Early Impact of Coronavirus Pandemic on Restaurant Industry*. https://restaurant.org/articles/news/study-details-impact-of-coronavirus-on-restaurants

Nicola, M., Alsafi, Z., Sohrabi, C., Kerwan, A., Al-Jabir, A., Iosifidis, C., & Agha, R. (2020). The socio-economic implications of the coronavirus pandemic (COVID-19): A review. *International Journal of Surgery*, *78*, 185–193. doi:10.1016/j.ijsu.2020.04.018 PMID:32305533

Oberholzer, M., & Febbo, P. (2020). *What we know today about coronavirus SARS-CoV-2 and where do we go from here*. Genetic Engineering and Biotechnology News. https://www.genengnews.com/insights/what-we-know-today-about-coronavirus-sars-cov-2-and-where-do-we-go-from-here/

Oliveira, T. C., Abranches, M. V., & Lana, R. M. (2020). Food security in Brazil in the context of the SARS-CoV-2 pandemic. *Cadernos de Saude Publica*, *36*(4), e00055220. doi:10.1590/0102-311X00055220 PMID:32267383

OziliP. K.ArunT. (2020). Spillover of COVID-19: impact on the Global Economy. SSRN 3562570. doi:10.2139/ssrn.3562570

Pan, X., Chen, D., Xia, Y., Wu, X., Li, T., & Ou, X. (2020). Asymptomatic cases in a family cluster with SARS-CoV-2 infection. *The Lancet. Infectious Diseases*, *20*, 410–411. doi:10.10161473-3099(20)30114-6 PMID:32087116

Peeri, N. C., Shrestha, N., Rahman, M. D. S., Zaki, R., Tan, Z., & Bibi, S. (2020). The SARS, MERS and novel coronavirus (COVID-19) epidemics, the newest and biggest global health threats. What lessons have we learned? *International Journal of Epidemiology*, *49*, 717–726. doi:10.1093/ije/dyaa033 PMID:32086938

Pezzotti, G., Ohgitani, E., Shin-Ya, M., Adachi, T. D., Marin, E., & Boschetto, F. (2020). Rapid inactivation of SARS-CoV-2 by silicon nitride, copper, and aluminum nitride. [Preprint]. *bioRxiv*. Advance online publication. doi:10.1101/2020.06.19.159970

Pigini, D., & Conti, M. (2017). NFC-based traceability in the food chain. *Sustainability*, *9*(10), 1910–1930. doi:10.3390u9101910

Pressman, P., Naidu, A. S., & Clemens, R. (2020). COVID-19 and Food Safety: Risk Management and Future Considerations. *Wolters Kluwer Health*, *55*(3), 125–128.

Pung, R., Chiew, C. J., Young, B. E., Chin, S., Chen, M. I., & Clapham, H. E. (2020). Investigation of three clusters of COVID-19 in Singapore: Implications for surveillance and response measures. *Lancet*, *395*(1029), 1039–1046. doi:10.1016/S0140-6736(20)30528-6 PMID:32192580

Quevedo-Leon, R., Bastías-Montes, J. M., Espinoza-Tellez, T., Ronceros, B., Balic, I., & Munoz, O. (2020). Inactivation of coronaviruses in food industry: The use of inorganic and organic disinfectants, ozone, and UV radiation. *Scientia Agropecuaria*, *11*(2), 257–266. doi:10.17268ci.agropecu.2020.02.14

Randazzo, W., Fabra, M. J., Falcó, I., López-Rubio, A., & Sánchez, G. (2018). Polymers and biopolymers with antiviral activity: Potential applications for improving food safety. *Comprehensive Reviews in Food Science and Food Safety*, *17*, 754–768. doi:10.1111/1541- 4337.12349 PMID:33350126

Ratnesar-Shumate, S., Williams, G., Green, B., Krause, M., Holland, B., & Wood, S. (2020). Simulated sunlight rapidly inactivates SARS-CoV-2 on surfaces. *The Journal of Infectious Diseases*, *222*, 214–222. doi:10.1093/infdis/jiaa274/5841129 PMID:32432672

Skara, T., & Rosnes, J. T. (2016). Emerging methods and principles in food contact surface decontamination/prevention environmental factors in infectious disease. In C. E. Leadley (Ed.), *Innovation and Future Trends in Food Manufacturing and Supply Chain Technologies* (pp. 151–172). Woodhead Publishing Series in Food Science. doi:10.1016/B978-1-78242-447-5.00006-X

Sportelli, M. C., Izzi, M., Kukushkina, E. A., Hossain, S. I., Picca, R. A., & Ditaranto, N. (2020). Can nanotechnology and materials science help the fight against SARS-CoV-2? *Nanomaterials (Basel, Switzerland), 10*(4), 802. doi:10.3390/nano1004 0802 PMID:32326343

Steardo, L., Steardo, L. Jr, Zorec, R., & Verkhratsky, A. (2020). Neuroinfection may attribute to pathophysiology and clinical manifestations of COVID-19. *Acta Physiologica (Oxford, England), 229*(3), e13473. Advance online publication. doi:10.1111/alph.13473 PMID:32223077

Tong, Z. D., Tang, A., Li, K. F., Li, P., Wang, H. L., & Yi, J. P. (2020). Potential presymptomatic transmission of SARS-CoV-2, Zhejiang Province, China. *Emerging Infectious Diseases, 26*, 1052–1054. doi:10.3201/eid2605.200198 PMID:32091386

Van Doremalen, N., Bushmaker, T., Morris, D. H., Holbrook, M. G., Gamble, A., & Williamson, B. N. (2020). Aerosol and surface stability of SARS-CoV-2 as compared with SARS-CoV-1. *The New England Journal of Medicine, 382*(16), 1564–1567. doi:10.1056/NEJMc2004973 PMID:32182409

Wang, W., Tang, J., & Wei, F. (2020). Updated understanding of the outbreak of 2019 novel coronavirus (2019-nCoV) in Wuhan, China. *Journal of Medical Virology, 92*(4), 441–447. doi:10.1002/jmv.25689 PMID:31994742

Warnes, S., Little, Z. R., & Keevil, C. W. (2015). Human Coronavirus 229E Remains Infectious on Common Touch Surface Materials. *mBio, 6*(6). Advance online publication. doi:10.1128/mBio.01697-15 PMID:26556276

Wei, W. E., Li, Z., Chiew, C. J., Yong, S. E., Toh, M. P., & Lee, V. J. (2020). Presymptomatic transmission of SARS-CoV-2. *Morbidity and Mortality Weekly Report, 69*, 411–415. doi:10.15585/mmwr.mm6914e1 PMID:32271722

Williams, G., & Price, M. (2020, March 18). Businesses Pleased with Support. *Otago Daily Times*. https://www.odt.co.nz/regions/queenstown/businesses-pleased-support

World Health Organization. (2014b). *Infection prevention and control of epidemic and pandemic Prone Acute respiratory infections in health care*. WHO Press.

World Health Organization. (2014a). *Zoonotic Disease: Emerging Public Health Threats in the Region*. http://www.emro.who.int/about-who /rc61/zoonotic-diseases.html

World Health Organization. (2020a). *Infection Prevention and Control during Health Care When Novel Coronavirus (nCoV) Infection is Suspected: Interim Guidance*. https://www.who.int/publicationsdetail/infection-prevention-and-control-during-health-care-when-novelcoronavirus-(ncov)-infection-is-suspected-20200125

World Health Organization. (2020b). *Water, Sanitation, Hygiene and Waste Management for COVID-19: Technical Brief.* https://www.who.int/publications-detail/watersanitation-hygiene-and-waste-management-for-COVID-19

World Health Organization. (2020c). *Coronavirus disease (COVID-19) pandemic.* http ://www.who.int/emergencies/diseases/novel-coronavirus-2019/

World Health Organization. (2020d). *Cleaning and disinfection of environmental surfaces in the context of COVID-19: Interim guidance.* https://www.who.int/ publications/i/item/cleaning-and-disinfection-of-environmental-surfaces-inthecontext-of-COVID-19

World Health Organization. (2021). *WHO Coronavirus (COVID-19) Dashboard.* https://covid19.who.int/

Wu, Y., Guo, C., Tang, L., Hong, Z., Zhou, J., & Dong, X. (2020). Prolonged persistence of SARS-CoV-2 viral RNA in faecal samples. *The Lancet. Gastroenterology & Hepatology, 5*(5), 434–435. doi:10.1016/S2468-1253(20)30083-2 PMID:32199469

Xu, J., Guo, S., Xie, D., & Yan, Y. (2020). Blockchain: A new safeguard for agri-foods. *Artificial Intelligence in Agriculture, 4*, 153–161. doi:10.1016/j.aiia.2020.08.002

Yang, C. (2020). Does hand hygiene reduce SARS-CoV-2 transmission? *Graefes Archive for Clinical and Experimental Ophthalmology, 258*, 1133–1134. doi:10.100700417-020-04652-5 PMID:32221693

Yazaki, Y., Mizuo, H., Takahashi, M., Nishizawa, T., Sasaki, N., Gotanda, Y., & Okamoto, H. (2003). Sporadic acute or fulminant hepatitis E in Hokkaido, Japan, may be food-borne, as suggested by the presence of hepatitis E virus in pig liver as food. *The Journal of General Virology, 84*(9), 2351–2357. doi:10.1099/vir.0.19242-0 PMID:12917455

Yépiz-Gómez, M. S., Gerba, C. P., & Bright, K. R. (2013). Survival of respiratory viruses on fresh produce. *Food and Environmental Virology, 5*, 150–156. doi:10.100712560-013-9114-4 PMID:23681671

Yiannas, F. (2018). A new era of food transparency powered by blockchain. *Innovations: Technology, Governance, Globalization, 12*(1–2), 46–56. doi:10.1162/ inov_a_00266

Yin, C., Ling, W., Hong, S., & Yan, X. (2020). Vitro inactivation of SARS-CoV-2 by commonly used disinfection products and methods. Preprints. 10.2 1203/rs.3.rs-43489/v1

Yoshimoto, J., Ono, C., Tsuchiya, Y., Kabuto, S., Kishi, M., & Matsuura, Y. (2020). *Virucidal effect of acetic acid and vinegar on SARS-CoV-2.* Preprints., doi:10.21203/rs.3.rs-37750/v1

Yu, P., Zhu, J., Zhang, Z., Han, Y., & Huang, L. (2020). A familial cluster of infection associated with the 2019 novel coronavirus indicating potential person-to-person transmission during the incubation period. *The Journal of Infectious Diseases, 221*, 1757–1761. doi:10.1093/infdis/jiaa077 PMID:32067043

Zhu, N., Zhang, D., Wang, W., Li, X., Yang, B., & Song, J. (2019). A Novel Coronavirus from Patients with Pneumonia in China. China Novel Coronavirus Investigating and Research Team. *The New England and Journal of Medicine, 382*(8), 727-733.

Zuber, S., & Brüssow, H. (2020). COVID-19: Challenges for virologists in the food industry. *Microbial Biotechnology*, 1–13. doi:10.1111/1751-7915.13638 PMID:32700430

Chapter 7
The Present and Future of Gastronomy Against Global Threats

Cigdem Sabbag
Adiyaman University, Turkey

ABSTRACT

In today's world, tourism is one of the world's leading economic sectors in revenue generation with the number of jobs and clients in addition to its cultural services. To get a share from this economic cake, many countries highlight their gastronomic values to enhance their attractiveness, competition, and sustainability of the destination. Countries promote their global, regional, and local cuisines by employing a variety of methods. In the last 20 years, adding a gastronomic component to historical touristic trips has accelerated the development of gastronomy by attracting not only tourists but also residents of the region. Despite these encouraging developments, climate change, desertification, biodiversity loss, increasing population pressure, and pandemic processes, all of which are global issues threatening food security, pose threats to gastronomy culture. In this context, immediate steps should be taken to ensure the survival of the newly emerging gastronomy community.

INTRODUCTION

It would not be incorrect to say that the kitchen is an indicator of almost all components of a geography. Climate, geology, natural vegetation, soil and water resources, culture, and traditions all influence the cuisine of a particular place (Gillespie, 2001). The basic kitchen materials of various countries are commodities grown in the geography and climatic settings of those countries. Societies then have established a food production culture that will give them with the energy they require in their daily lives by utilizing the materials available in their cultures. The eating habits of agricultural communities that rely heavily on physical strength and the food habits of nomadic societies, for example, differ significantly (Sabbağ, 2012). Travelers have visited other locations because of these diverse tastes since ancient times. Lucian from Samsat (2nd Century) expressed his embarrassment at not knowing the food customs of that

DOI: 10.4018/978-1-7998-9148-2.ch007

region at a dinner invitation in a different environment than the one he lived in (Wilkins & Hill, 2009). Gastronomy is also a nice example of globalization because it facilitates intercultural connection by facilitating cultural exchange everywhere now (Hwang et al., 2005). As travelers try dishes from other cultures on the location, the level of intercultural social interaction improves. Except for ceremonies, eating was once regarded a basic need, but it has now evolved into a source of pleasure, amusement, and show (Çaycı, 2019). The identification of local, regional, and national identities through gastronomy, on the other hand, is threatened by the globalization (Richards, 2003). Local food cultures, which are material cultural components, have been endangered aftereffect of globalization, particularly due to global food chains, and societies' lifestyle and, most crucially, eating habits have modified.

It would not be inappropriate to suggest that gastronomy would undergo a transformation and change due to climate change, which is one of the worldwide challenges that has become increasingly apparent in recent years and is expected to become much more severe in the future. Climate change, according to researchers, will harm the nutritional content of the food as well as yield (Hummel et al., 2018; Berbegal et al., 2019). Extreme temperatures caused by climate change are said to have an impact on animal growth, meat, milk, egg yield, and quality, as well as having a detrimental impact on their taste (Koyuncu, 2017). Climate change is also putting small indigenous civilizations, which are already endangered, in jeopardy. Climate extremes, such as record-breaking heat, drought, and the burning of forests, many of which are natural food supplies, are threatening to undermine local food systems, which are the backbone of gastronomy (Zavaleta-Cortijo et al., 2020). Protecting the local food system from threats (production, supply chain, etc.) will also ensure the survival of the local food culture. Covid-19, for example, has caused change and transformation in the economic, political, social, and cultural fields in all countries worldwide (Kurtdaş, 2020). Although the Covid-19 epidemic did not result in significant production losses in agricultural and animal production on a country-by-country basis, the control over the export and import of certain foods for transportation between countries and their own food safety has raised concerns about future crises in global food imports and exports. According to the OECD report, over 60 countries prohibited the export of agricultural and food products during the early stages of Covid-19 (OECD, 2020).

Moreover, due to the pandemic, consumers' behavior in the tourism field has changed, forcing them to maintain restricted travel, comply to tight hygiene, and maintain social distance (Korkmaz & Başaran, 2021). Because of Covid-19, tourists prioritize current health measures, as well as taste, location, and price, when choosing restaurants for eating and drinking. According to Kim et al. (2021), the importance of cleanliness during Covid-19 in small or large-scale restaurants has become much more important than before in the customer's restaurant selection. In addition to these, foods in the gastronomy sector, as well as cleaning and hygiene, should contain immune-boosting ingredients. Rapid urbanization and population increase endangers valued local cultural characteristics like gastronomy, which have been acquired over time (Aksoy & Sezgi, 2015). The present global population of 7.7 billion people is expected to rise to 9.7 billion by 2050 (United Nations, 2019). Food safety will be undermined in the future due to a lack of supplies for appropriate nourishment for an ever-increasing population, and countries' gastronomy will feel the consequences.

BACKGROUND

The current and future impact of global environmental problems on gastronomy is outlined in this review. Globalization, climate change, biodiversity loss, Covid-19, population pressure, and migration, all of which endanger human life and the environment, have an impact on gastronomy alone or with a combination. According to global future forecasts (Calıcıoğlu et al. 2019, Kubiszewski et al. 2020, Saghai, 2021), the extent and prevalence of the said problems will worsen in the future. Therefore, the current and future impact of each problem has been evaluated using current sources, and its impact on gastronomy has been reviewed in the main part of the text.

The global threats discussed in this research are issues that pose a serious threat to food safety (Ristaino et al., 2021; Rasul, 2021), and because gastronomy is a food phenomenon, it is directly affected by these problems. In today's world, when hunger and malnutrition are still problems, thorough identification and prevention of dangers to gastronomy, which is an important part of tourism, is vital (FAO hunger). Tourism employed 334 million people globally, accounting for 10.6% of all jobs prior to the pandemic, and generated US$9.2 trillion, or 10.4% of global gross domestic product (WTTC, 2022). As a result, exhibiting dangers to gastronomy can be deemed a priority issue for tourism. Moreover, the UN's Sustainable Development Goals are also closely linked to ensuring the sustainability of gastronomy in tourism, which involves 1.7 billion people with 1.4 billion travelers and 334 million employees. Zero Hunger (Goal 2), Good Health and Well-being (Goal 3), Decent Work and Economic Growth (Goal 8), Responsible Consumption and Production (Goal 12), Clean Water and Sanitation (Goal 6), Climate Action (Goal 13), Life below Water (Goal 14), and Life on Land (Goal 15) are some of the UN's goals that are directly related to gastronomy. In addition, due to quarantines and travel prohibitions, tourism is one of the first industries to be negatively impacted by the pandemic, which began in 2019 (Nicola et al., 2020). Within the context of the forecast that the pandemic will continue with different SARS-CoV-2 mutants (Skegg et al. 2021), it is vital that tourism be prepared for those kinds of processes. To summarize, gastronomy is a phenomenon that is highly vulnerable to global issues, making it important to strengthen its resilience.

Covid-19 is, obviously, the book's main focus. Global issues, on the other hand, are said to be one of the factors that contributed to the outbreak. In the analysis of this circumstance, Mishra et al., (2021) expressed the drivers of Covid-19 as follows "Despite their differences in origin, the major factors that contribute to the emergence and spread of zoonotic pandemics include habitat fragmentation, deforestation, biodiversity loss, intensive agriculture, livestock farming, uncontrolled urbanization, pollution, climate change, and bush meat hunting and trading...". Another study that supports the idea that global environmental and socio-economic issues are triggered Covid-19 can be found in Harvard's web page (Harvard, n.d.) as follows: "As the planet warms, animals large and small, on land and at sea, are flocking to the poles to escape the heat. This means that animals are coming into contact with other animals they would not normally come into contact with, creating an opportunity for pathogens to enter new hosts". Large-scale and rapid environmental changes over the last few decades have been suggested as a contributor in the emergence of the Covid-19 pandemic, increasing the likelihood of zoonotic disease emergence and dissemination, reducing food security (Gorji & Gorji, 2021). Lawler et al. (2021) have presented strong arguments that the Covid-19 pandemic is closely associated to ecosystem health and biodiversity loss. Skegg et al. (2021) argue in the Lancet journal that unless countries collaborate more to avert global problems, pandemics and other transboundary disasters with similar implications, such as those caused by climate change, would remain a constant threat.

MAIN FOCUS OF THE CHAPTER

The current and future effects of the pandemic, which no one could have predicted before 2019, on food safety and thus gastronomy, as well as the negative environmental effects, particularly climate change and biodiversity loss, which are becoming increasingly felt around the world, and the socio-economic pressures brought on by population growth and migration. Their probable consequences have indeed been sought to be revealed through the study of a wide number of sources from around the world.

MATERIALS AND METHODS

The terms "Gastronomy," "Globalization," "Climate Change," "Biodiversity Loss," "Covid-19," "Population Growth," and "Migration" were searched in the Web of Science and Scopus databases between April 2021 and January 2022. Within the scope of the aforementioned keywords, the threats confronting gastronomy were reviewed. The global and environmental challenges that anticipate gastronomy in the future were evaluated in view of the information gathered from the sources, and suggestions for the sustainability of gastronomic activities were intended to be proposed (Figure 1).

GLOBAL THREATS TO GASTRONOMY

Globalization and Gastronomy

With changes in the conception of tourism, or the strategy of engaging into the daily and traditional life of the places visited, travelers explore a variety of social and cultural concepts before deciding on a destination. Gastronomy, in this sense, plays a crucial role in tourism destination competitiveness, distinction, and branding (Haven-Tang & Jones, 2005; López-Guzmán & Sánchez-Cañizares, 2012). Local dishes are effective in destination selection because tourists can get information about the food of the place, they are visiting directly from tourists who've already eaten there, thanks to social media sites like TripAdvisor and Hotwire, which arose because of communication technology advancements in the last 20 years. The genuineness, taste and place of the dishes affect the satisfaction of the travelers (Björk & Kauppinen-Räisänen, 2016). Gastronomy, on the other hand, aids to the tourism growth of a destination and the formation of a particular identity (Stavrianea et al., 2017). Gastronomy also contributes to the positive side of globalization by facilitating international dialogue through cultural exchange (Hwang et al., 2005). As travelers sample the cuisines of various cultures, the level of intercultural social interaction grows. Until recently, food was regarded solely as a necessity, but in today's tourism concept, it has evolved into a source of pleasure, entertainment, and side show factor (Çaycı, 2019).

Globalization is a complicated phenomenon with economic, social, political, technological, and cultural components (Mir et al., 2014). It can have an impact on many parts of life, from music to dressing, behavior, and traditional food (Garvey, 2010). As a result, the globalization process poses a threat to local, regional, and national gastronomy (Richards, 2003). Globalization has resulted in a homogenization of food culture, as well as changes in the production and consumption of local foods (Kolawole Olayiwola, 2007). In developing countries where globalization and rapid urbanization are taking place, the media has an impact on local people's culinary tastes and preferences by influencing the original

Figure 1. Publication dates of references reviewed

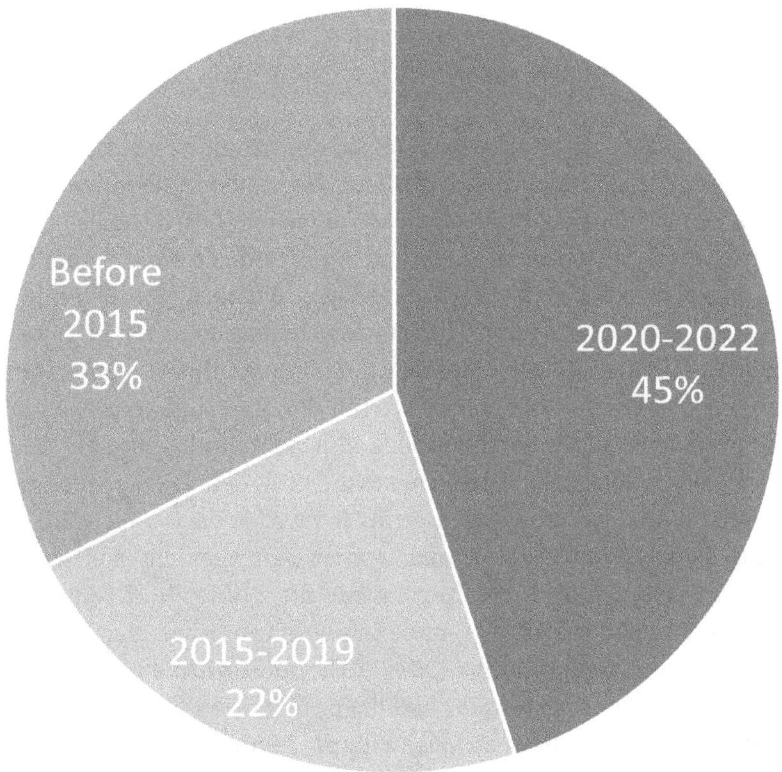

taste and preparation of traditional dishes, and even reducing access to these culinary delights (Mendez & Popkin, 2004). International food chains adapt local favorites and sell them to customers after making numerous modifications, as can be seen in falafel, a well-known Middle Eastern traditional meal (Ram, 2004). Many consumers remark that, like falafel, there are significant flavor differences between modified and traditional dishes (Sudiyarto, et al., 2018). As a result, it's important to keep in mind that these modifications have a negative impact on local cuisines, which have evolved over thousands of years of experience.

Food intake is often conservative, which means that those with a well-developed palate may be skeptical of novel flavors. Some travelers, on the other hand, are delighted with the homogeneity of gastronomy since it enables them to eat in a cost-effective, predictable, and safe manner anywhere in the globe. The vast majority, however, opposes this process since it leads gastronomy to become homogenized and lose its identity due to fast food's unhealthy and artificial standardization (Richards, 2003). Global tourism has grown inevitably by the advancements in communication, technology, and transportation (Dwyer, 2015). In the 1960s, global passenger numbers increased from ten to twenty million to 1.5 billion in 2019 (UNWTO, 2020). Given the current global population of 8 billion people, tourism is estimated to service close to 20% of the world's population. Intercultural interaction has a positive and negative impact on the material and spiritual cultural elements of societies triggered by high population movement. Global food chains have put local food culture, which is a material cultural element, in jeopardy, and as a result, society's lifestyles and, most significantly, eating habits have shifted. On a global scale, the expansion of fast-food restaurant franchises has changed and continues to modify people's eating habits. For example, in the 2000s, McDonald's, the world's largest food company with more than 60 years in business, served 43 million people each day with around 27,000 restaurants, whereas in 2019, before the pandemic, it served 69 million customers with 38,695 restaurants in 119 countries (McDonald's Reports Fourth Quarter and Full Year 2020).

Gastronomic cultural change can be viewed as positive in some respects, as it essentially leads to the formation of new tastes. The colonial period effect, for example, increased the use of hot spices in England. Most countries are moving to Mediterranean cuisine-based meals because of tourism and awareness, which have existed since ancient times and are regarded a healthy diet due to its wide range of food choices (Martínez-González et al., 2017). Sabbağ & Mankan (2012), on the other hand, claim that globalization caused food consumption shifting toward fast food rather than new exotic preferences in several countries. They also discovered that, even though this tendency is especially prevalent in children and young people, responsible parties are not doing enough to prevent malnutrition. In the developed world, the proportion of homemade foods in a household's diet has declined dramatically over the last few decades (Crossley et al., 2018). Due to the growth in average income, changes in eating patterns such as eating out more and/or skipping meals in favor of easy-to-prepare quick foods at home have occurred, owing to the increase in leisure time, women's engagement in business life, and extended working hours (Jabs & Devine, 2006). In addition, while the table is traditionally the gathering place for all family members, industrialization has resulted in a shift toward eating alone, which has increased the consumption of ready-made food (Gürhan, 2017). The most visible evidence of this is the rise in the selling of fast-cooked items in supermarkets and the proliferation of fast-food restaurants. Fast food restaurants are still operating in numerous nations, with productions tailored to each country's culture. Transnational chain corporations, for example, avoid using pork by adjusting the products they offer in Muslim countries, and since Hindu traditions prohibit the consumption of beef, original recipes are altered to include chicken, sheep, and goat meat (Ahmad, 2013). In formulating foods that respond to local tastes, these multinational corporations have been found to adopt global approaches in the standardization of local products, such as fast cooking, high fat, and high carbohydrate (Orkun, 2009). The high energy, saturated fat, salt concentration, and lack of fruit and vegetable content of the foods created following these adjustments resulted in several health issues, including obesity, particularly among youngsters, who are the group that eats these goods the most (Paeratakul et al., 2003). Barriers to the cross-border movement of goods, services, and capital have been reduced by globalization (Shetty, 2003). Thanks to technological developments, the international trade in new foods has led to a shift in kitchens. The recognition and cultivation of the new products not usually grown locally in countries have begun. It is well known that increased human mobility due to reasons like labor, migration, and tourism which speed up the process of globalization also helps to change the cuisine. Tourism has aided in the production of global food items while also strengthening local culinary traditions as a tool of globalization (Hall & Mitchell, 2003). Exotic plants brought to local habitats because of increasing global trade have an impact on food culture change. The greater use of avocado and kiwi in world cuisines over the last 20 years is one of the most visible examples of this change. Avocado output in Turkey has exceeded 13.000 tons in the last five years (2015-2019), and kiwi production has exceeded 267,000 tons (FAO, n.d.), according to data from the Turkish Ministry of Agriculture. Numerous studies have shown that the losses of globalization in terms of local food culture outnumber the positives. The research on the negative features of fast food in the book "fast food nation," which was published in 2001 and became one of the best-selling books at the time (Schlosser, 2001), is the most notable study on the subject. In many nations, a movement to preserve food culture has begun because of the negative effects of globalization on local culinary culture. The slow-food movement, which began in Italy in the 1980s, was founded on the premise of protecting local food culture against fast-food products. Slow food has grown into a global movement including millions of individuals in over 160 countries who are trying to assure that everyone has access to healthy, clean, and balanced food. Many countries are taking numerous steps to disclose and preserve their endangered national and regional culinary cultures. Slow food allows guests to enjoy local cuisine rather than globalized (modified) food (Slow Food, n.d.). Countries, in a way, tend to localize in opposition to globalization. National and regional customs, religious practices, tastes, and gastronomic cultures, according to Hall & Mitchell (2003), are powerful variables that clearly limit food globalization. However, it is unclear how long these local movements will be able to survive global food pressure.

Many countries have attempted to preserve their current cultures to protect their culinary traditions from the threat of globalization and avoid extinction. These activities necessitate a significant amount of money and labor. In this situation, the truth that local foods' production and consumption costs are higher than fast food is seen as a significant disadvantage. Local meals have become more expensive in recent years (Lehtinen, 2012). it is believed that local items generated from this unit area cannot compete with cash crops such as maize. Furthermore, USDA (2020) reported that the increase in labor costs in recent years has affected the organic product price by 1/3. This high input cause farmers to leave the product in the field which contributed negatively to the availability and prices of products. So, this kind of actions reduces organically produced crops use in kitchens (Astil, 2020).

Climate Change and Gastronomy

The key physical variables in the evolution of local cuisine are climate, soil, and water supplies. The cuisine of countries is made up of the diversity of plants and animals that emerge in response to physical elements, as well as the expertise of food production and preparation handed down through generations. Climate change harms biodiversity, which is the driving force behind the extinction of plant and animal species that are fundamental to the ecosystem (IPCC, 2019). Extreme temperature differences during plant development, irregularities in storms and precipitation, frosts, floods, and drought all have a detrimental effect on the morphological, physiological, biochemical, and yield characteristics of plants (Jangid & Dwivedi, 2016). Hurricane Harvey, which struck Texas in 2017, was projected to have caused a monetary loss of 125 billion dollars (Cho, 2019). Extreme cold weather events caused a 1.3% yield loss in grains and 2.6% loss in non-grain products, according to a study done in 28 European Union member countries between 1961 and 2018, whereas heat waves produced losses of up to 7%. (Cross, 2021). According to Russo et al. (2015), with the RCP8.5 scenario, the number of heatwaves will rise in the future. The increase in atmospheric CO_2 levels from 337 to 415 parts per million in the last 40 years has caused complications in agricultural output and food storage, along with reducing the quality of food produced (Co2 earth, n.d.). The cooling demand of pistachio, which ranges between 500-1000 hours under 7.2°C in the Southeastern Anatolia Region of Turkey, which is regulated by the continental climate, will be threatened by the predicted 2°C temperature increase (Tekin et al., 2001). In addition to raising prices, this predicament will necessitate the cultivation of plants other than pistachio in the region. This negative trend will also lead to a decrease in the use of pistachio, which is the basis of pastry desserts in Turkish cuisine and will jeopardize the production of geographically registered desserts such as baklava.

Climate change will inevitably affect the materials used in the kitchen by reducing agricultural production and biodiversity. Vegetables with high water requirements, as well as cereals like corn, will suffer in terms of yield and nutritional quality mostly as consequence of natural stress caused by drought and irregular rainfall. A rise in food prices soon could be the result of these negative events, as well as the cessation of local crop production and perhaps farmers' withdrawals from agricultural areas altogether. Drought-related land abandonment has been documented in Chihuahua's arid northwestern region in the 1930s, 1950s, and 1990s (Sandoval, 2003). According to reports, after 2010, the drought was as effective as political conflicts in causing Syrians to flee to neighboring nations. Farmers in Syria who were farming on the relatively fertile Mesopotamian regions lost a substantial amount of their revenue between 2006 and 2010 because of the drought, which was the worst since 1980 (DW, 2020).

Food quality is directly linked to geography, it is widely assumed that food cultivated in the geographical conditions to which it belongs is of greater quality. Climate change will almost certainly have an impact on food quality. In viticulture, for example, the flavor and aroma of grapes in red and white wines are regulated by geographical factors (Harrington, 2005). Likewise, so the acid, sugar, and lycopene content of the tomato, which originated in the Americas, is influenced by high temperatures, not only the appearance but also the aroma, as well as the taste and flavor, cannot reach the desired level (Dumas et al., 2003; Gruda, 2005). According to Sugiura et al. (2013), the flavor and textural features of apple grown in Japan have changed by global warming, based on 30-40 years of records. According to a study conducted by Melbourne University (2015), the texture and taste of carrots, which are widely used in Australian cuisine, will deteriorate significantly, cabbage may become more bitter, eggplant quality and shape deformations will occur, and canola oil will lose about a quarter of its nutritional value as climate changes. Climate change will have an impact on more than just the taste and flavor quality of foods. Crops grown in different regional soils may be lost because they are unable to adapt to climate change or will require climate-controlled growing environments with extremely high inputs to be cultivated. In this case, countries will look for new products that are suitable for changing climatic conditions, and the gastronomy map will most likely change. Mostly as consequence, flavors that have been around for centuries will be replaced by new foods. Wild herbs, for example, are cultivated in a completely natural environment and contribute to a variety of flavors in the creation of local dishes. In some locations, these herbs are among the principal gourmet items, and various events, are held to highlight the natural flora richness, such as the Herb Harvest Fall Festival featuring the foods of South America. Belsinger, (n.d.), on the other hand, become more difficult to find in their natural environment and may even disappear with climate change. According to an interview with locals in Uganda's Teso-Karamoja region, 100 different wild herbs were used to prepare dishes from the forest area. It has been claimed that consuming these plants ensures food safety and cultural taste continuity. However, it is claimed that this food richness will be adversely affected by plant decline and extinction, and that this will irreversibly change the Teso-Karamoja weed gastronomy culture (Ojelel et al., 2019). Luczaj et al. (2020) revealed that weed use is decreasing today compared to the 19th century in their study on the prevalence of weed use in kitchens in Poland, Italy, Spain, Estonia, and Sweden. The reason for this decrease was attributed to a decrease in plant culture due to an ecosystem change by the authors. It is undeniable that global warming and drought will affect countries' crop production. Alike potato production, which plays an important role in people's carbohydrate consumption and is relatively inexpensive, mainly takes place in cool and humid areas, such as Northern Europe (Thornton & Cramer, 2012) however temperature increase in these regions will most probably hamper potato production.

Climate change has an impact not only on plant species but also on animal husbandry. Climate change has resulted in decreases in the amount and quality of animal production, increased sensitivity to diseases and pests, animal losses, and regressions in the conversion of feed into the product, as seen in the case of silage corn (Koyuncu, 2017). These conditions also have a negative impact on the quality of milk used in cheese production (Melbourne University, 2015). Most nomadic countries' culinary cultures are based on meat and dairy products, particularly, Kenya, Iran, Somalia, Morocco, Algeria, Kazakhstan, Kyrgyzstan, and Afghanistan. Because the pastures where herds breed will be adversely affected by climate change, meat and milk production will decrease directly, and due to meeting demand, prices will rise and their use in meals will decrease. New Zealand's mutton consumption, for example, has decreased from 25.5 kg per capita in 2000 to slightly under 4 kg per capita in 2020, according to the Data Agency Statista. High prices, according to Beef and Lamb New Zealand's CEO Rod Slater, are a major factor in the reduction in domestic red meat consumption (Dunkley, 2021).

Regardless of the cause of food price increases, many studies show that price has a significant impact on eating and drinking habits. It has been demonstrated, for example, that a 1% increase in grain prices from 136 different countries results in a 0.61% decrease in consumption (Green et al., 2013). Likewise, the increase in food costs in Iran between 2004 and 2012 ended in households spending 48% of their income on food, and they changed their food consumption to cereals, which are cheaper calorie sources, in 2012 (Ghahremanzadeh & Ziaei, 2014). In this case, it is expected that, instead of animal products, alternative products with high-calorie content and low prices will predominate in the local cuisine. Cases illustrated that climate change will cause transformation and change by influencing traditional tastes in gastronomy culture.

Fish, which our ancestors consumed for thousands of years, is not only a source of animal protein but also a unique food that should be consumed at a young age because it contains EPA (eicosapentaenoic acid) and DHA (docosahexaenoic acid) fatty acids, which play a role in brain development. These nutritional advantages, as well as the fertile waters, have increased fish consumption in recent years, particularly in developing countries (De Silva & Soto, 2009). Seafood now, which is consumed by 3.3 billion people, accounts for 20% of animal protein consumption per capita. On the other side, seafood, which is a protein source that is expected to decline by climate change, will have an adverse influence on developing nations' fish consumption culture, economy, and diet. Water temperature rise is one of the negative effects of climate change. Global sea surface temperatures are expected to rise by 0.4 – 1.1°C by 2025, according to the IPCC (2019). In today's world, where yearly sea-based food production has risen to 155 million tons (Ritchie & Roser, 2019), unfavorable sea-level rise will represent a significant danger to the food security of coastal countries as well as seafood-importing countries.

Climate change, on the other hand, is expected to increase the population of some sea creatures. Since 1924, 149 tropical marine species have spread to the Mediterranean, according to research (Raitsos et al., 2010). This study predicted that conditions caused by the invasion of alien species threaten the diversity of traditional seafood used in local cuisine. Changes in seafood production will put undue pressure on costs, limiting the chance of all segments of society consuming seafood, particularly omega 3 fatty acids (Gephart et al., 2020; Farmery et al., 2021). Furthermore, indigenous communities' fisheries in affected areas information about fishing heritage will be lost, resulting in the decline and maybe loss of local gastronomic delicacies day by day. Due to a rise in the number of unexpected weather incidents, in the Ondo and Ekiti Regions of Niger, 61.6% of total aquaculture yield was lost in just one flood (Adebo & Ayelari, 2011). Most ethnic groups who rely on fishing for a living in coastal locations are among the most vulnerable communities to climate change, and it has been suggested that they are attempting to flee the problem by migrating driven by climate change (Maldonado et al., 2013). The Shishmaref Community, which lives in the Bering Strait in Alaska, are well-known for changing their living habitat throughout time due to adverse weather conditions (Marino, 2012). Climate change not only deprives vulnerable populations of their income and food resources but also endangers their cultural identity (IPCC, 2019).

Covid-19 and Gastronomy

WHO labeled the coronavirus pandemic a global public health threat on March 11, 2020. Since the outbreak of Covid-19 in July 2021 (WHO, 2020e), more than 184 million people have been infected and 4 million have died worldwide (Statista, n.d.) despite states taking urgent measures such as travel restrictions, social distancing, mask use, and quarantine (WHO, 2020e). It is estimated that Covid-19 has a 4.5% impact on global gross product across all sectors, resulting in a loss of 3.94 trillion dollars (Szmigiera, 2021). According to UNWTO (2020), 100 million people are at risk of losing their jobs as a result of Covid-19, with a financial loss of between 910 billion and 1.2 trillion dollars. In this situation, it may be said that tourism, with 1.2 billion clients, is the most affected sector because of Covid-19 travel limitations.

Tourism is threatened not only by Covid-19, but also by a variety of other factors such as drought, climate change, terrorism, and economic ups and downs (Hoogendoorn & Fitchett, 2018; Saint Akadiri et al., 2020). As a result, tourism is exceedingly vulnerable to problems arising from environmental, economic, and health concerns. Gastronomy is an essential component of today's tourism, thus Covid-19 has also influenced it in this manner (Harms et al., 2021). This effect will be amplified by the inclusion of Covid-19's negative effects on labor, transportation, travel, and marketing (CIHEAM, 2020). Because of the curfews imposed in Covid-19 and businesses' inability to serve clients on-site, the tendency toward takeout has risen. Consumption of ready meals and takeaways changes in eating and drinking habits have a touch on gastronomic preferences, both directly and indirectly. For example, the Italian restaurant brand Carluccio discharged around 1000 employees, although 30 restaurants were purchased by another restaurant chain, Boparan Restaurant Group, in 2020 due to financial troubles. There are 2.2 million restaurants on the verge of closing around the world, according to reports (Hughes, 2020). As a result, countries have taken a variety of measures to address the pandemic, particularly in the tourism sector, which is heavily impacted by Covid-19. For example, under the administration of Turkey's Ministry of Culture and Tourism, the Safe Tourism Certification Program, which will operate as an indicator of the protection of tourists' health, was developed and made mandatory in lodging facilities with 30 or more rooms as of January 1, 2021. Because of the constrained mobility and social distance embedded in touristic activities, the tourism sector is extremely vulnerable to pandemic-related measures such as quarantine, controlled travel, and food consumption in closed areas (Gössling et al., 2020). As a result of the restrictive measures adopted during diseases and epidemics, the food and beverage business faces dangers such as failure to provide enough food supply, loss of customers, and business bankruptcy (Lagiso, 2020; Shivakoti, 2021). In Ecuador, for example, the overall loss in tourism revenues in 2020 has been estimated to be 11 billion dollars, or 10.9% of the country's total income (Herrera, 2021). The pandemic has impacted the whole food and beverage supply chain, particularly manufacturers, distributors, retailers, and customers (Aday & Aday, 2020). However, this effect was caused by a lack of expertise about how to respond to the epidemic, or in other words, a lack of strategies in the supply chain during the pandemic (Cohen, 2020). This underlines those countries should develop methods to secure the food supply chain's continuity in the event of a pandemic. In this regard, risk management strategies such as "collaborative management," "proactive business continuity planning," and "financial sustainability" have been recommended for the food chain in pandemic-like processes (Kumar et al., 2021). There is also a broad belief that the food business and supply chain affected by the epidemic must address four critical challenges. These are a focus on healthy diets to maintain the immune system, hygienic food distribution to reduce viral spread, preventing food security issues in importing countries, and preserving the long-term viability of food systems (Galanakis, 2020).

With the pandemic, the importance of healthy meals that stimulate the immune system has been recognized once more (Aman & Masood, 2020). During the pandemic, consumption of nutritious and immune-boosting foods such as vegetables and fruits with high vitamin, mineral, and antioxidant content increased relative to the pre-pandemic period (Başaran & Purut, 2021). For example, it was determined that during the pandemic in the United States, more customers opted to plant-based foods, and their sales soared by 90% compared to March 2019 (SPINS, 2020).

While traditional gastronomy acts as a mechanism for bringing people together, sharing, and socialization, all of these have influenced community relations that have nearly vanished because of the pandemic, and individualization has been fostered by the fast-food system. Consequently, the pandemic has altered consumer behavior and severely disrupted social interactions (Barman et al., 2021). The rise in food-related concerns, particularly concerning hygiene, has influenced consumer food attitudes and habits. Because of the possibility of transmitting the virus, people began to be concerned about offering or receiving food from one another. As a result, numerous functions that are part of the civilizations' food culture, such as catering, invitation, and gift-giving, were harmed. Although the influence of pandemic on gastronomy is slight, it can be claimed that it has the potential to endanger the continuation of traditional food culture, and the level of damage will increase as identical problems occur more frequently in the coming years.

Access to gourmet products has been limited because of restaurant closures during the pandemic. The trend toward short-sold items (Çelik & Ergönül, 2020), which had previously been questioned for sanitary grounds, has considerably diminished during the Covid-19, resulting in significant damage to small-scale caterers. Despite the fact that there is no definite link between Covid-19 and food (De Faria Coelho-Ravagnani et al., 2021), food packaging is always regarded as safe in terms of cleanliness. As a result, for health reasons, the desire for multi-packaged hygienic items made from local materials has as increased (Bakalis et al., 2020). After the epidemic, there will undoubtedly be substantial changes in the food and beverage business. Kitchen managers, for example, believe that customers would prefer individualized consumption over open buffet, that menu plans will modify, and that the trend toward more local and healthier dishes will expand (Bucak & Yiğit, 2021). Bakalis et al., (2020) published research titled "Perspectives from CO+RE: How Covid-19 transformed our food systems and food security paradigms," which will be advised for food consumption trends following Covid-19.

From a wide perspective, the pandemic has influenced dietary habits in both positive and negative ways. People who were forced to stay at home due to restrictions at the start of the epidemic began to eat foods they had never eaten before, possibly by cooking them themselves rather of purchasing ready-made food. The fact that most recipe videos were seen during the epidemic and that the most bread recipes were searched on Google in 2020 is the clearest evidence of this transition. In another case, via this transformation, many people had the opportunity to visit the kitchen, which they had been neglecting due to their hectic work schedule, and the return to the healthier Mediterranean diet, which had been ignored before the pandemic, increased (Rodríguez-Pérez et al., 2020). As a result, the consumption of traditional foods has increased rather than fast-food products when most family members are at home.

This contributes to the preservation of gastronomic culture in a favorable way. On the other side, it has been observed that snack foods such as chocolate, sweet goods, pastries, and beverages are being consumed more often in many countries (Bracale & Vaccaro, 2020; Coulthard et al., 2021). There is ample evidence that excessive consumption of easy-to-bake pastries, along with inactivity at home, leads to a rise in BMI. In one of them, it was reported that overweight/obesity increased by 25% among youngsters in China during the epidemic (Yang et al., 2020).

Another aspect endangering cuisine is the reduction in agricultural labor, albeit for a while, as a result of Covid-19. Although there is no imminent threat to production, it has become evident that it is vital to be prepared in such outbreaks where food safety is imperiled. Although the pandemic is not severe enough to produce a global agricultural disaster, it is expected to cause a drop in production of between 0.4 and 2 percent in some nations, such as China (Zhang et al., 2020). According to reports, countries that rely on imports, particularly due to transportation bans, are the ones most affected by the pandemic in terms of food supply, not only in terms of finding food, but also in terms of access to food due to increased transportation prices (Freightos, 2022). During the pandemic, there have been inconsistencies in the cultivation of basic foods such as corn, sunflower, wheat, barley, canola, and open field vegetables (Poudel et al., 2020). Given that the products in question have been used in gastronomic kitchens, it is possible to argue that crises directly threaten not just food safety but also the future of the country's cuisines.

Population Growth, Migration and Gastronomy

Historically, the rate of change in local cuisine was slow because it was not affected by causes such as overcrowding, migration, and urbanization that exist today (Wahlqvist, 2007). Migration, population increase, globalization, climate change, and political disputes will all intensify cultural interactions in the next decades, causing tensions both inside and between countries (FAO, 2017). Migration from rural to urban areas has been fueled by the aging of the rural population, high input costs, price fluctuations, extreme climatic events, and the inability of the youthful population to find a decent social environment. In addition, internal turmoil in the countries resulted in overseas migration and a decline in the number of young people working in agriculture (Jentsch & Simard, 2016). Societies migrate voluntarily for various causes such as conflicts, natural disasters, or for better living conditions. Material and spiritual-cultural aspects are also transported during these migrations (Yılmaz, 2014). Food culture can be transferred to different living areas and created in a contemporary paradigm (Tellström, 2016). Movements of people will continue today and in the future. The number of international migrants in 2019 is anticipated to be around 272 million (3.5 percent of the global population), with approximately two-thirds of them being labor migrants (IOM, 2020). All these factors endanger agricultural production. Despite the growing population, the scarcity of food supplies necessitated the deployment of technological advances in food production. In this situation, the use of chemicals and technology in food processing leads to flavor modifications. These technical tools, which are utilized in national and international cuisines, have an impact on food cultures by modifying taste, flavor, and appearance.

The feature cuisine of a location is based on the contributions of many people who have passed through that territory. The foods deemed fundamental in Mediterranean cuisine are commodities from other regions that have been adopted over time. Tomatoes, potatoes, and peppers, for example, came from America; citrus fruits came from Asia; while eggplant, spinach, and artichokes came from the Arabs. This process is still going on today, with refugee influx between the Mediterranean's coastlines

(Oussedik, 2012). Rather than participating with the community, they migrated to, immigrant societies frequently transmit their cultures to those countries in a conservative manner. Immigrant societies also bring food-related weddings, funeral, and Eid ceremonies with them wherever they migrate. As a result, migrations affect local cuisines, which are material cultural components. Regional food become more diverse because of immigration, but it also loses its authenticity. The most notable example is doner kebab, which was introduced to Germany by Turks in the 1960s and is now one of Germany's most popular fast foods (Möhring, 2010). Similarly, Italians, who have been one of Australia's major ethnic communities since the 1800s, have influenced British-originated food culture by substituting more fish, salad, and olive oil (Sala & Baldassar, 2017). In Turkey, where the closest immigration phenomena are observed, around 4 million Syrian immigrants have meals like Turkish cuisine, but the style they are prepared, the names, and the materials used differ, for example, being oilier with more use of tomato paste (Avşar & Seçim, 2020). Syrians have opened restaurants and dessert stores in several Turkish provinces, offering their distinct gastronomic culture. With the establishment of such conservative ethnic eateries, these cuisines became more well-known (Eser, 2018). Again, since the middle of the nineteenth century, because of intense immigration from Asian countries to America, the culinary customs of the country have been brought to America. Immigrants from Asian countries have now built thousands of restaurants in America, each with its distinct food culture. A new kitchen trend, fusion kitchens, has evolved as by blending of different cuisines. This can be evaluated as a positive trend in terms of keeping these countries' traditions alive and transmitting their cultural identities.

SOLUTIONS AND RECOMMENDATIONS

Keeping a meal alive in the location where it originated is a requirement of cultural heritage. Protecting the water, soil, vegetation, and even the temperature of the place in issue is a top responsibility in this regard. Furthermore, initiatives such as preserving the socioeconomic opportunities of the society that produces the food, such as decreasing migration from the countryside to the city, will assure the taste's preservation. The fact that living conditions in every country, developed or developing, are becoming increasingly difficult in terms of time and money, unavoidably lowers people's ability to devote time to food as they did in previous centuries. During the chaos of life, especially young generations regard food as a necessity rather than a pleasure, and they fulfill this need as quickly and cheaply as possible. Furthermore, not just with tourists' visits, but also with increased refugee and migrant movements, particularly strong intercultural changes and interactions are observed. Climate change, population pressure, and the likelihood of shifting demographic structures are all factors to consider while taking measures against all these events that place positive and negative impact on gastronomy. For example, because the ability to grow local plants will be reduced in places where annual temperatures rise due to global warming, the viability of cultivating these plants should be examined first. As a response, the following topics can be considered as potential measures:

- Digital recording and archiving of gastronomic values in order keep traditional data against globalization
- Making plans to ensure the production of geographically specific foods, such as protecting endemic species' habitats in forests and meadows due to increasing biodiversity loss

- Keeping soil and water resources in decent health (prevention of pollution) since land degradation along with pollution threatens natural resources,
- Beginning in primary school, activities to produce gastronomic products needs be initiated. For example, these trainings could begin with student engagement how to recognize natural plants used in the preparation of local food in nature. This approach will provide natural and cultural awareness to adolescents,
- Providing some tax benefits and incentives to restaurants that serve gastronomic meals since increasing prices of foods that are used in traditional kitchen likely to decrease profit of food serving enterprises which put pressure to abandon cooking gastronomic foods,
- Experts inspecting culinary meals in restaurants on a regular basis and assigning ratings such as quality stars which will contribute to attracting tourists on a global and local scale.,
- Food used in gastronomic products is provided with economic support in rural areas, particularly in women's home gardens, as well as the establishment of a chain that will allow the products produced to reach restaurants. Two benefits will indeed be obtained by means of this. The first is to protect local culture from globalization, while the second is to prevent problems like migration by providing economic input to women.
- State support for cultural events and gastronomic product activities (festivals, seminars, workshops). These initiatives will demonstrate that environmental protection and local production will result in delight for all classes of society.

FUTURE RESEARCH DIRECTIONS

- Conducting study on a local scale in regions where gourmet values are unknown or have yet to be fully unveiled due to their lack of closeness to well-known tourist destinations that will enhance awareness.
- Implementation of gastronomic cultural heritage research on a local to regional scale, employing data from historical archives and archaeological excavations that would integrate various interdisciplinary study on gastronomy.
- Interviewing elderly people to record local gastronomic products which would provide conservation of heritage.
- Trying to figure out how efficiently gourmet cuisine matches today's nutritional, health, and flavor needs which will ease directing young generations to local food consumption

Studies on how gastronomy would adapt and prepare in the event of a future climate changes/crisis are recognized as crucial for future generations to benefit from gastronomic diversity and cultural heritage protection. Consequently, decision-makers will be able to develop actions and policies based on sound information to preserve and sustain gastronomic values.

POST-COVID-19 PERSPECTIVES OF GASTRONOMY

Covid 19 has had an impact on a wide range of sectors, causing social, political, economic, and environmental concerns all around the world. One of these sectors, gastronomy, is a vulnerable one, afflicted by economic, environmental, and social issues. Covid-19, which impacts all countries throughout the world, has had a severe impact on the gastronomy sector due to consumer loss and uncertainty (Harms et al., 2021). Consumer behavior had changed over the course of the pandemic, and it was assumed that consumers' willingness to visit restaurants had decreased as a result of their fear of getting down with Covid-19 (Skenderi, 2021).

Food crises are expected to occur in the future, along with other global problems such as the effects of the post-Covid 19 pandemic process, population growth, climatic changes, and migrations, so it is critical to take global measures for sustainable food security, which is among the sustainable development goals of the UN. As consumers' priority for safe (hygienic) food rises during the Covid-19, food and beverage companies must strengthen their food control practices even more than before the pandemic to earn customers' trust (De Souza et al., 2022). During the Covid-19 epidemic, technological applications and tools have become extremely critical and widespread for delivering information, services, and products to customers (Garibaldi and Pozzi, 2020). Applications such as online workshops, trainings, non - contact food delivery, the use of disposable dinnerware in restaurants, an increase in online food deliveries by restaurants, and packaging that preserves food quality are anticipated to remain after Covid-19. At the same time, following Covid 19, it is projected that the trend toward immune-boosting foods will continue to grow.

Furthermore, because environmental issues such as climate change and biodiversity loss may play a role in the formation and spread of zoonotic pandemics like the Covid 19, strategic plans that include environmental sustainability have been developed to combat such epidemics. If environmental issues are not addressed, the likelihood of analogous health issues arising and spreading in the future will increase. It is essential to note that no problem occurs on its own. It is fair to argue that Covid-19 is both a cause and an effect in this context.

CONCLUSION

Globalization, climate change, pandemics, population increase, and migration all have an adverse effect on gastronomy at diverse scales. Local actions, such as environmental conservation and biodiversity restoration, must be implemented promptly to ensure the continuity of gastronomy, which has been developed over many years, or cultural heritage and healthy nutrition will be seriously undermined. The standardization of food culture, as well as the rise of international fast-food networks, will cause local food cultures to decline and vanish quickly as a result of globalization. Climate change, which is one of the global gastronomy concerns, will have an adverse effect on not only the taste and quality of foods, but also on plant and animal production. All of this will have a negative impact on gastronomic culture. As a result, documenting and conserving existing food cultures should be among the top priorities for transferring cultural legacy to future generations, as reversing climatic change becomes exceedingly difficult with each passing day. Gastronomic products including healthful foods are expected to be in high demand, particularly during the pandemic. However, as the population increases, biodiversity declines, GMO products, additive-laden food, and economic barriers become more prevalent, all segments of society will have trouble accessing healthy culinary products equally. As a result, access to such healthful items will be hampered by social and economic inequity. The maintenance and long-term viability of gastronomy will be jeopardized, as there will be items that only people with a particular level of economic power will seek out in search of fresh and safe cuisine.

REFERENCES

Aday, S., & Aday, M. S. (2020). Impact of Covid-19 on the food supply chain. *Food Quality and Safety*, *4*(4), 167–180. doi:10.1093/fqsafe/fyaa024

Adebo, G. M., & Ayelari, T. A. (2011). Climate change and vulnerability of fish farmers in Southwestern Nigeria. *African Journal of Agricultural Research*, *6*(18), 4230–4238. doi:10.5897/AJAR10.948

Ahmad, A. (2013). *New age globalization: Meaning and metaphors*. Springer.

Aksoy, M., & Sezgi, G. (2015). Gastronomy Tourism and Southeastern Anatolia Region Gastronomic Elements. *Journal of Tourism and Gastronomy Studies*, *3*(3), 79–89.

Aman, F., & Masood, S. (2020). How Nutrition can help to fight against Covıd-19 Pandemic. *Pakistan Journal of Medical Sciences*, *36*(S4), 121. doi:10.12669/pjms.36.COVID19-S4.2776

Astil, G. (2020, March 2). *Food Loss: Why Food Stays on the Farm or off the Market*. Economic Research Service. https://www.ers.usda.gov/amber-waves/2020/march/food-loss-why-food-stays-on-the-farm-or-off-the-market/

Avşar, Ö. Y., & Seçim, Y. (2020). The Process of Adaptation of the Culinary Culture of Syrian Immigrants; The Case of Konya. *Journal of Recreation and Tourism Research*, *7*(2), 209–221. doi:10.31771/jrtr.2020.61

Bakalis, S., Valdramidis, V. P., Argyropoulos, D., Ahrne, L., Chen, J., Cullen, P., & … . (2020). Perspectives from CO+ RE: How COVID-19 changed our food systems and food security paradigms. *Current Research in Food Science*, *3*, 166. doi:10.1016/j.crfs.2020.05.003 PMID:32908972

Barman, A., Das, R., & De, P. K. (2021). Impact of COVID-19 in food supply chain: Disruptions and recovery strategy. *Current Research in Behavioral Sciences*, *2*, 100017. doi:10.1016/j.crbeha.2021.100017

Başaran, B., & Purut, H. P. (2021). The Impact of the Covıd-19 Pandemic on the Frequency of Food Consumption. *Journal of Tourism and Gastronomy Studies*, *9*(1), 47–66. doi:10.21325/jotags.2021.776

Belsinger, S. (n.d.). *Herb harvest fall festival featuring plants of South America*. Fine Gardening. https://www.finegardening.com/article/herb-harvest-fall-festival-featuring-plants-of-south-america

Berbegal, C., Fragasso, M., Russo, P., Bimbo, F., Grieco, F., Spano, G., & Capozzi, V. (2019). Climate changes and food quality: The potential of microbial activities as mitigating strategies in the wine sector. *Fermentation (Basel, Switzerland)*, *5*(4), 85. doi:10.3390/fermentation5040085

Björk, P., & Kauppinen-Räisänen, H. (2016). Local food: A source for destination attraction. *International Journal of Contemporary Hospitality Management*, *28*(1), 177–194. doi:10.1108/IJCHM-05-2014-0214

Bracale, R., & Vaccaro, C. M. (2020). Changes in food choice following restrictive measures due to Covıd-19. *Nutrition, Metabolism, and Cardiovascular Diseases*, *30*, 1423–1426. doi:10.1016/j.numecd.2020.05.027 PMID:32600957

Bucak, T., & Yiğit, S. (2021). The future of the chef occupation and the food and beverage sector after the Covıd-19 outbreak: Opinions of Turkish chefs. *International Journal of Hospitality Management, 92*, 102682. doi:10.1016/j.ijhm.2020.102682 PMID:33071426

Co2 earth. (n.d.). *Página inicial de CO2 da Terra* [Earth CO2 homepage.] https://www.co2.earth

Calıcıoglu, O., Flammini, A., Bracco, S., Bellù, L., & Sims, R. (2019). The future challenges of food and agriculture: An integrated analysis of trends and solutions. *Sustainability, 11*(1), 222. doi:10.3390u11010222

Çaycı, A. E. (2019). *The role of social media in the transformation of globalized food culture: Instagram gourmets* [Doctoral Thesis]. Istanbul University Institute of Social Sciences.

Çelik, K., & Ergönül, B. (2020). Determination of Microbiological Attributes of Unpackaged and Packaged Raisin Samples Sold in Manisa Markets. *Acta Scientific Nutritional Health, 4*(5), 10–13.

Cho, R. (2019, June 20). *How Climate Change Impacts the Economy, How Climate Change Impacts the Economy*. State of the Planet. https://news.climate.columbia.edu/2019/06/20/climate-change-economy-impacts/

CIHEAM. (2020). *The Covıd-19 Pandemıc: Threats on Food Securıty In The Medıterranean Regıon.* https://www.ciheam.org/wp-content/uploads/2020/07/COVID-rapport-FINAL.pdf

Cohen, M. J. (2020). Does the Covıd-19 outbreak mark the onset of a sustainable consumption transition? *Sustainability: Science Practice and Policy, 16*(1), 1–3. doi:10.1080/15487733.2020.1740472

Coulthard, H., Sharps, M., Cunliffe, L., & Van den Tol, A. (2021). Eating in the lockdown during the Covid 19 pandemic; self-reported changes in eating behaviour, and associations with BMI, eating style, coping and health anxiety. *Appetite, 161*, 105082. PMID:33476651

Cross, T. D. (2021). *Frequent extreme weather events are increasing crop losses in Europe.* Sustainability Times. https://www.sustainability-times.com/environmental-protection/frequent-extreme-weather-events-are-increasing-crop-losses-in-europe/

Crossley, T. F., Griffith, R., Jin, W. M., & Lechene, V. (2018). *A structural analysis of the decline of home-cooked food* [Conference session]. LISER Conference, Luxembourg. https://www.bc.edu/content/dam/bc1/schools/mcas/economics/pdf/seminars/CGJL.pdf

De Faria Coelho-Ravagnani, C., Corgosinho, F. C., Sanches, F. L. F. Z., Prado, C. M. M., Laviano, A., & Mota, J. F. (2021). Dietary recommendations during the Covıd-19 pandemic. *Nutrition Reviews, 79*(4), 382–393. doi:10.1093/nutrit/nuaa067 PMID:32653930

De Silva, S. S., & Soto, D. (2009). Climate change and aquaculture: Potential impacts, adaptation and mitigation. *FAO Fisheries and Aquaculture Technical Paper, 530*, 151–212.

De Souza, T. S. P., Miyahira, R. F., Matheus, J. R. V., de Brito Nogueira, T. B., Maragoni-Santos, C., Barros, F. F. C., Costa Antunes, A. E., & Fai, A. E. C. (2022). Food services in times of uncertainty: Remodeling operations, changing trends, and looking into perspectives after the COVID-19 pandemic. Trends in Food Science & Technology. *Trends in Food Science & Technology, 120*, 301–307. doi:10.1016/j.tifs.2022.01.005 PMID:35035090

Dumas, Y., Dadomo, M., Di Lucca, G., & Grolier, P. (2003). Effects of environmental factors and agricultural techniques on antioxidant content of tomatoes. *Journal of the Science of Food and Agriculture, 83*(5), 369–382. doi:10.1002/jsfa.1370

Dunkley, D. (2021, February 14). *A nation of meat-eaters: How Kiwis swapped lamb for chicken*. Stuff. https://www.stuff.co.nz/business/300224037/a-nation-of-meateaters-how-kiwis-swapped-lamb-for-chicken

DW. (2020). *How climate change paved the way to war in Syria*. DW Made for minds. https://www.dw.com/en/how-climate-change-paved-the-way-to-war-in-syria/a-56711650

Dwyer, L. (2015). Globalization of tourism: Drivers and outcomes. *Tourism Recreation Research, 40*(3), 326–339. doi:10.1080/02508281.2015.1075723

Eser, B. (2018). Where are the migrant restaurants on the gastronomic field? A theoretical introduction to immigrant neighborhood, refugee cuisine and spatial segregation. *ViraVerita E-Journal: Interdisciplinary Encounters, 8*, 172–195.

FAO. (2017). The future of food and agriculture – Trends and challenges. FAO.

FAO. (n.d.). http://www.fao.org/faostat/en/?#data/QC

Farmery, A. K., Alexander, K., Anderson, K., Blanchard, J. L., Carter, C. G., & Evans, K. (2021). Food for all: Designing sustainable and secure future seafood systems. *Reviews in Fish Biology and Fisheries*, 1–21. doi:10.100711160-021-09663-x(0123456789) PMID:34092936

Freightos. (2022, March 3). *Shipping & Freight Cost Increases, Freight Capacity, and Shipping Container Shortage*. https://www.freightos.com/freight-resources/coronavirus-updates/

Galanakis, C. M. (2020). The food systems in the era of the coronavirus (COVID-19) pandemic crisis. *Foods, 9*(4), 523. doi:10.3390/foods9040523 PMID:32331259

Garibaldi, R., & Pozzi, A. (2020). Gastronomy tourism and Covid-19: technologies for overcoming current and future restrictions. In F. Burini (Ed.), *Tourism facing a pandemic: From Crisis to recovery*. https://aisberg.unibg.it/handle/10446/160698#.XxlqKp4zbIX

Garvey, N. L. (2010). *À table: The impacts of globalization on French cuisine* [Doctoral dissertation]. Texas A & M University.

Gephart, J. A., Golden, C. D., Asche, F., Belton, B., Brugere, C., & Froehlich, H. E. (2020). Scenarios for global aquaculture and its role in human nutrition. *Reviews in Fisheries Science & Aquaculture*, 1–17. doi:10.1080/23308249.2020.1782342

Ghahremanzadeh, M., & Ziaei, M. B. (2014). Food Price Change and its welfare impact on Iranian Households. *International Journal of Agricultural Management and Development, 4*, 313–323. doi:10.22004/ag.econ.234396

Gillespie, C. (2001). *European gastronomy into the 21st century*. Butterworth-Heinemann.

Gorji, S., & Gorji, A. (2021). COVID-19 pandemic: The possible influence of the long-term ignorance about climate change. *Environmental Science and Pollution Research International, 28*(13), 15575–15579. doi:10.100711356-020-12167-z PMID:33403640

Gössling, S., Scott, D., & Hall, C. M. (2020). Pandemics, tourism, and global change: A rapid assessment of COVID-19. *Journal of Sustainable Tourism, 29*(1), 1–20. doi:10.1080/09669582.2020.1758708

Green, R., Cornelsen, L., Dangour, A. D., Turner, R., Shankar, B., Mazzocchi, M., & Smith, R. D. (2013). The effect of rising food prices on food consumption: Systematic review with meta-regression. *British Medical Journal, 346*. Advance online publication. doi:10.1136/bmj.f3703 PMID:23775799

Gruda, N. (2005). Impact of environmental factors on product quality of greenhouse vegetables for fresh consumption. *Critical Reviews in Plant Sciences, 24*(3), 227–247. doi:10.1080/07352680591008628

Gürhan, N. (2017). A Sociological Analysis on Social Change and Food Culture: Mardin Sample. *Journal of International Social Research, 10*(54), 561–570.

Hall, M., & Mitchell, R. (2003). Tourism as a force for gastronomic globalization and localization. In A. M. Hjalager & G. Richards (Eds.), *Tourism and gastronomy*. Routledge.

Harms, R., Alfert, C., Cheng, C. F., & Kraus, S. (2021). Effectuation and causation configurations for business model innovation: Addressing COVID-19 in the gastronomy industry. *International Journal of Hospitality Management, 95*, 102896. doi:10.1016/j.ijhm.2021.102896

Harrington, R. J. (2005). Defining gastronomic identity: The impact of environment and culture on prevailing components, texture and flavors in wine and food. *Journal of Culinary Science & Technology, 4*(2-3), 129–152. doi:10.1300/J385v04n02_10

Harvard. (n.d.). *Coronavírus, Mudanças Climáticas e Meio Ambiente Uma Conversa sobre COVID-19 com Dr. Aaron Bernstein, Diretor de Harvard Chan C-CHANGE* [Coronavirus, Climate Change and Environment A Conversation on COVID-19 with Dr. Aaron Bernstein, Director of Harvard Chan C-CHANGE]. https://www.hsph.harvard.edu/c-change/subtopics/coronavirus-and-climate-change/

Haven-Tang, C., & Jones, E. (2005). Using local food and drink to differentiate tourism destinations through a sense of place: A story from Wales-dining at Monmouthshire's great table. *Journal of Culinary Science & Technology, 4*(4), 69–86. doi:10.1300/J385v04n04_07

Herrera, D. B. (2021). Estimation of the tourism crisis in Ecuador by covıd-19 with a view to a Post-Pandemic Reactivation. *Journal of Contemporary Issues in Business and Government, 27*(4), 91–110.

Hoogendoorn, G., & Fitchett, J. M. (2018). Tourism and climate change: A review of threats and adaptation strategies for Africa. *Current Issues in Tourism, 21*(7), 742–759. doi:10.1080/13683500.2016.1188893

Hughes, L. (2020). *The restaurant chains which collapsed, restructured and cut jobs in 2020*. InYourArea. https://www.inyourarea.co.uk/news/the-restaurant-chains-which-collapsed-restructured-and-slashed-jobs-in-2020/

Hummel, M., Hallahan, B. F., Brychkova, G., Ramirez-Villegas, J., Guwela, V., Chataika, B., & ... . (2018). Reduction in nutritional quality and growing area suitability of common bean under climate change induced drought stress in Africa. *Scientific Reports*, *8*(1), 1–11. doi:10.103841598-018-33952-4 PMID:30385766

Hwang, L. J. J., van Westering, J., & Chen, H. H. (2005). Exploration of the linkages between the gastronomy and heritage of Tainan city, Taiwan. In *Advances in hospitality and leisure*. Emerald Group Publishing Limited.

International Organization for Migration. (2020). *World Migration Report 2020.* https://publications.iom.int/system/files/pdf/wmr_2020.pdf

IPCC. (2019). *IPCC Special Report on the Ocean and Cryosphere in a Changing Climate. IPCC.*

Jabs, J., & Devine, C. M. (2006). Time scarcity and food choices: An overview. *Appetite*, *47*(2), 196–204. doi:10.1016/j.appet.2006.02.014 PMID:16698116

Jangid, K. K., & Dwivedi, P. (2016). Physiological responses of drought stress in tomato: A review. *International Journal of Agriculture Environment and Biotechnology*, *9*(1), 53–61. doi:10.5958/2230-732X.2016.00009.7

Jentsch, B., & Simard, M. (2016). *International Migration and Rural Areas, Cross-National Comparative Perspectives - Studies in Migration and Diaspora*. Taylor and Francis.

Kim, K., Bonn, M. A., & Cho, M. (2021). Clean safety message framing as survival strategies for small independent restaurants during the Covid-19 pandemic. *Journal of Hospitality and Tourism Management*, *46*, 423–431. doi:10.1016/j.jhtm.2021.01.016

Kolawole Olayiwola, A. S. (2007). Impact of globalization on food consumption, health and nutrition in Nigeria. In FAO (Ed.), Globalization of Food Systems in Developing Countries: impact on food security and nutrition. FAO.

Korkmaz, M., & Başaran, G. (2021). The domestic transformation of coffee consumption habits in the Covid-19 pandemic process. *Journal of Original Studies*, *2*(1), 43–56.

Koyuncu, M. (2017). Global Climate Change and Animal Husbandry. *Selcuk Journal of Agriculture and Food Sciences*, *31*(2), 98–106. doi:10.15316/SJAFS.2017.26

Kubiszewski, I., Costanza, R., Anderson, S., & Sutton, P. (2020). The future value of ecosystem services: Global scenarios and national implications. In Environmental Assessments. In K. N. Ninan (Ed.), *Environmental Assessments: Scenarios, Modelling and Policy* (pp. 81–108). Edward Elgar Publishing. doi:10.4337/9781788976879.00016

Kumar, A., Mangla, S. K., Kumar, P., & Song, M. (2021). Mitigate risks in perishable food supply chains: Learning from COVID-19. *Technological Forecasting and Social Change*, *166*, 120643. doi:10.1016/j.techfore.2021.120643

Kurtdaş, M. Ç. (2020). Some evaluations on the social effects of Covid-19. *Journal of City and Civilization,* 530-545.

Lagiso, D. (2020). Challenges and opportunities of Covid-19 in agricultural economy: The case of Ethiopia. *International Journal of African and Asian Studies*, *65*, 24–29. https://iiste.org/Journals/index.php/JAAS/article/view/52989

Lawler, O. K., Allan, H. L., Baxter, P. W., Castagnino, R., Tor, M. C., & Dann, L. E. (2021). The COVID-19 pandemic is intricately linked to biodiversity loss and ecosystem health. *The Lancet. Planetary Health*, *5*(11), e840–e850. doi:10.1016/S2542-5196(21)00258-8 PMID:34774124

Lehtinen, U. (2012). Sustainability and local food procurement: A case study of finnish public catering. *British Food Journal*, *114*(8), 1053–1071. doi:10.1108/00070701211252048

López-Guzmán, T., & Sánchez-Cañizares, S. (2012). Gastronomy, tourism and destination differentiation: A case study in Spain. *Revue d'Economie Financiere*, *1*, 63–72.

Luczaj, L., Pieroni, A., Tardío, J., Pardo-de-Santayana, M., Sõukand, R., Svanberg, I., & Kalle, R. (2012). Wild food plant use in 21 st century Europe, the disapperance of old traditions and the search for new ciusines involving wild edibles. *Acta Societatis Botanicorum Poloniae*, *81*(4), 359–370. doi:10.5586/asbp.2012.031

Maldonado, J. K., Shearer, C., Bronen, R., Peterson, K., & Lazrus, H. (2013). The impact of climate change on tribal communities in the US: displacement, relocation, and human rights. In J. K. Maldonado, C. Shearer, R. Bronen, K. Peterson, & H. Lazrus (Eds.), *Climate change and indigenous peoples in the United States* (pp. 93–106). Springer.

Marino, E. (2012). The long history of environmental migration: Assessing vulnerability construction and obstacles to successful relocation in Shishmaref, Alaska. *Global Environmental Change*, *22*(2), 374–381. doi:10.1016/j.gloenvcha.2011.09.016

Martínez-González, M. Á., Hershey, M. S., Zazpe, I., & Trichopoulou, A. (2017). Transferability of the Mediterranean diet to non-Mediterranean countries. What is and what is not the Mediterranean diet. *Nutrients*, *9*(11), 1226. doi:10.3390/nu9111226 PMID:29117146

McDonald's Reports Fourth Quarter & Full Year. (2019). *Results and Quarterly Cash Dividend*. McDonald's Newsroom. McDonald's Corporation.

Melbourne University Report. (2015). *Appetite for Change Global Warming Impacts on Food and Farming Regions in Australia*, Melbourne Sustainable Society Institute.

Mendez, M., & Popkin, B. (2004). Globalization, urbanization, and nutritional change in the developing world. *The Electronic Journal of Agricultural and Development Economics*, *1*(2), 220–241.

Mir, U. R., Hassan, S. M., & Qadri, M. M. (2014). Understanding globalization and its future: An analysis. *Pakistan Journal of Social Sciences*, *34*(2), 607–624. http://pjss.bzu.edu.pk/website/journal/article/5ea0309b5c9a2/page

Mishra, J., Mishra, P., & Arora, N. K. (2021). Linkages between environmental issues and zoonotic diseases: With reference to COVID-19 pandemic. *Environmental Sustainability*, *4*(3), 455–467. doi:10.100742398-021-00165-x

Möhring, M. (2010). Döner Kebab and West German Consumer (Multi-) Cultures. *Cross/Cultures, 129*, 151-166.

Nicola, M., Alsafi, Z., Sohrabi, C., Kerwan, A., Al-Jabir, A., Iosifidis, C., Agha, M., & Agha, R. (2020). The socio-economic implications of the coronavirus pandemic (COVID-19): A review. *International Journal of Surgery*, *78*, 185–193. doi:10.1016/j.ijsu.2020.04.018 PMID:32305533

OECD. (2020). *COVID-19 and international trade: Issues and actions*. https://www.oecd.org/coronavirus/policy-responses/covid-19-and-international-trade-issues-and-actions-494da2fa/

Ojelel, S., Mucunguzi, P., Katuura, E., Kakudidi, E. K., Namaganda, M., & Kalema, J. (2019). Wild edible plants used by communities in and around selected forest reserves of Teso-Karamoja region, Uganda. *Journal of Ethnobiology and Ethnomedicine*, *15*(1), 1–14. doi:10.118613002-018-0278-8 PMID:30626418

Orkun, N. D. (2009). *Food culture changed by globalization: Istanbul-Beyoğlu 2002-2009* [Doctoral Thesis]. Marmara University Institute of Social Sciences.

Oussedik, S. (2012). *Food and cuisine: Part of the migration process*. Institut Europeu de la Mediterrània.

Paeratakul, S., Ferdinand, D. P., Champagne, C. M., Ryan, D. H., & Bray, G. A. (2003). Fast-food consumption among US adults and children: Dietary and nutrient intake profile. *Journal of the American Dietetic Association*, *103*(10), 1332–1338. doi:10.1016/S0002-8223(03)01086-1 PMID:14520253

Poudel, P. B., Poudel, M. R., Gautam, A., Phuyal, S., Tiwari, C. K., Bashyal, N., & Bashyal, S. (2020). COVID-19 and its global impact on food and agriculture. *Journal of Biology and Today's World*, *9*(5), 221–225.

Raitsos, D. E., Beaugrand, G., Georgopoulos, D., Zenetos, A., Pancucci-Papadopoulou, A. M., Theocharis, A., & Papathanassiou, E. (2010). Global climate change amplifies the entry of tropical species into the Eastern Mediterranean Sea. *Limnology and Oceanography*, *55*(4), 1478–1484. doi:10.4319/lo.2010.55.4.1478

Ram, U. (2004). Glocommodification: How the global consumes the local-McDonald's in Israel. *Current Sociology*, *52*(1), 11–31. doi:10.1177/0011392104039311

Rasul, G. (2021). Twin challenges of COVID-19 pandemic and climate change for agriculture and food security in South Asia. *Environmental Challenges*, *2*, 100027. doi:10.1016/j.envc.2021.100027

Richards, G. (2003). Gastronomy: an essential ingredient in tourism production and consumption? In Tourism and gastronomy. Routledge.

Ristaino, J. B., Anderson, P. K., Bebber, D. P., Brauman, K. A., Cunniffe, N. J., & Fedoroff, N. V. (2021). The persistent threat of emerging plant disease pandemics to global food security. *Proceedings of the National Academy of Sciences of the United States of America*, *118*(23), e2022239118. doi:10.1073/pnas.2022239118 PMID:34021073

Ritchie, H., & Roser, M. (2019). *Seafood Production*. https://ourworldindata.org/seafood-production

Rodríguez-Pérez, C., Molina-Montes, E., Verardo, V., Artacho, R., García-Villanova, B., Guerra-Hernández, E. J., & Ruíz-López, M. D. (2020). Changes in dietary behaviors during the COVID-19 outbreak confinement in the Spanish COVIDiet study. *Nutrients, 12*(6), 1730. doi:10.3390/nu12061730 PMID:32531892

Russo, S., Sillmann, J., & Fischer, E. M. (2015). Top ten European heatwaves since 1950 and their occurrence in the coming decades. *Environmental Research Letters, 10*(12), 124003. doi:10.1088/1748-9326/10/12/124003

Sabbağ, Ç. (2012). The seasonal BMI changes of rural women living in Anatolia. *International Journal of Environmental Research and Public Health, 9*(4), 1159–1170. doi:10.3390/ijerph9041159 PMID:22690188

Sabbağ, Ç., & Mankan, E. (2012). Crosschecking Eastern and Western universities students food consumption attitudes in Turkey. *International Journal of Academic Research, 4*(2), 23–27.

Saghai, Y. (2021). Subversive future seeks like-minded model: On the mismatch between visions of food sovereignty futures and quantified scenarios of global food futures. *Ethics & International Affairs, 35*(1), 51–67.

Saint Akadiri, S., Alola, A. A., Olasehinde-Williams, G., & Etokakpan, M. U. (2020). The role of electricity consumption, globalization and economic growth in carbon dioxide emissions and its implications for environmental sustainability targets. *The Science of the Total Environment, 708*, 134653. doi:10.1016/j.scitotenv.2019.134653 PMID:31806296

Sala, E., & Baldassar, L. (2017). Leaving Family to Return to Family: Roots Migration Among Second-Generation Italian-Australians. *Ethos (Berkeley, Calif.), 45*(3), 386–408. doi:10.1111/etho.12173

Sandoval, C. A. E. (2003). Vulnerability due to extremes: Droughts, Enzo Levi Lecture, 2002. *Ingeneria Hidraulica en Mexico, 18*(2), 133–155.

Schlosser, E. (2001). *Fast Food Nation the Dark Side of the All-American Meal.* Mariner Books Houghton Mifflin Harcourt.

Shetty, P. (2003, January 22-24). *Impact of globalization on food and agriculture from the farm to the plate* [Conference session]. Impacts of Globalization on Agricultural Production and Marketing with Focus on Food Quality, Tokyo.

Shivakoti, A. (2021). Impact of COVID-19 on tourism in Nepal. *The Gaze: Journal of Tourism and Hospitality, 12*(1), 1–22. doi:10.3126/gaze.v12i1.35674

Skegg, D., Gluckman, P., Boulton, G., Hackmann, H., Karim, S. S. A., Piot, P., & Woopen, C. (2021). Future scenarios for the COVID-19 pandemic. *Lancet, 397*(10276), 777–778. doi:10.1016/S0140-6736(21)00424-4 PMID:33607000

Skenderi, R. (2021). *Dissecting the Normal: The Effects of Covid-19 Pandemic in the Gastronomy Sector of Kosovo* [Thesis]. Rochester Institute of Technology. https://scholarworks.rit.edu/theses/10946

Slow Food. (n.d.). *Sobre nós.* https://www.slowfood.com/about-us/

SPINS. (2020, May 26). *New Data Shows Plant-Based Food Outpacing Total Food Sales During COVID-19.* Plant Based Foods Association. https://www.spins.com/new-data-shows-plant-based-food-outpacing-total-food-sales-during-covid-19

Statista. (n.d.). *COVID-19/Coronavirus: Facts and Figures.* https://www.statista.com/page/covid-19-coronavirus

Stavrianea, A., Dipidis, C., & Siomkos, G. (2017). Gastronomy tourism: An examination of the "Greek Breakfast Initiative" potential. In N. Tsounis & A. Vlachvei (Eds.), *Advances in Applied Economic Research* (pp. 841–848). Springer.

Sudiyarto, S., Yuliati, N., Uhertiana, I., & Pratiwi, L. F. L. (2018). Exploring youngest consumers perceptions of traditional foods and fast foods product. In J. Island (Ed.), *AIP Publishing LLC.* AIP Conference Proceedings. doi:10.1063/1.5061859

Sugiura, T., Ogawa, H., Fukuda, N., & Moriguchi, T. (2013). Changes in the taste and textural attributes of apples in response to climate change. *Scientific Reports, 3*(1), 1–7. doi:10.1038rep02418 PMID:23948740

Szmigiera, M. (2021, February 21). *Impact of the coronavirus pandemic on the global economy - Statistics & Facts.* Statista. https://www.statista.com/topics/6139/covid-19-impact-on-the-global-economy/

Tekin, H., Arpacı, S., Atlı, S., Açar, İ., Karadağ, S., Yükçeken, Y., & Yaman, A. (2001). *Cultivation of Pistachios. T.C. Ministry of Agriculture and Rural Affairs, General Directorate of Agricultural Research.* Pistachio Research Institute.

Tellström, R. (2016, November 28-29). *Inheriting Your Ancestor's Diasporic Food Culture–A Planned Project How Swedish Fourth and Fifth Generation Emigrants in USA Relate to their Ancestor's Former Swedish Food Culture, People Moving with Food: Food, Migration and Multiculturalism* [Conference session]. NAFS conference. https://www.diva-portal.org/smash/get/diva2:1119911/FULLTEXT01.pdf

Thornton, P. & Cramer, L. (2012). *Impacts of climate change on the agricultural and aquatic systems and natural resources within the CGIAR's mandate* [CCAFS Working Paper 23]. CGIAR Research Program on Climate Change, Agriculture and Food Security.

United Nations. (2019). *World Population Prospects 2019.* https://www.un.org/development/desa/en/news/population/world-population-prospects-2019.html

Wahlqvist, M. L. (2007). Regional food culture and development. *Asia Pacific Journal of Clinical Nutrition, 16*(1), 2–7. PMID:17392068

Wilkins, J., & Hill, S. (2009). *Food in the ancient world.* John Wiley & Sons.

World Health Organization. (2020e). *Responding to COVID-19 Community Transmission.* https://www.who.int/publications/i/item/responding-to-community-spread-of-covid-19

World Tourism Organization. (2020). *World Tourism Barometer.* WTO.

World Travel & Tourism Council. (2022). https://wttc.org/

Yang, S., Guo, B., Ao, L., Yang, C., Zhang, L., Zhou, J., & Jia, P. (2020). Obesity and activity patterns before and during COVID-19 lockdown among youths in China. *Clinical Obesity*, *10*(6), e12416. doi:10.1111/cob.12416 PMID:33009706

Yılmaz, A. (2014). International Migration: Its types, reasons and effects. *Electronic Turkish Studies*, *9*(2), 1685–1704. doi:10.7827/TurkishStudies.6274

Zavaleta-Cortijo, C., Ford, J. D., Arotoma-Rojas, I., Lwasa, S., Lancha-Rucoba, G., García, P. J., & ... . (2020). Climate change and COVID-19: Reinforcing Indigenous food systems. *The Lancet. Planetary Health*, *4*(9), e381–e382. doi:10.1016/S2542-5196(20)30173-X PMID:32777205

Zhang, S., Wang, S., Yuan, L., Liu, X., & Gong, B. (2020). The impact of epidemics on agricultural production and forecast of COVID-19. *China Agricultural Economic Review*, *12*(3), 409–425. doi:10.1108/CAER-04-2020-0055

KEY TERMS AND DEFINITIONS

Biodiversity: It covers the diversity of all living species found in terrestrial and marine ecosystems.

Climate Change/Crisis: The phenomena in which the ecosystem is damaged by the occurrence of extreme weather events as a result of changes in global average climatic events.

Gastronomy: It is the science that studies the preparation, cooking, and presentation of local flavors, as well as their cultural impact.

Globalizing Food: It refers to the global standardization and homogenization of food cultures.

Local Cuisine: It is the food, drink, and cuisine traditions, beliefs, and practices associated with a specific region's food, drink, and cuisine.

Migration: As a result of the irreversible deterioration of the socioeconomic structure in the living territory, locals are fleeing to new living spaces inside or beyond the country in search of better living conditions.

Pandemic: It refers to contagious epidemic diseases that have spread throughout a large part of the world.

APPENDIX: ABBREVIATIONS

BMI: Body Mass Index
CO_2: Carbon dioxide
Covid-19: Coronavirus Disease-2019
DW: Deutsche Welle is Germany's international broadcasting organization
FAO: Food and Agriculture Organization of the United Nations
GMO: Genetically Modified Food
IOM: International Organization for Migration
IPCC: Intergovernmental Panel on Climate Change
OECD: Organization for Economic Co-operation and Development
RCP: Representative Concentration Pathways
SE: Southeastern Anatolian Region/Turkey
UN: United Nations
UNWTO: The World Tourism Organization
USDA: United States Department of Agriculture
WHO: World Health Organization

Chapter 8
An Investigation of Key Success Factors for Kebab as an Example of Local Food Tourism in Bonab:
Customer Perspectives

Hamed Rezapouraghdam
https://orcid.org/0000-0001-8190-6565
Eastern Mediterranean University, Turkey

Osman M. Karatepe
Eastern Mediterranean University, Turkey

Hasan Kilic
Eastern Mediterranean University, Turkey

ABSTRACT

In today's competitive market environment, restaurateurs realize that they should differentiate their products and services by offering unique tastes and experiences. This is also true for restaurants that promote and deliver local foods to customers. Local food which is considered as a cultural identity in every destination can create a positive image about a city and even help the local authorities to establish a city brand. Despite this realization, there is limited research about local food tourism and its services, especially in non-Western countries. In light of this, the study uses Bonab city as a popular food tourism destination in Iran and aims to ascertain the factors that motivate customers to buy kebabs in Bonab city. Utilizing a qualitative research approach, interviews were conducted with customers who consumed kebabs in this city. The findings delineate useful managerial implications for practice.

DOI: 10.4018/978-1-7998-9148-2.ch008

INTRODUCTION

Tourism is undeniably one of the forerunners of the experience economy (Rezapouraghdam et al., 2021). However, the contribution of tourism in destinations extends from economics to a plethora of socio-cultural possibilities for both locals and visitors as well (Toker & Rezapouraghdam, 2021). Meanwhile, of the momentous markers of socio-cultural distinction, food is considered a substantial symbol of the tourism industry (Kim et al., 2009). Particularly, in the age of globalization, local food and locality are frequently associated with tourism sustainability that emphasizes regional identity and conservation (Ellis et al., 2018). In sum, meals are viewed as a marker of culture identification, sense of belonging to a certain community or ethnic group, with some aspects of food preparation and consumption serving as visible badges of identity (Nor et al., 2012).

Tourism, due to its enchanting economic benefits and its influence on the well-being of individuals (Alipour et al., 2020), is a preferred phenomenon in most societies. According to Telfer and Wall (2000) eating out constitutes almost one-third of all expenditures which travelers spend on their holidays. In different studies, the consumption of local food has been suggested as a major reason for traveling to a specific destination and consequently a determinant factor of economic enhancement in destinations (e.g., Kivela & Crotts, 2006; Telfer & Wall, 2000). However, like other places, the homogenizing force of globalization (Mak et al., 2012) has also influenced Iranian food preferences and has created intense competition in the restaurant industry in this nation.

According to Momtaz et al. (2013), there are more than 20 different international food chains (e.g., pizzas, burgers, sandwiches, steaks) available in Iran which, besides many other types of restaurants (e.g., Chinese, Japanese), provide services to customers. The study of local food tourism has been recommended in the recent literature to understand various aspects of this phenomenon in destinations (Toker & Rezapouraghdam, 2021). Obviously, in the complex and competitive market environment, understanding and analyzing the factors affecting customers' choice of a specific sort of food can have several implications for theory and practice. For instance, it can help practitioners to have a detailed idea about the taste of the target market and serve them better. Moreover, it can contribute to the enhancement of the local food preservation which is the cultural treasure and image of destinations. However, in spite of the highly important position of food consumption in tourism destinations, there are very scarce numbers of studies in this area (e.g., Kim et al., 2009; Mak et al., 2012; Pamukcu et al., 2021; Stalmirska, 2021).

This study scrutinizes to fill the aforementioned gaps in the food tourism literature by identifying the factors affecting the preference of Bonab kebab which is a famous local food in Iran. Bonab kebab restaurants are the most memorable image of this city in travelers' minds when traveling to this destination.

LITERATURE REVIEW

Based on Nummedal and Hall's study (2006), the food that is produced in local places and the local indigenous have the specialty of making that food is called local food. Rozin and Vollmecke (1986) define food preference as the choice of a particular type of food item over the others. The review of the literature revealed that in many studies scholars tried to identify the elements which influenced customers' food selection (e.g., Randall & Sanjur, 1981; Wright et al., 2001). According to Longue (1991), culture has been introduced as the major factor affecting food preferences.

The determinant role of the culture on food preferences also has been emphasized as a filter through which the appropriateness of the food is confirmed (Long, 2004; Prescott et al., 2002). Moreover, Fields (2002) noted that motivational factors are a significant element in the process of food choice of people. That is, food can be a physical motivator since food consumption by itself is basically a physiological need. Food also can be categorized as a cultural motivator because it is representative of a particular culture's values. Additionally, food can be an interpersonal motivator resulting from the fact that by eating it out, individuals build or renew social relations with other members of the community. The food also may function as a status motivator for consumers who find eating out as an element differentiating them from others (Fields, 2002).

Introduced by Atkinson (1957), the expectancy-value theory is one of the most significant schools of thought that explains the essence of achievement motivation. This theory proposes that individuals form attitudes that boost the probability of pleasant outcomes and reduce the risk of unpleasant ones (Dunn et al., 2008). There are a number of studies in the food literature that have been framed based on this theory (Armitage & Conner, 1999; Towler & Shepherd, 1992). The expectancy-value theory indicates the importance of identifying the factors that influence the attitudes and beliefs of people about the consumption of a specific type of food. Apart from what has been discussed above, one should note that available options are also so crucial in forming human attitudes and behaviors.

Referring to restaurants and considering their resources, the resource-based theory seems a relevant framework as well. In various studies, the resource-based view has been used to illustrate companies' competitive advantage. Barney (1991) argues that in order for firms to be able to gain competitive advantages over their competitors their resources should be rare and valuable. In addition, the resources should also be inimitable and non-substitutable so that the owners of firms sustainably may retain their advantages (Newbert, 2008). These resources may include all of the financial, physical, human, and organizational assets by which a firm pursues its goals (Barney, 1991). According to Barney (1991), the financial resources of firms are their incomes, debts, and so forth while the human resources of the firms comprise the experience, knowledge, wisdom, aptitudes, attitudes, and the other personality traits of firms' employees. The most significant physical resources of firms also are considered their manufacturing tools and machines. Finally, the organizational assets of firms include but not limited to relationships and trust as well as the organizational culture.

BONAB KEBAB

Koobideh Kebab is among the most famous traditional foods in Iran and Bonab Kebab is a localized type of Iranian Koobideh Kebab. Therefore, when discussing local foods in the study area (Bonab city), the first food which comes at the top of the customers' list is Bonab kebab. Bonab kebab is a long and hand-minced meat kebab mounted on an iron skewer and grilled on a mangal burning with charcoal. This kebab was named after Bonab, a small city located in the Northern West of Iran. The characteristics which have distinguished Bonab kebabs from other sorts of kebabs are the traditional methods of its preparation and its cooking styles. The fresh ground beef or lamb mixed with onion is chopped by a huge cleaver on a chunk of tree trunk until it reaches the desired juicy and sticky state. Within the process of the preparation, only a suitable amount of salt and in some cases saffron is added to Bonab kebabs (Hakimi, 2013). Local yogurt, fresh basil, onion, sumac, roasted peppers, and tomatoes as well as fresh lemon are the life-long accompaniments of Bonab kebabs served at restaurants with freshly baked bread or "Sangak".

Study Setting

Bonab, a city located in East Azerbaijan province of Iran where meets Maragheh from the east, is famous for its big Koobideh Kebabs. Bonab kebab, as a former luxury meal of the Qajar dynasty and the traditional meal of Iranians in the current era, is the best brand of kebab all over the country. The outstanding image that Bonab resembles in minds of everyone is its authentic scenery created by the series of kebab restaurants and butcheries located side by side in every street and bazaar. It might sound bizarre hearing one can have difficulty finding any fast food or other sorts of restaurants than those serving kebab. The history of kebab is not only rooted in the background of the people of this city but also the city's cultural context and urban context are also tied to this traditional food (Keshavarzi & Akbari, 2012). This city has been able to successfully develop food tourism in the region and attracts a considerable number of visitors annually.

According to Bonab's Head of Cultural Heritage, Tourism and Handicraft Organization, there will be efforts on behalf of this organization to list Bonab kebab as one of the world's cultural heritage in UNESCO. Moreover, the municipality of Bonab decided to propose a town-twinning agreement to Adana's municipality (Turkey) because of the commonality those two cities have in kebab popularity in 2014 (IRNA, 2022). Bonab kebab does not only own pleasant taste and appearance but is also so nutritious and healthy. Based on a past study, it has been asserted that Bonab kebabs have acceptable nutritional value and hygienic quality (Nemati et al., 2008). Kebab is the image of Bonab city and even there are many kebab restaurants in other cities of Iran. Moreover, the geo-strategic position of this small city as a corridor linking western provinces together and constantly serves travelers throughout the year.

METHODOLOGY

A qualitative research approach was applied in this study since this exploratory method fits best for the studies which try to discover and explain a phenomenon such as the factors affecting Bonab kebab preference among customers. Semi-structured interviews were employed in the study to access a deep understanding of participants' experiences (Rezapouraghdam et al., 2018a).

Purposive sampling was utilized in this study since its initial objective is to generate understanding rather than the generalization of the findings (Saunders et al., 2000). Thus, being a native to the area and co-authoring this research, one of the authors during his frequent visits to restaurants located in Bonab collected the data. After explaining the purpose of the study and getting the consent of customers, the interviews started, and each lasted between 10-30 minutes. In total, 28 customers with various background and age groups participated in this research (see Table 1). This sample size is deemed appropriate according to the literature (e.g., Alsaawi, 2014; Boddy, 2016; Marshall et al., 2013).

Data Analysis and Findings

We followed a four-step procedure: collecting the data (recording the interviews), transcribing the collected data using Microsoft word document, reducing the unrelated data, coding the data, and analyzing

Table 1. Participants' profile

Variable	Number	Percentage
Age		
20-39	11	40.0
40-59	9	32.0
Over 59	8	28.0
Gender		
Male	21	75.0
Female	7	25.0
Originality		
Local (Bonab)	13	46.0
Non-local (Visitor)	15	54.0
Occupation		
Employee (Governmental)	7	25.0
Employee (Private/ Self-employed)	13	46.0
Others (Educator/ Household/Unemployed)	8	29.0
Marital Status		
Single	12	43.0
Married	16	57.0

the data (Rezapouraghdam et al., 2018a). As a result of reading and analyzing the material, the transcribed interviews were subjected to qualitative analysis to look for recurrent themes and codes (i.e., thematic and content analysis) (Alipour et al., 2021). In other words, the acquired data (i.e., the content of the interviews) was reviewed many times and analyzed for recurrent storylines in order to develop codes (Alipour et al., 2017). The findings of the data provided valuable themes (summing up a piece of text) which were outlined in Table 2. Below are some of the examples of how the themes in Table 2 emerged:

Need for Mastery and Professionalism to Make Kebabs (Theme 1)

The majority of the respondents were in a common belief that the preparation of Bonab kebabs needs mastering and a state of professionalism. They confirmed that they had tried kebabs in other places. However, none of these kebabs was comparable in taste and appearance with the ones they had at Bonab Kebab restaurants.

Respondent number 5 stated: "It is not easy to prepare the meat in a way it sticks on skewers. The size of ingredients and even the heat of mangal all are things that armatures cannot balance. I tried to make kebabs in our garden several times but the result was not like these".

Respondent number 11 mentioned that: "making kebabs looks easy but when you try it, you understand that it needs its level of professionalism and experience which takes time and practice for one to achieve that".

Nutritious and Premium Quality (Theme 6)

Most of the respondents have a consensus on the idea that this food item in comparison to other available outdoor foods is healthier and more nutritious. They believed that the ingredients of Bonab kebab are natural and good for the body.

Respondent number 2 declared: "I was born in a family that from childhood my father always was preaching to us: eat kebab to get stronger. Now I am a 57-years old man and at least once a week I eat Bonab kebab. This has been a part of my lifestyle".

Respondent number 19 said: "comparing kebab with other foods that can be prepared at restaurants quickly, I prefer this (Bonab kebab) because at least I am sure about the ingredients. This makes me confident to choose it. Fresh meat, onion, and pepper, all are good for the body".

Table 2. The major themes affecting customers' choice of Bonab Kebab restaurants

Number Category and description
1. Cooking shish kebab needs specific mastering and professionalism
2. Shish kebab needs to be cooked on barbeque, not stove
3. Preparation of shish kebab makes so much smoke and smell
4. Shish kebab suits well with hot fresh furnace bread available at Bonab restaurants
5. The taste of Bonab shish kebab is different from homemade ones
6. The ingredients of shish kebab are nutritious
7. Bonab kebabs are served with local healthy desserts (yogurt, pepper, buttermilk)
8. Shish kebabs are considered a traditional food
9. The prices of Bonab shish kebabs are reasonable
10. The time between ordering and getting Bonab shish kebabs is short
11. Employees at Bonab shish kebab restaurants are so friendly and warm
12. Since the whole process of kebab preparation can be seen, its hygienic quality is high
13. The size of each Bonab Kebab is bigger than the other types of kebabs
14. These Kebab shops are open till late at night and serve customers at any time they come
15. The freshness of the meat and bread are always guaranteed at these restaurants
16. There are many Bonab Kebab restaurants available around the city
17. Tea service (free) after meal and boiling water and fresh bread (free) for the ones who do not eat at Bonab restaurants

Table 3 presents the main motives for customers visiting Bonab Kebab restaurants.

DISCUSSION AND CONCLUSION

Considering the scarcity of evidence in the field of food preference in Iran, this study is a considerable effort in the identification of food preference factors of Bonab kebab restaurants. This study provides a

Table 3. Respondents' visitation motivation

Number	Factor
1.	Eating Bonab kebab as a main meal after work
2.	Eating Bonab kebab while passing from this city (travel)
3.	Buying Bonab kebabs when having a sudden guest (take home)
4.	Buying Bonab kebabs when there is no prepared home meal
5.	Eating Bonab kebab with friends/relatives as a hobby
6.	Celebrating social/cultural/religious events (e.g., anniversary, birthday).
7.	Traveling to Bonab at weekends for eating kebabs (from nearby cities)

deep analysis of the motivational factors of Kebab preference as well as the context where these preferences occur most.

The findings of this study asserted that Bonab city and its kebab have the potential capacity to become an appealing tourism attraction not only for the inbound travelers but also for outbound travelers. The findings of the current study contribute to the sociological aspects of Bonab kebab as a popular local food in the region in some ways. Broadly speaking, it provides a deep understanding of the factors which are influential in choosing Bonab kebab for customers. Therefore, restaurant owners can promote the characteristics of their products (kebabs) in order to attract the potential buyers and retain the current ones. For example, the health and nutrition features of kebabs can be highlighted in advertisements. The restaurant owners can also use the findings to increase the quality of their services to customers. For example, it has been identified that customers consider fresh bread which is baked at restaurants and served beside Kebabs as a positive attribute. Therefore, restaurants can try to provide such services to their customers.

The inclusion of free services for travelers who pass through Bonab city can make Bonab kebab restaurants' customers more delighted. These free services as identified in this study are boiling water, tea, and in some cases fresh bread. As Bonab is a significant connection hub in the region, there are many daily travelers passing through this city. Thus, Bonab kebab restaurants can benefit from this and leave positive memories in their customers' minds.

The findings of the current study also clarify a number of valuable facts about the local food consumption, marketing, and attraction that can be used in the food sector. The local indigenous and travelers apart from the taste and nutrition criteria preferred Bonab kebab because of the transparency of the production process. Unlike most other types of restaurants, at Bonab kebab restaurants the whole process of food preparation can be seen by customers. Even customers can purchase the raw materials (fresh meat) from butcheries and deliver them to Bonab kebab restaurants to prepare kebabs. In other words, there is nothing hidden between customers and chefs and this factor ensures the healthiness of Bonab kebabs. Therefore, in other places, restaurant managers can benefit from this finding and prepare a condition through which customers may ensure about the hygiene of the food preparation. If such facilities are missing, managers can use technological tools to connect their customers with the chefs and kitchen in one way or another.

Due to the wide range of culinary traditions, cultural diversity and natural environmental distinctions in Iran, foods in this nation are extremely diverse. There are more than two thousand different items of local and traditional foods in Iran many Iranians are not aware of them (Iranwonders, 2022). This enormous food diversification definitely can be utilized as a competitive advantage for promoting gastronomy tourism in Iran both at national and international levels. It should be noted that in consequence of the globalization trend, consumption of fast food items can be considered as a threat and an opportunity for the local food. That is, if a wide range of nutritious and healthy local foods of every region may receive enough attention (e.g., advertisement, availability) by different stakeholders (restaurant owners, tourism sectors, governmental/non-governmental organizations, local individuals, local and national media), then local foods will be preserved, the local image will be enhanced, and the economic benefits will be generated. This is very important for the sustainable development of tourism destinations.

Local foods, as cultural resources of any community, need to be treated in a sustainable manner to guarantee the satisfaction of all stakeholders. During these days the pivotal role of sustainability in organizations and tourism destinations has attracted substantial attention in academia (e.g., Akhshik et al., 2021; Karatepe et al., 2021a, 2021b; Rezapouraghdam et al., 2018; Rezapouraghdam et al., 2019). Therefore, future studies should observe how Bonab kebab restaurants deal with this issue and how the local food tourism in Bonab city is in line with the sustainable development guidelines. This is important since tourism is a complex phenomenon and, if not managed sustainably, can lead to many socio-cultural and environmental problems for destinations (Rezapouraghdam & Akhsik, 2021).

Finally, it should be noted that data for this study were collected before the emergence of the Coronavirus pandemic. As the pandemic has severely hit the tourism and hospitality industry (Rezapouraghdam & Karatepe, 2020) including restaurants, future studies can focus on how the pandemic affected individuals' food preferences, especially kebab items. Moreover, as the findings of this study asserted, professional and experienced kebab makers who work at restaurants are valuable assets and resources for the sector. Therefore, it is recommended that future studies consider how the well-being of the employees at these restaurants is considered by the employers. Particularly, future research can observe the negative consequences of the COVID-19 pandemic on restaurant employees' psychological well-being (e.g., insecurity, burnout) (Abubakar et al., 2021; Karatepe et al., 2020).

REFERENCES

Abubakar, A. M., Rezapouraghdam, H., Behravesh, E., & Megeirhi, H. A. (2021). Burnout or boreout: A meta-analytic review and synthesis of burnout and boreout literature in hospitality and tourism. *Journal of Hospitality Marketing & Management*, 1–46. doi:10.1080/19368623.2022.1996304

Akhshik, A., Ozturen, A., & Rezapouraghdam, H. (2021). A passionate travel to mind green turtles-Unpacking the complexity of visitors' green behavior. *International Journal of Tourism Research*, *23*(3), 301–318. doi:10.1002/jtr.2407

Alipour, H., Olya, H. G., Hassanzadeh, B., & Rezapouraghdam, H. (2017). Second home tourism impact and governance: Evidence from the Caspian Sea region of Iran. *Ocean and Coastal Management*, *136*, 165–176. doi:10.1016/j.ocecoaman.2016.12.006

Alipour, H., Rezapouraghdam, H., & Akhshik, A. (2021). Heritage redemption and the curse of tourism: The case of world's last inhabited troglodyte village. *Tourism Planning & Development*, *18*(1), 68–85. doi:10.1080/21568316.2020.1775114

Alipour, H., Rezapouraghdam, H., & Esmaeili, B. (2020). Public health and tourism: A personalist approach to community well-being: A narrative review. *Iranian Journal of Public Health*, *49*(3), 447–453. doi:10.18502/ijph.v49i3.3127

Alsaawi, A. (2014). A critical review of qualitative interviews. *European Journal of Business and Social Sciences*, *3*(4), 149–156. doi:10.2139srn.2819536

Armitage, C. J., & Conner, M. (1999). Distinguishing perceptions of control from self-efficacy: Predicting consumption of a low-fat diet using the theory of planned behavior. *Journal of Applied Social Psychology*, *29*, 72–90. doi:10.1111/j.1559-1816.1999.tb01375.x

Atkinson, J. W. (1957). Motivational determinants of risk-taking behavior. *Psychological Review*, *64*(6), 359–372. doi:10.1037/h0043445 PMID:13505972

Barney, J. B. (1991). Firm resources and sustained competitive advantage. *Journal of Management*, *17*(1), 99–120. doi:10.1177/014920639101700108

Boddy, C. R. (2016). Sample size for qualitative research. *Qualitative Market Research*, *19*(4), 426–432. doi:10.1108/QMR-06-2016-0053

Dunn, K. I., Mohr, P. B., Wilson, C. J., & Wittert, G. A. (2008). Beliefs about fast food in Australia: A qualitative analysis. *Appetite*, *51*(2), 331–334. doi:10.1016/j.appet.2008.03.003 PMID:18430490

Ellis, A., Park, E., Kim, S., & Yeoman, I. (2018). What is food tourism? *Tourism Management*, *68*, 250–263. doi:10.1016/j.tourman.2018.03.025

Fields, K. (2002). Demand for the gastronomy tourism product: Motivational factors. In A. M. Hjalager & G. Richards (Eds.), *Tourism and gastronomy* (pp. 37–50). Routledge.

Hakimi, M. A. (2013, August 31). *Kebab, the symbol of Iranian hospitality*. www.hotelmagazine.ir

Iranwonders. (2022). https://iranwonders.com/fa/articles-fa/165-iran-cuisine

IRNA. (2022). https://www.irna.ir/news/81130204

Karatepe, O. M., Rezapouraghdam, H., & Hassannia, R. (2020). Job insecurity, work engagement and their effects on hotel employees' non-green and nonattendance behaviors. *International Journal of Hospitality Management*, *87*, 102472. doi:10.1016/j.ijhm.2020.102472

Karatepe, O. M., Rezapouraghdam, H., & Hassannia, R. (2021a). Sense of calling, emotional exhaustion and their effects on hotel employees' green and non-green work outcomes. *International Journal of Contemporary Hospitality Management*, *33*(10), 3705–3728.

Karatepe, O. M., Rezapouraghdam, H., & Hassannia, R. (2021b). Does employee engagement mediate the influence of psychological contract breach on pro-environmental behaviors and intent to remain with the organization in the hotel industry? *Journal of Hospitality Marketing & Management*, *30*(3), 326–353. doi:10.1080/19368623.2020.1812142

Keshavarzi, P., & Akbari, M. (2012). *The role of food tourism in urban and regional development: the case of Bonab city* [Conference session]. 1st National Conference on Tourism, Geography and Environmental Sustainability, Hamadan, Iran.

Kim, Y. G., Eves, A., & Scarles, C. (2009). Building a model of local food consumption on trips and holidays: A grounded theory approach. *International Journal of Hospitality Management, 28*, 423–431. doi:10.1016/j.ijhm.2008.11.005

Kivela, J., & Crotts, J. C. (2006). Tourism and gastronomy: Gastronomy's influence on how tourists experience a destination. *Journal of Hospitality & Tourism Research (Washington, D.C.), 30*(3), 354–377. doi:10.1177/1096348006286797

Long, L. M. (2004). *Culinary tourism*. The University Press of Kentucky.

Longue, A. W. (1991). *The psychology of eating and drinking: An introduction*. WH Freeman.

Mak, A. H., Lumbers, M., & Eves, A. (2012). Globalization and food consumption in tourism. *Annals of Tourism Research, 39*, 171–196. doi:10.1016/j.annals.2011.05.010

Marshall, B., Cardon, P., Poddar, A., & Fontenot, R. (2013). Does sample size matter in qualitative research? A review of qualitative interviews in IS research. *Journal of Computer Information Systems, 54*(1), 11–22. doi:10.1080/08874417.2013.11645667

Momtaz, N. J., Alizadeh, S., & Vaghefi, M. S. (2013). A new model for assessment fast food customer behavior case study: An Iranian fast-food restaurant. *British Food Journal, 115*(4), 601–613. doi:10.1108/00070701311317874

Nemati, M., Ghorbanpour, H., Razavieh, S. V., & Hoseini, M. (2008). Chemical composition and microbiological quality of the Bonab kebabs sold in Tabriz market. *Journal of Food Safety, 28*(3), 315–323. doi:10.1111/j.1745-4565.2008.00112.x

Newbert, S. L. (2008). Value, rareness, competitive advantage, and performance: A conceptual-level empirical investigation of the resource-based view of the firm. *Strategic Management Journal, 29*(7), 745–768. doi:10.1002mj.686

Nor, N. M., Sharif, M. S. M., Zahari, M. S. M., Salleh, H. M., Isha, N., & Muhammad, R. (2012). The transmission modes of Malay traditional food knowledge within generations. *Procedia: Social and Behavioral Sciences, 50*, 79–88. doi:10.1016/j.sbspro.2012.08.017

Nummedal, M., & Hall, M. (2006). Local food and tourism: An investigation of the New Zealand South Island's bed and breakfast section's use and perception of local food. *Tourism Review International, 9*, 365–378. doi:10.3727/154427206776330571

Pamukcu, H., Sarac, O., Aytugar, S., & Sandikci, M. (2021). The effects of local food and local products with geographical indication on the development of tourism gastronomy. *Sustainability, 13*(12), 6692. doi:10.3390u13126692

Prescott, J., Young, O., O'Neill, L., Yau, N. J. N., & Stevens, R. (2002). Motives for food choice: A comparison of consumers from Japan, Taiwan, Malaysia and New Zealand. *Food Quality and Preference, 13*, 489–495. doi:10.1016/S0950-3293(02)00010-1

Randall, E., & Sanjur, D. (1981). Food preferences: Their conceptualization and relationship to consumption. *Ecology of Food and Nutrition, 11*(3), 151–161. doi:10.1080/03670244.1981.9990671

Rezapouraghdam, H., & Akhshik, A. (2021). Tracing the complexity-sustainability nexus in a small Mediterranean island: Implications for hospitality and tourism education. *Worldwide Hospitality and Tourism Themes*, *13*(4), 476–487.

Rezapouraghdam, H., Akhshik, A., & Ramkissoon, H. (2021). Application of machine learning to predict visitors' green behavior in marine protected areas: Evidence from Cyprus. *Journal of Sustainable Tourism*. Advance online publication. doi:10.1080/09669582.2021.1887878

Rezapouraghdam, H., Alipour, H., & Arasli, H. (2019). Workplace spirituality and organization sustainability: A theoretical perspective on hospitality employees' sustainable behavior. *Environment, Development and Sustainability*, *21*(4), 1583–1601.

Rezapouraghdam, H., Alipour, H., & Darvishmotevali, M. (2018). Employee workplace spirituality and pro-environmental behavior in the hotel industry. *Journal of Sustainable Tourism*, *26*(5), 740–758. doi :10.1080/09669582.2017.1409229

Rezapouraghdam, H., Behravesh, E., Ari, E., & Doh, W. L. (2018a). Cost-benefit analysis of educational tourism in North Cyprus: A qualitative study of the socio-economic impacts. *e-Review of Tourism Research, 15*(6), 457-479. https://ertr-ojs-tamu.tdl.org/ertr/index.php/ertr/article/view/259

Rezapouraghdam, H., & Karatepe, O. M. (2020). Applying health belief model to unveil employees' workplace COVID-19 protective behaviors: Insights for the hospitality industry. *International Journal of Mental Health Promotion*, *22*(4), 233–247. doi:10.32604/IJMHP.2020.013214

Rozin, P., & Vollmecke, T. A. (1986). Food likes and dislikes. *Annual Review of Nutrition*, *6*, 433–456. PMID:3524623

Saunders, M., Lewis, P., & Thornhill, A. (2000). *Research methods for business students*. Pearson Education Limited.

Stalmirska, A. M. (2021). Local food in tourism destination development: The supply-side perspectives. *Tourism Planning & Development*. Advance online publication. doi:10.1080/21568316.2021.1928739

Telfer, D. J., & Wall, G. (2000). Strengthening backward economic linkages: Local food purchasing by three Indonesian hotels. *Tourism Geographies*, *2*(4), 421–447. doi:10.1080/146166800750035521

Toker, B., & Rezapouraghdam, H. (2021). Intangible cultural heritage and management of educational tourism: An experiential learning approach. In *Handbook of Research on Human Capital and People Management in the Tourism Industry*. IGI Global., doi:10.4018/978-1-7998-4318-4

Towler, G., & Shepherd, R. (1992). Application of Fishbein and Ajzen's expectancy-value model to understanding fat intake. *Appetite*, *18*, 15–27. doi:10.1016/0195-6663(92)90207-M PMID:1562199

Wright, L. T., Nancarrow, C., & Kwok, P. M. H. (2001). Food taste preferences and cultural influences on consumption. *British Food Journal*, *103*(5), 348–357. doi:10.1108/00070700110396321

Chapter 9
Tourism and Gastronomy in the "New Normal":
Building the Case for the Rise of Consumer Spiritual Quest to Feed the Body, the Mind, and the Soul

Mafalda Nogueira
https://orcid.org/0000-0002-0302-0483
IPAM, Portugal

ABSTRACT

This chapter aims to provide a critical reflection on how individuals, families, tourists, in general, are looking for alternative forms of touristic and gastronomic experiences, very much rooted on the spiritual quest for peace, security, hope, self-awareness, and above all, emotional wellbeing. Hence, it discusses the rise of new forms of tourists, in the so called 'new normal', who completely changed their mindset regarding holiday habits and eagerly seek to feed their body, mind, and soul, rethinking and replacing their previous touristic and gastronomic choices. Additionally, it argues that the pandemic is giving the tourism and hospitality industry an opportunity to reinvent their value propositions and design their offerings based on these new emerging features of post-COVID tourists by creating innovative touristic packages and crafting new marketing messages very much focused on the spiritual attributes of their offerings.

INTRODUCTION

Following the COVID outbreak, people all over the world are facing stressful challenges in their lives, caused by substantial threats to their physical health, but also and not surprisingly, caused by mental issues that emerged from a cacophony of reasons - fear to get infected, fear of losing a loved one, anxiety, job losses, work from home, children' home school, home isolation and, for some, total absence of physical contacts with families and friends. All these aspects unquestionably led to emotional break-

DOI: 10.4018/978-1-7998-9148-2.ch009

downs and depressions (Bhalla et al., 2021; Kala, 2021). Contrary to the physical consequences of the disease, these effects on peoples' wellbeing will remain present for a long time (Prime et al., 2020). In the socio-economic domain, the pandemic led to bankruptcies, unemployment, financial crisis, and general economic slowdowns worldwide, and its harmful effects in almost every sector are being subject of daily inquiry by academic and non-academic realms. The amount of research developed around the topic, since March 2020, is overwhelming. The negative impacts caused in the tourism, gastronomic and hospitality sector are no exception. In fact, COVID-19 impacts on local, national and international tourism are astounding (Sigala, 2020). According to the United Nations World Tourism Organization (UNWTO), the tourism sector worldwide observed the largest decline in history (UNWTO, 2020b). Indeed, as pointed out in March 2021, "the COVID-19 pandemic brought global tourism to a near-complete standstill" (UNWTO, 2020c).

By looking at a plethora of different sources, namely academic and nonacademic research, secondary data provided by the UNWTO (2020b, 2020c, 2020a, 2021d, 2021c), media, news, traveler and tourism blogs, this chapter aims to provide a critical reflection on how individuals, families, tourists, in general, are looking for alternative forms of touristic and gastronomic experiences, very much rooted on the spiritual quest for peace, security, hope, self-awareness, and above all, emotional wellbeing. Hence, this chapter does not intend to describe a systematic literature review on spiritual tourism or food spirituality in post pandemic times. Aware of the highly subjective and controversial dimensions of spirituality, the author intends to share with interested and intrigued readers, a critical reflection on the rise of new forms of consumers, in the so called 'new normal', who completely changed their mindset regarding holiday habits and eagerly seek to feed their body, mind and soul, rethinking and replacing their previous touristic and gastronomic choices. The concept of 'new normal' has been recurrently and broadly adopted in the past, in academic and public spheres, to translate periods of adaptation after disruptive events, forcing people, organizations and societies to readjust behaviors, trajectories and ways of living. In the milieu of the pandemic situation, and particularly, within the tourist and gastronomic contexts, new normality stands for the need to rearrange and apply behaviors in hotels, restaurants, travelling infrastructures, required by health protocols and government regulations aiming at preventing COVID 19 transmission. Social distancing, wear of masks, hands washing, and more recently vaccination certificates and proofs of COVID testing were compulsory in many touristic facilities, including hotels, restaurants, transportation means. Following these guidelines and complying with mandatory rules imposed by the protocols, allowed the tourism sector to revive, and safely return to their operations (Engel, 2021; Pramezwary et al., 2021). Hence this chapter draws attention for the set of possibilities and opportunities for tourism, hospitality and gastronomic sectors, to critical think about their current business models, create innovative touristic packages and craft new marketing messages very much focused on spiritual attributes.

SPIRITUALITY, TOURISM AND GASTRONOMY: WHERE IS THE LINK?

The Need to Contextualize Spirituality

Spirituality is emerging as a novel research area in a plethora of interdisciplinary research crossing well-being and health (Chirico, 2016; Garssen et al., 2021), management (Makkar & Singh, 2021; Rocha & Pinheiro, 2021), tourism (Garcês et al., 2018; Stausberg, 2014), food and gastronomy (Bastenegar, 2019; Michopoulou & Jauniškis, 2020) to name a few. One common tenet across these studies, is the difficulty of comprehensively defining its concept, given its highly subjective and debatable nature (Speck, 2005). Despite the lack of consensus in defining spirituality, the term has its origins on the Latin word *spiritus* meaning air, breath, breathing. In semantic and linguistic terms, the word translates a "correlation of the two main fields of human reality, the field of the human soul and the field of the spirit, which interpenetrate and complement each other" (Kielkiewicz & Dalzell, 2015, p.53). Historically linked to religion, and unquestionably a reflection of religious beliefs and practices for many people worldwide (Makkar & Singh, 2021; Parsian & Dunning, 2009), the term has been used, explored and debated in religion and non-religion settings (Beck, 1986), and its historical link has gradually been removed (Chirico, 2016; Gall et al., 2011).

The World Health Organization also plays a crucial role in providing guidelines regarding the linkages between spirituality, health and personal beliefs (see for example: World Health Organization, 1998, 2012), suggesting that "Spirituality addresses questions such as meaning of life and purpose in life and is not necessarily limited to any specific types of beliefs or practices" (World Health Organization, 1998, p. 7). As for Beck (1986), spirituality is an inner state of being and a force within the heart of a person. Hence, the common ground is the consideration that spirituality helps people to deal with major life stressors (Del Castillo, 2020) and deals with human beliefs, instituting itself as a universal topic, whether rooted in religious or non-religious convictions, independently of cultures, civilizations, origins and timeframes (Michopoulou & Jauniškis, 2020).

Spirituality, in fact, has also been discussed as a keystone of wellbeing and quality of life. Lück & Aquino (2021) reminds us the work of Hettler (2006), suggesting that wellbeing comprises six dimensions of life balance: intellectual, physical, emotional, occupational, social and spiritual wellness. Quality of life, in turn, is also discussed by Cherblanc et al. (2021) as encompassing physical and psychological wellbeing, social relationships, environment, personal beliefs, religion and spirituality. To wrap up so many different perspectives on what spirituality might be, Ross's (1995) definition seems to be extremely useful to contextualize this chapter viewpoint. The author suggests that defining spirituality encompasses three components: i) the need to find meaning, purpose and fulfilment in life, ii) the need for hope and will to live, and lastly, iii) the need for belief and faith in self, others and God.

Understanding Spirituality in the Context of Touristic Choices

The idea that particular forms of tourism, such as nature-based tourism, transformational tourism, eudaimonia tourism, spiritual tourism, have the potential to transform people, to improve well-being and mental health and facilitate individuals' internal transformations (Gezon, 2018; Little, 2012), is not new. Such forms are suggested as alternative forms of therapies for people to recover from stress and mental fatigue, aiming at reconciling body, mind and spirit (Smith, 2003; Smith & Kelly, 2006). Research already validated this argument years before the pandemic situation (see for example: Ferguson & Tamburello, 2015; Gezon, 2018; Little, 2012; Pomfret, 2012). Spirituality in the context of touristic choices is not circumscribed, in this chapter, to religious beliefs or pilgrimages to religious places. On the contrary, it opens the discussion to more holistic perspectives on how non-religious touristic choices are gaining momentum in the quest for peace, well-being and spiritual wellness (Sharma & Nayak, 2019; Sharpley & Jepson, 2011; Sheldon, 2020; Tuo et al., 2014), as discussed above. Thus, the viewpoint of Sheldon (2020) is here embraced to argument that travelling might act as a tool for inner transformation. It is important to bring some light to the meaning of transformation in this context. The concept of transformational tourism aims to translate the role of countryside nature and wild nature activities, in tackling contemporary individuals' concerns towards their health, well-being and body shapes and sizes, with the ultimate goal of taking care of the self (Jepson, 2015; Little, 2012). It is worth mentioning that the so-called nature-based tourism (NBT) has received a lot of attention in recent years, whether positioned as adventure tourism, ecotourism, geotourism, wildlife tourism, natural tourism, marine tourism. To put it simply NBT is defined as the "nexus between tourism, outdoor recreational activities, and nature areas" (Fossgard & Fredman, 2019, p. 2). In the context of nature-based touristic choices, several studies revealed that tourists engage in personal and philosophical reflections while challenged by trekking and other nature-related activities (Aquino & Hyde, 2018), that tourism experiences linked to nature, hiking and mountaineering constitute personal spiritual journeys (Pomfret, 2012), that non-religious spiritual tourism connected and inspired by nature provides healing effects (Ferguson & Tamburello, 2015) and spiritual fulfilment (Willis, 2015) and also that nature-based practices rooted in ancient cultural heritages such as forest bathing, that is to say, spending time among trees (e.g. in Japan, Germany, China, Poland), besides contributing to physical wellness, are also directly linked to spirituality (Bhalla et al., 2021; Lück & Aquino, 2021).

The main argument across this research frame, is that nature generates some kind of healing power (Buzzell & Chalquist, 2009), by answering to the individual's quest for solitude, silence, inner peace and deep restoration (Sharpley & Jepson, 2011; Sheldon, 2020). In sum, tourism and travelling in nature-based settings and landscapes, provides us with the opportunity to reflect and self-realize on personal worldviews (Lück & Aquino, 2021) and as suggested by Jepson (2015), stimulates the generation of emotional effects, that we may call spiritual experiences.

Understanding Spirituality in the Context of Gastronomic Experiences

Regarding the linkages between spirituality and food/gastronomy, a lot of research has also been done in recent years (Bone, 2005; Coveney, 1999; Dias et al., 2021; Dias & Nogueira, 2018; Michopoulou & Jauniškis, 2020). Researchers claim that food preparation and serving is, in fact, an act of love portraying strong emotional and spiritual connections between household and family members. Hence, individuals express their spirituality through these rituals and social practices of acquiring, preparing, cooking, serving and sharing food (Bone, 2005; Caldwell, 2007; Classen, 2007; Mcguire, 2003; Parkins & Craig, 2015). It is no surprise then, that academics and practitioners have started to evince a keen interest on the interplay between food, gastronomy and spirituality.

Within the abovementioned relationship between spirituality and wellbeing, one finds fertile ground when connecting food as a third dimension of analysis. Referring to the expression 'breaking bread' as a symbolic aphorism to translate any form of communal eating, but also as a shared experience quite connotated to the sacred, Bone (2005, p. 309) uses the concept of *everyday spirituality* to illustrate "the idea that taken-for-granted routines and patterns in daily life can be perceived as spiritual.". The author further adds that social practices around eating, not only feed body and soul, but also provide social interaction, encourage harmony and love for life and more importantly "an opportunity for spiritual renewal" and to connect with a variety of philosophical beliefs (p.316). Other authors follow this line of reasoning, claiming that food and gastronomic experiences may contribute to restore emotional and physical wellbeing in the sense that it provides pleasure, positive sensorial experiences and a sense of self and family caring, thus encompassing deep spiritual and psychological implications (Caldwell, 2007; Michopoulou & Jauniškis, 2020).

The role played by spirituality in food preparation and sharing has also been subject of inquiry by this chapter's author and colleagues (Dias et al., 2021; Dias & Nogueira, 2018, 2019). Previous reflections on the topic highlighted the need to bridge spirituality with the slow movement, where we are invited to slow down the pace at which we live and downshift to a slow connected life (see for example: Honoré, 2010, 2013). Honoré explains that the movement began with the Slow Food Movement (SFM) in Italy back in the 1980s, within which unsustainable agri-food systems are heavily contested. SFM was founded in 1986 by Carlo Petrini, a food activist, to dispute the rapid proliferation of fast food chains or as simply put by Ritzer (2001b) to resist the 'McDonaldization' of food. Slow Food (SF) supporters advocate a multidisciplinary approach to sustainable food production and consumption towards more balanced and slower paces of life, contesting modern, industrialized and capitalistic food systems. According to the SF philosophy, informed consumers make better decisions and become co-producers of sustainable food. Hence, always cherishing the gastronomic pleasure, the movement aims to protect our right to consume good food while conserving regional, traditional and artisan cuisines, products and techniques, respecting food heritage and ancient rituals, promoting fair trade, and ultimately, while protecting the Planet's biodiversity in harmony with the ecosystems (Morrisey, 2010; Petrini, 2001, 2006, 2007, 2013; Ritzer, 2001a). To achieve such endeavor, the SFM proposes that truly sustainable food emerges from three interconnected principles: good, clean and fair (Petrini, 2006, 2007; Ritzer, 2001a).

SF philosophy and principles, have been delicately connected to spirituality. Using Sonnenfeld's words, a culinary historian, Laudan (2004) encourages us to contemplate the table as an 'altar' "that offers the template for the preservation of human rights and the environment." (p. 135). She also brings to light Batali's perspective, a well-know, chef, writer, restaurateur, who claimed that SF movement is "far more spiritual, nay, religious, than any club (or religion, for that matter)" (p. 135). Acknowledging the relevance of this apparently subtle link between SF and spirituality, and based on a Delphi panel of 24 experts from the public, private and academic domains, Dias and Nogueira (2019) developed a more overarching and embracing Slow Food model, inviting us to reflect on how social food practices are heavily rooted on philosophical and spiritual beliefs and proposing the inclusion of a fourth principle – the agri-food philosophy unfolded in three dimensions: spirituality, socialization and knowledge. The authors claim that embracing spirituality in food practices and rituals contributes to our understanding on the meaning of life, balance, peace of mind and harmony through food. As argued elsewhere (Dias et al., 2021) a double significance is attributed to food spirituality: firstly, it encompasses the metaphorical quest for *Food for the Soul*, that is, "the sensation of happiness, peace, harmony and balance with the nature resulting from contact with the earth in the moments of production and preparing food" (p.5). And secondly, it sheds light to the raising concerns of consumers regarding their feelings of safety in their food purchasing decision-making processes. Consumers, consciously and deliberately, search for natural and spiritual food attributes, such as genuine taste, texture, smell and appearance, by which they can clean and purify the body.

CONSUMERS' QUEST FOR SPIRITUALITY WITHIN TOURISTIC AND GASTRONOMIC EXPERIENCIES IN THE 'NEW NORMAL'

When COVID-19 Meets Spirituality

This chapter started by raising attention to the fact that, following the COVID outbreak, people worldwide were confronted with exasperating situations in their lives, leading to countless health-related issues, whether mental and/or physical, caused by the disease itself, but foremost, caused by fear of the unknown virus and its long-term consequences. Hence, the influence of the pandemic on peoples' well-being and mental health has been a hot topic among researchers from different fields, in the past few months. For the most part, it is unquestionable that peoples' lives worldwide faced massive challenges. Abounding studies confirm a positive association between the Covid-19 pandemic and all sorts of well-being and mental health issues such as trauma, fear, depression, panic, anxiety, burnout (Arslan & Yildirim, 2021; Del Castillo, 2020; Kumar & Nayar, 2021; Prime et al., 2020). To overcome and reduce the impact of the aforesaid issues, several scholars are revealing that people are eagerly searching for inner peace, self-control, life balance, self-consciousness and well-being, through spirituality (Arslan & Yildirim, 2021; Bhalla et al., 2021; Del Castillo, 2020; Kala, 2021).

Del Castillo (2020), for example, provided an overview of themes and insights from recent studies on health, spirituality and Covid-19, showing that the pandemic significantly impacted peoples' normal way of living, forcing them to deal with unexpected and stressful situations that were dealt with spiritual practices. The main argument is that spiritual well-being constitutes a cornerstone of holistic care, beyond physical and mental health. In this sense, it is suggested that "belief in the sacred or divine and the consequent spiritual experience leads to positive psychological states of peace, healing, contentment, hope and joy" (Del Castillo, 2020, p. 254). Following the same line of inquiry, Arslan and Yildirim (2021) explored spirituality as an underlying psychological mechanism that helps to mitigate COVID-19 detrimental stressors.

Different forms of spiritual practices were studied, namely meditation, yoga, prayer, reading holy books, and results revealed that such practices contributed to achieve states of healing, by embodying a positive mindset, reconstructing hope (Kala, 2021) and acting as resources to cope and manage stressful environments (Del Castillo, 2020). Similarly, Cherblanc et al (2021) demonstrated that hope, optimism, awe and meaning of life are dimensions of spirituality, that taken together provide "a source of spiritual force that is useful in coping with the stress caused by the pandemic" (p.13).

Touristic and Gastronomic Consumer Demands for Spirituality in the New Normal

As bluntly recognized in the introduction of this chapter, the United Nations World Tourism Organization (UNWTO), pointed earlier this year, that the COVID-19 pandemic brought global tourism close to a near-disaster position. However, as the global vaccination rollout advanced and the number of infected people showed signs of reduction worldwide, at the beginning of the second semester of 2021, international tourism enjoyed signs of rebound. Hence, since June 2021 many travel destinations were reopened mostly in Europe and the Americas and travel restrictions were lifted, or better said, relaxed, to vaccinated travelers (UNWTO, 2021e). Hence, in close coordination and collaboration with the World Health Organization (WHO), UNWTO is constantly delivering warning calls to the tourism and hospitality

Figure 1. Tourism

industry, claiming that its renaissance and recovery plans must be deeply rooted in promoting consumers' local and global health and sense of safety and wellbeing. Hereof, Chhabra (2020, p.78) raises a key question to the tourism and hospitality industry: "how to build and sustain healthy and robust consumer sentiments so that when the world is ready to travel, they serve as a push motivation factors and generate demand for health/wellness trips to authentic domestic and eventually international destinations?".

Answering Chhabra's question goes hand in hand with the above discussion on the link between spirituality and touristic/gastronomic choices. As noted earlier in this chapter, certain patterns of gastronomic and touristic experiences might act as key drivers of physical and mental well-being when consumers connect with spiritual, loving and caring practices. The novelty here, that is worth reflecting upon, is that tourism post-COVID-19 may have the potential to heal or reduce states of depression, anxiety and other mental health issues caused by it (Bhalla et al., 2021) and, tourism in natural and peaceful settings may actively contribute to rebound the sector, compared with other forms of tourism (Quendler et al., 2021). Many authors are providing useful recommendations for the tourism industry, based on consumers' current preferences for spiritual quest and engagement in new touristic configurations (Bhalla et al., 2021; Chhabra, 2020; Kala, 2021). Using NVivo to analyze the plethora of research that has been developed on the linkages between tourism, gastronomy and spirituality, in the past few months, the following word cloud shows the main issues, words, and questions raised by its authors. One can clearly identify, on one hand, words related to spirituality such as 'sacred', 'pilgrimages', 'minds', 'religion', 'consciousness', 'restoration', 'self'. On the other hand, words related to individuals and consumer choices and feelings are also present, namely, 'landscapes', 'connection', 'emotions', 'food', 'gastronomy', 'nature', 'table', 'caring', 'trips'.

Some of these studies explored consumers' preferences towards touristic sites, during the COVID-19 pandemic, and after the lock-down periods, in the so called 'new normal'. On this matter, research shows that a growing number of individuals and families, in the 'new normal', are craving to heal their bodies and minds after the peak pandemic period namely through the opportunity to spend more time closer to nature (Cao, 2020; Qiu et al., 2021), participate in spiritual retreats offering yoga and meditation programs as restorative therapeutic interventions (Chhabra, 2020) and engage in rural tourism and farm holidays embedded in natural settings with sunlight and fresh air (Quendler et al., 2021), to name a few.

Lück and Aquino' study (2021) focused precisely on consumers' search for holidays during the pandemic and revealed that although faced with rules and restrictions, consumers' preferences laid in domestic nature-based tourism activities. Findings demonstrated that regardless of duration or distance from home, nature-based touristic options contributed to build upon individual wellbeing. Many other studies reached similar results. Bhalla et al (2021) attempted to understand people's sentiments and willingness for travelling post-COVID-19 and unveiled that participants are prone to nature-based tourism, referring to beautiful landscapes, mountains, sunlight and all sorts of natural spaces, to recover from the hindrances and negative effects of COVID-19. The narratives show that participants are much more inclined to choose nature-based touristic solutions, to feed their needs of spiritual elevation and healing, after spending long periods of lock-down, stress, fear, despair and isolation. Travel motivations were, therefore, inclined to spiritual tourism, not necessarily religious touristic sites, but foremost linked to the positive effects of nature embeddedness, such as releasing psychological pressures, calming their minds, finding inner connection, and renovating their sense of healing, peace and positivity. As stated by one participant: "Nature makes me positive. I feel spiritually uplifted, more conscious, and I believe a trip post-COVID-19 will surely help me to make some tough decisions in my life" (Bhalla et al., 2021, p.6).

Moreover, as people were deprived from gatherings with families and friends, there was a generalized feeling of emptiness, as if social life was nothing but a void. Hence, when lock-downs started to open up, people moved fast forward to travel to outdoor sites and share meals with their families and friends before new lock-downs arose. In the context of the impact of COVID on food choices, a lot of research demonstrated that consumers started searching for healthier food choices across the globe (Marty et al., 2021; Ogundijo et al., 2021), but the main point to be made here is that the pandemic period forced families to engage in new dynamics at home, like cooking together, trying new recipes, sharing it with friends and family. Thus, happiness and wellbeing were also generated from this new form of continuous conviviality with loved ones, twenty-four/seven, something that never happened before in peoples' life. Aligned with the lessons learned from the Slow Food movement on conviviality and commensality, this chapter raises attention to the potential of the SF framework to restore, renew, and ultimately, rethink our food choices, and the importance of food in social, cultural and health settings, during and after the COVID pandemic. And even after long periods of lockdown, consumers are craving for gastronomic experiences in a sharing environment, with their loved ones, families and friends all together, with no limit of people per table, room or space. As beautifully acknowledged by Fourat, Fournier and Lepiller (2021) commensality at the table is as reassurance of respect for social bonds, respect for food rituals and an opportunity to restoration since "eating together normalizes the abnormal and helps people find meaning in an unusual situation" (p.206).

SOLUTIONS AND RECOMMENDATIONS

Following UNWTO incessant warning calls to the tourism and hospitality industry to rethink its recovery plans, the main recommendation to be made is to design offerings with a strong foci on current tourists/ consumers' preferences and needs for spiritual quest, very much rooted in promoting a sense of safety and wellbeing. Hence, the pathway to recovery could be anchored in two pillars: crafting value propositions towards spiritual healing and investing in new forms of consumer engagement trough digital formats.

The first pillar – crafting value propositions towards spiritual healing – departs from the conclusions retrieved by the studies scrutinized above that, in the post pandemic period, tourists are eagerly seeking for products and services that offer healing. In the context of the supply side of tourism, gastronomic and hospitality, it will be necessary to explore and understand tourists' motivations to, and expectations of, travelling and rethink and redesign strategies to reposition destination appeals and attributes towards emotional wellbeing and spiritual peace. It follows that value propositions need to be crafted in alignment with the recognition that the most attractive alternatives for tourists are those who appeal for their spiritual quest to feed their body, mind and soul.

Regarding the second pillar – investing in new forms of consumer engagement through digital formats – one cannot ignore that during lock-downs, consumers worldwide relied on the internet to work, shop, and ultimately, to stay connected with family, friends and the outside world. One trend that became clear was the possibility to engage consumers in digital travelling, where new forms of virtual tourists emerged. Travelling blogs, virtual museums, virtual travelling to beautiful landscapes, virtual reality experiences for e-food and e-wine destinations, online cooking tutorials and even digital stay-temple visits gained momentum, as people craved for travelling, even if only by means of the virtual world. Tourism firms have now an opportunity to think ahead the potential of this type of engagement. The digitalization of tourism might act as a pull factor towards future physical touristic choices. Furthermore, these digital platforms can also serve as spaces to frame effective messages to mitigate perceived risk and fear of travelling and rebuild consumers' trust. By informing consumers on the safety schemes put into action in touristic sites in accordance with the World Health Organization guidelines to travelers, tourism firms also contribute to build consumers' confidence and increase their willingness to travel. There is no doubt that the demand side of the tourism market is showing very optimistic signs already, but it is also true that people search for information that assures them, they are being offered with safe ways of travelling, from start to finish, i.e., during the trip, during the stay and during visits.

CONCLUSION AND FUTURE RESEARCH DIRECTIONS

To sum up, this chapter drew attention to the (not so!) subtle linkages between touristic and gastronomic choices, spirituality and the COVID-19 crisis. The main argument addressed in these pages is that the pandemic is giving the tourism and hospitality industry an opportunity to reinvent their value propositions and design their offerings based on the post COVID tourists, which reveal different attitudes, needs and wants, regarding their tourism demands. This chapter outlined different perspectives about consumers' behavior in the post-pandemic phase, that might help to revive and support the tourism and hospitality sector: domestic tourism away from big metropolis to avoid over tourism, nature-related tourism, rural tourism, spiritual tourism or any other forms that allow for some spiritual quest (Bhalla et al., 2021; Chhabra, 2020). All these forms share a common attribute utterly valued by current consumers, who have suffered, and are still suffering, from post pandemic stress: the possibilities and opportunities to invest in touristic and gastronomic experiences that might trigger inner peace, emotional wellbeing, safety (Lück & Aquino, 2021), commensality and happiness at the table (Dias et al., 2021; Fourat et al., 2021). All these aspects taken together, comprise a mix of opportunities for tourism, hospitality and gastronomic sectors, to critically think about their current business models, create innovative touristic packages and craft new marketing messages very much focused on spiritual attributes, since the 'new normal' is witnessing the rise of new consumers of touristic products who genuinely quest to heal, both mentally and physically, in post pandemic times.

The spiritual side of food and gastronomic consumption is an emerging research topic, since the literature and the number of articles published regarding both food and spirituality conjointly, is scarce and barely existent in the hospitality industry (Michopoulou & Jauniškis, 2020). Moreover, despite blissful news that Covid-19 infections are decreasing as populations get vaccinated and collectively develop immunity, Covid-19 negative, and long-term, effects on mental well-being are still worrisome. On that account, post pandemic times within the so-called 'new normal' are an opportunity to further investigate the role of spirituality in the present on-going struggle (Del Castillo, 2020). People are eager to win back their sense of freedom, happiness and well-being. After so many months of isolation and compulsory lock-downs, traveling, tourism and gastronomic experiences are at the top of consumers' priorities, but all in all, this eagerness brought about new forms of consumption very much rooted on physical, mental and spiritual well-being. Hence, to better inform practitioners on how to develop successful business models based on these premises, new research is needed to understand in-depth what consumers want and feel regarding their touristic choices. Implied in this call for new research is the author's strong believe that the current tourism and hospitality research could benefit from engaging in more qualitative research designs and move away from positivist approaches employed in the majority of studies in this field. In-depth understanding of consumers' feelings in this new normal cannot be simply tackled with quantitative measurement tools.

To conclude, the author entrusts the readers with some *food for thought* about her own empirical observation of current generations that, as remarked by Classen (2007), are gradually losing respect and concern for food preparation and eating rituals and "with this loss, however, also comes a loss of respect for life and spirituality in their crucial components" (p.316.). Given the scant research acknowledged before, this chapter hopes to engage and inspire other academics and practitioners to embrace philosophical and spiritual approaches to better understand consumers' needs and wants embedded in their *everyday spirituality*, which ultimately translates their touristic and gastronomic demands, towards more balanced, slower, happier, and above all, spiritual choices. At the end of the day, we all wish to finally enjoy holidays with a sense of freedom and wellbeing, bearing in mind that even the word *holy-days* is pledging us to reflect on the spiritual side of it!

REFERENCES

Aquino, R. S., & Hyde, K. F. (2018). Unearthing the geotourism experience: Geotourist perspectives at Mount Pinatubo, Philippines. *Tourist Studies*, *18*(1), 41–62. doi:10.1177/1468797617717465

Arslan, G., & Yildirim, M. (2021). Meaning-based coping and spirituality during the COVID-19 pandemic : Mediating effects on subjective well-being. *Frontiers in Psychology*, *12*. Advance online publication. doi:10.3389/fpsyg.2021.646572 PMID:33935906

Bastenegar, M. (2019). Spiritual understanding and experience in the creative tourism of gastronomy. *International Journal of Tourism and Spirituality*, *3*(2), 43–67.

Beck, C. (1986). Education for spirituality. *Interchange*, *17*(2), 148–156.

Bhalla, R., Chowdhary, N., & Ranjan, A. (2021). Spiritual tourism for psychotherapeutic healing post COVID-19. *Journal of Travel & Tourism Marketing*, *0*(0), 1–13. doi:10.1080/10548408.2021.1930630

Bone, J. (2005). Breaking bread : Spirituality, food and early childhood education. *International Journal of Children's Spirituality*, *10*(3), 307–317. doi:10.1080/13644360500347607

Buzzell, L., & Chalquist, C. (2009). Psyche and nature in a circle of healing. In Ecotherapy: Healing with Nature in Mind. Catapult.

Caldwell, M. L. (2007). Feeding the body nourishing the soul. *Food, Culture, & Society*, *10*(1), 43–71.

Cao, X. (2020). COVID-19: Immunopathology and its implications for therapy. *Nature Reviews. Immunology*, *20*(5), 269–270. doi:10.103841577-020-0308-3 PMID:32273594

Cherblanc, J., Bergeron, C., Maltais, D., Cadell, S., Gauthier, G., & Labra, O. (2021). Predictive factors of spiritual quality of life during the COVID - 19 Pandemic : A multivariate analysis. *Journal of Religion and Health*, *60*(3), 1475–1493. doi:10.100710943-021-01233-6 PMID:33759072

Chhabra, D. (2020). Transformational wellness tourism system Model in the pandemic era. *International Journal of Health Management and Tourism*, *5*(2), 76–101. doi:10.31201/ijhmt.788628

Chirico, F. (2016). Spiritual well-being in the 21st century : It is time to review the current WHO ' s health definition. *Journal of Health and Social Sciences*, *1*(1), 11–16. doi:10.19204/2016prt2

Classen, A. (2007). The symbolic function of food as iconic representation of culture and spirituality in Wolfram von Eschenbach's Parzival. *Orbis Litterarum*, *62*(4), 315–335.

Coveney, J. (1999). The science and spirituality of nutrition. *Critical Public Health*, *9*(1), 23–37. doi:10.1080/09581599908409217

Del Castillo, F. (2020). Health, spirituality and Covid-19: Themes and insights. *Journal of Public Health Health*, *43*(2), 254–255. doi:10.1093/pubmed/fdaa185 PMID:33044544

Dias, R. R., & Nogueira, M. (2018). *Agri-Food trends in Portugal for 2027 according to the principles of slow food.* Academic Press.

Dias, R. R., & Nogueira, M. (2019). Agri-food trends for Portugal based on the slow food principles: a new approach. *13th International European Forum on System Dynamics and Innovation in Food Networks*.

Dias, R. R., Nogueira, M., & Azinheira, F. (2021). Spirituality in agri-food sector : A complementary vision of slow food principles. *International Journal of Food System Dynamics. Proceedings in System Dynamics and Innovation in Food Networks*, *2021*, 4–13.

Engel, Ü. (2021). COVID-19 and "New Normal" tourism: Reconstructing tourism. *Journal of Tourism and Development*, *35*, 217–226. https://proa.ua.pt/index.php/rtd/article/view/24652

Ferguson, T. W., & Tamburello, J. A. (2015). The natural environment as a spiritual resource: A theory of regional variation in religious adherence. *Sociology of Religion*, *76*(3), 295–314.

Fossgard, K., & Fredman, P. (2019). Dimensions in the nature-based tourism experiencescape: An explorative analysis. *Journal of Outdoor Recreation and Tourism*, *28*, 100219. doi:10.1016/j.jort.2019.04.001

Fourat, E., Fournier, T., & Lepiller, O. (2021). Snatched commensality: To eat or not to eat together in times of Covid-19 in France. *Food & Foodways*, *29*(2), 204–212. doi:10.1080/07409710.2021.1901383

Gall, T. L., Malette, J., & Guirguis-Younger, M. (2011). Spirituality and religiousness : A diversity of definitions. *Journal of Spirituality in Mental Health*, *13*(3), 158–181. doi:10.1080/19349637.2011.593404

Garcês, S., Pocinho, M., & de Jesus, S. N. (2018). Review of optimism, creativity and spirituality in tourism research. *Tourism and Hospitality Management*, *24*(1), 107–117. doi:10.20867/thm.24.1.6

Garssen, B., Visser, A., Pool, G., & Garssen, B. (2021). Does spirituality or religion positively affect mental health? *The International Journal for the Psychology of Religion*, *31*(1), 4–20. doi:10.1080/10508619.2020.1729570

Gezon, L. L. (2018). Global scouts: Youth engagement with spirituality and wellness through travel, Lake Atitlán, Guatemala. *Journal of Tourism and Cultural Change*, *16*(4), 365–378. doi:10.1080/14766825.2017.1310217

Honoré, C. (2010). *In praise of slow: How a worldwide movement is challenging the cult of speed*. Hachette.

Honoré, C. (2013). *The slow fix: solve problems, work smarter and live better in a world addicted to speed*. Knopf Canada.

Jepson, D. (2015). The lure of the countryside: the spiritual dimension of rural spaces of leisure. In S. Gammon & S. Elkington (Eds.), *Landscapes of leisure* (pp. 202–219). Palgrave Macmillan. doi:10.1057/9781137428530_15

Kala, D. (2021). Thank you, God. You saved us' - examining tourists' intention to visit religious destinations in the post COVID. *Current Issues in Tourism*, *0*(0), 1–7. doi:10.1080/13683500.2021.1876643

Kielkiewicz, K., & Dalzell, T. (2015). Towards comprehension of spirituality through its semantics. *A Journal of Religion, Education and the Arts, 9*(January).

Kumar, A., & Nayar, K. R. (2021). COVID 19 and its mental health consequences. *Journal of Mental Health (Abingdon, England)*, *30*(1), 1–2. doi:10.1080/09638237.2020.1757052 PMID:32339041

Laudan, R. (2004). Slow Food: The french terroir strategy, and culinary modernism. *Food, Culture, & Society*, *7*(2), 133–144. doi:10.2752/155280104786577833

Little, J. (2012). Transformational tourism, nature and wellbeing: New perspectives on fitness and the body. *Sociologia Ruralis*, *52*(3), 257–271.

Lück, M., & Aquino, R. S. (2021). Domestic nature-based tourism and wellbeing - A roadmap for the new domestic nature. In J. Wilks, D. Pendergast, P. Leggat, & D. Morgan (Eds.), *Tourist health, safety and wellbeing in the new normal*. Springer., doi:10.1007/978-981-16-5415-2

Makkar, S., & Singh, A. K. (2021). Development of a spirituality measurement scale. *Current Psychology (New Brunswick, N.J.)*, *40*(3), 1490–1497.

Marty, L., de Lauzon-Guillain, B., Labesse, M., & Nicklaus, S. (2021). Food choice motives and the nutritional quality of diet during the COVID-19 lockdown in France. *Appetite*, *157*, 105005. PMID:33068666

Mcguire, M. (2003). Why bodies matter: A sociological reflection on spirituality and materiality. *Journal of Christian Spirituality*, *3*(1), 1–18.

Michopoulou, E., & Jauniškis, P. (2020). Exploring the relationship between food and spirituality: A literature review. *International Journal of Hospitality Management*, *87*(February), 102494. doi:10.1016/j.ijhm.2020.102494

Morrisey, L. (2010). Slow Food: An Interview with Carlo Petrini. *Development*, *53*(3), 447–448. doi:10.1057/dev.2010.51 PMID:20056678

Ogundijo, D. A., Tas, A. A., & Onarinde, B. A. (2021). Exploring the impact of COVID-19 pandemic on eating and purchasing behaviours of people living in England. *Nutrients*, *13*(5). Advance online publication. doi:10.3390/nu13051499 PMID:33946799

Parkins, W., & Craig, G. (2015). Culture and the politics of alternative food networks. *Food, Culture, & Society*, *12*(1), 77–103. doi:10.2752/155280109X368679

Parsian, N., & Dunning, T. (2009). Developing and validating a questionnaire to measure spirituality : A psychometric process. *Global Journal of Health Science*, *1*(1), 2–11.

Petrini, C. (2001). *Slow food: the case for taste*. Columbia University Press.

Petrini, C. (2006). *Slow Food Revolution*. Rizzoli.

Petrini, C. (2007). *Slow food nation*. Rizzoli.

Petrini, C. (2013). *Slow food nation: Why our food should be good, clean, and fair*. Rizzoli Publications.

Pomfret, G. (2012). Personal emotional journeys associated with adventure activities on packaged mountaineering holidays. *Tourism Management Perspectives*, *4*, 145–154.

Pramezwary, A., Pramono, R., Situmorang, J. M. H., Salim, S. L., & Christina, F. (2021). Impulsive buying of five star hotel Jakarta: effect new normal era mediated of price and promotion. *Academy of Strategic Management Journal, 20*(2), 1–16.

Prime, H., Wade, M., & Browne, D. T. (2020). Risk and resilience in family well-being during the COVID-19 pandemic. *The American Psychologist*, *75*(5), 631–643. doi:10.1037/amp0000660 PMID:32437181

Qiu, M., Sha, J., & Scott, N. (2021). Restoration of visitors through nature-based tourism: A systematic review, conceptual framework, and future research directions. *International Journal of Environmental Research and Public Health*, *18*(5), 1–19. doi:10.3390/ijerph18052299 PMID:33652652

Quendler, E., Magnini, V., & Driouech, N. (2021). Strategic positioning of the farm holiday's post-pandemic competitive advantages: Fresh air and sunlight. *Zeitschrift für Tourismuswissenschaft*, *13*(1), 22–51. https://econpapers.repec.org/RePEc:bpj:touwis:v:13:y:2021:i:1:p:22-51:n:1

Ritzer, G. (2001a). *Explorations in the sociology of consumption: fast food, credit cards and casinos*. SAGE Publications.

Ritzer, G. (2001b). Slow food versus McDonald's'. In C. Petrini & B. Watson (Eds.), Slow food: Collected thoughts on taste, tradition and the honest pleasures of food. Chelsea Green Publishing.

Rocha, R. G., & Pinheiro, P. G. (2021). Organizational spirituality: Concept and perspectives. *Journal of Business Ethics*, *171*(2), 241–252. doi:10.100710551-020-04463-y

Ross, L. (1995). The spiritual dimension: Its importance to patients' health, well-being and quality of life and its implications for nursing practice. *International Journal of Nursing Studies*, *32*(5), 457–468. doi:10.1016/0020-7489(95)00007-K PMID:8550306

Sharma, P., & Nayak, J. K. (2019). Do tourists' emotional experiences influence images and intentions in yoga tourism? *Tourism Review*, *74*(3). Advance online publication. doi:10.1108/TR-05-2018-0060

Sharpley, R., & Jepson, D. (2011). Rural tourism: A spiritual experience? *Annals of Tourism Research*, *38*(1), 52–71. doi:10.1016/j.annals.2010.05.002

Sheldon, P. J. (2020). Designing tourism experiences for inner transformation. *Annals of Tourism Research*, *83*, 102935.

Sigala, M. (2020). Tourism and COVID-19: Impacts and implications for advancing and resetting industry and research. *Journal of Business Research*, *117*, 312–321. doi:10.1016/j.jbusres.2020.06.015 PMID:32546875

Smith, M. (2003). Holistic holidays: Tourism and the reconciliation of body, mind and spirit. *Tourism Recreation Research*, *28*(1), 103–108. doi:10.1080/02508281.2003.11081392

Smith, M., & Kelly, C. (2006). Holistic tourism: Journeys of the self? *Tourism Recreation Research*, *31*(1), 15–24. doi:10.1080/02508281.2006.11081243

Speck, B. W. (2005). *What is spirituality?* Academic Press.

Stausberg, M. (2014). Religion and spirituality in tourism. Academic Press.

Tuo, Y., Bai, C., & Chen, Z. (2014). Travel for meaning: the present and future of tourist wellbeing. *Seventh International Joint Conference on Computational Sciences and Optimization (IEEE)*, 479–483. 10.1109/CSO.2014.95

Willis, C. (2015). The contribution of cultural ecosystem services to understanding the tourism–nature–wellbeing nexus. *Journal of Outdoor Recreation and Tourism*, *10*, 38–43.

World Health Organization. (1998). *WHOQOL and spirituality, religiousness and personal beliefs (SRPB)*. https://apps.who.int/iris/bitstream/handle/10665/70897/WHO_MSA_MHP_98.2_eng.pdf?sequence=1&isAllowed=y

World Health Organization. (2012). *Spirituality, religiousness and personal Beliefs (SRPB) field-test instrument*. https://apps.who.int/iris/bitstream/handle/10665/77777/WHO_MSD_MER_Rev.2012.04_eng.pdf?sequence=1&isAllowed=y

World Tourism Organization. (2020a). *Conceptual guidance on tourism statistics in the COVID-19 Context*. UNWTO.

World Tourism Organization. (2020b). How are countries supporting tourism recovery? *UNWTO Briefing Note – Tourism and COVID-19, 1*, 1–28.

World Tourism Organization. (2020c). *Understanding domestic tourism and seizing its opportunities* (Issue 3). WTO.

World Tourism Organization. (2021a). *How COVID-19 is changing the world : a statistical perspective Volume III*. WTO.

World Tourism Organization. (2021b). *One year ago today: WHO defined the COVID-19 outbreak as a pandemic*. https://www.unwto.org/one-year-ago-today-who-defined-the-covid-19-outbreak-as-a-pandemic

World Tourism Organization. (2021c). *Vaccines and Reopen Borders Driving Tourism's Recovery*. https://webunwto.s3.eu-west-1.amazonaws.com/s3fs-public/2021-10/211004-unwto-barometer-en.pdf?wLLYFuBNpS3FXCLZzMvreo8oOXC2UIbW

Chapter 10
The Portuguese Hospitality and Restaurant Industry:
Tax Challenges Post COVID-19

Susana Cristina Rodrigues Aldeia
https://orcid.org/0000-0002-2611-7683
ISAG - European Business School, Portugal & Research Center in Business Sciences and Tourism (CICET-FCVC), Portugal

Paulo Macedo
ISAG - European Business School, Portugal & Research Center in Business Sciences and Tourism (CICET-FCVC), Portugal

Liliana Ivone Pereira
IPCA, Portugal

Joaquim Jorge Mota
IPCA, Portugal

ABSTRACT

The main goal of this paper is to understand the evolution of taxation in the accommodation and restaurant industry. In particular, it has three objectives: (1) to study how it has been occurring the taxation of this sector in Portugal, (2) to know if Portuguese legislators adopted tax measures specific to this economic area, and (3) to understand what post-COVID-19 perspectives are in the sector of the fiscal domain. For this purpose, business and tax law of Portugal were investigated. The results show that the Portuguese tax legislator hasn't been paying attention to the lodging and restaurant sector. There were some changes in the income and consumption tax domain in the last 10 years underpinned by the will to increase investment and to control tax avoidance, but hardly were introduced tax measures to help this business area in the COVID-19 pandemic. Although there was another kind of economic financial support given by the Portuguese government, in the fiscal domain, the IVAucher measure was introduced.

DOI: 10.4018/978-1-7998-9148-2.ch010

INTRODUCTION

Portugal is characterized by the dependency in the touristic activity, from which the accommodation and restaurant activities cannot be dissociated. Thus, tourism is a strategic economic activity for the country's economic and social development, particularly for employment and export growth (Turismo de Portugal, 2017). Tourism business area constitutes a mechanism to economic development of several countries (Fernández & Rivero, 2010). It is comprehensible that the restrictions imposed by the need of control of the pandemic situation evolution of COVID-19 have been embarrassing the business driving. The limitation to the functioning of the establishments, and the adaptation to the pandemic rules imposed by the Government have been conditioning the management decisions. In demanding moments such as the ones the sector has been going through, the relevance of public measures to the stimulus to the development of the activity should not neglect, nevertheless, its enjoyment sometimes can be faced as excessive dependency to the financial support.

According to the sectorial statistical synthesis prepared by the *Gabinete de Estratégia do Ministério da Economia e Transição Digital*, the accommodation and restaurant sectors assume a significant weight between the non-financial companies' area[1]. Considering the same source:

1. 9% of the total number of entities, with successive growth since 2007.
2. 9.4% of the total number of employees, with successive growth since 2007.
3. 3.9% of the total turnover, with successive growth since 2007.
4. 6.6% of total gross value added, with successive growth since 2007.
5. 8.6% of total gross capital formation, with successive growth since 2007.

Considering the sector's relevance, important economic decisions have been made internationally as an answer to the pandemic crisis. European Union approved the *NextGenerationEU* instrument[2], and it has been essential to the combat of the harmful consequences of the COVID-19 crisis. This instrument has the goal of the economies' transformation of the European Union's states, through measures such as the increase of employment and the reinforcing of decarbonization and digitization. The *NextGenerationEU* integrates the Recovery and Resilience Mechanism, making available financial resources to finance reforms and investments in the EU Member States. This financial support integrates loans and grants, and it can be applied until 2026.

Taking into consideration the privileged economic relationship between the Iberian countries, Portugal and Spain, the crisis allowed the development of tax agreements relevant to the employees, and it includes the touristic sector workers. The status of the cross-border worker was approved, providing benefits to that kind of employees, as (1) the access to the public services in both countries, (2) the social security recognition in both countries, (3) the launch of joint programs to attract investment and workers to cross-border areas.

It is why questions related to COVID-19 and tourism sector has been researched by Portuguese scholars. Oliveira (2020) highlighted the most relevant measures to tackle the effects of pandemic crisis on credit contracts. Machado, et al. (2020) analyse the communication strategies used by the sector to combat the negative effects of Covid-19 available on social networks, the research analyses Portugal and Brazil. Lopes et al. (2021) examine the tourism employees characteristics that make them more vulnerable or immune to the unemployment. Marques et al. (2021) research the tourism preference in the summer of 2020. Almeida and Silva (2020) evaluate the COVID-19 impact on tourism sustainability.

Therefore, it is possible to conclude that although there are some studies about the issue, the theme is still underexplored and to the best of our knowledge there are any studies that related tax benefits and Portuguese tourism companies. There is a need to better understand the evolution of taxation in the accommodation and restaurant industry in Portugal. In specific, it has three objectives (1) to study how it has been occurring the taxation of this sector in Portugal, (2) to know if Portuguese legislator adopted tax measures specific to this economic area, and (3) to understand what post-COVID-19 perspectives are in the sector of the fiscal domain. This research addresses this gap by studying the legal norms that approached the fiscal regulation on the Portuguese tax law.

As tax matters are part of the economic dimension, managers need to attend to fiscal questions to make better economic decisions. This paper helps to understand the evolution and tendencies of taxation in Portugal's catering and lodging activity. Also, it identifies the primary needs to be observed to help the sector face the pandemic crisis challenges.

This paper is composed of six sections. First, the introduction, in which the framing of the subject is done and where it is explained why it is important and updated to study this concept. Second, it makes a literature review explaining the concept of extra taxation. In the third section, it is exposed the paper's methodology. In the fourth section, it is done a framework of the sector' companies in the Portuguese Accounting and Tax Law. The following chapter explains the perspectives post-COVID-19 in the tax domain, and finally, the sixth presents discussion and conclusions.

FRAMEWORK

Article 7 of Portuguese General Tax Law - *Lei Geral Tributária* (GTL) determines that the taxation has as goals to favour the employment, socially relevant investment, competitiveness and the Portuguese economy internationalization. To this effect, the Portuguese legislator can determine tax benefits to specific activity sectors, since there is economic, social or environmental justification.

Buffon and Bassani (2013) refer that this possibility for the State to use taxation as an instrument of intervention in society, encouraging or discouraging behaviour, is called extra taxation. It constitutes a mechanism to control the economy and society (Riker et al., 2018). Through the extra taxation, the State can foment exemptions and tax benefits to promote de economic-social development of a specific region or economic activity that has collective interest (Buffon & Bassani, 2013). Tax benefits and incentives to hotel and restaurant activity are justified because this sector is one of the most affected by the pandemic crises, and Government need to take the post-pandemic to reinforce this sector, that it is one of the most relevant to the country's development.

Sofronov (2017) considers that although tourism area foments the direct and indirect employees, promote the development in rural zones, and increase the tax collection, it presents some threats. There are internal pressures so that States are not exposed to the outside, to commercial barriers and people movement. In that sense, the accommodation and food sector constitutes a pillar in the tourism area, so it needs to be supported to can continue to be a motor to the economic development. Portuguese legislator can utilize the extra taxation to incentive companies and consumers in the immediate environmental conservation and education to long-term. According to Riker et al. (2018) the extra taxation can generate positive fruits, well directed to sustainability questions.

In addition, institutions have an important role in the tax regulations, as determined by the institutional theory. The institutional theory's application to organizational studies allows a higher comprehension of the implicit patterns and of the diversity in organizations (Pereira, 2012). The idea of the institutions as human creations that define the limits that influence individual actions is irrefutable (North, 1990). The organizations shape themselves due to the informal and formal restrictions, being the first one underpinned on the myths, costumes, culture, and being the second one based on the legislations applied, for example. In this way, the institutional theory provides essential contributions to the entity's management because the organizational performance results not only from human action but also politics and cultural interactions (Fachin & Mendonça, 2003). In this sense, it is important to understand how (1) the organizations evolve as answer to the incentives, strategies and individual choices, and (2) in what way the institutions affect the political, economic and tax systems (Peci, 2006).

A recent view of the institutional approach emphasizes the relevance of the institutions to normal societies' functioning, as the legal and tax system. Considering that the companies' accounting and taxation areas sustain the information supplying for the different stakeholders (Callén & Gadea, 2004) and that the produced and provided financial information is influenced by the organizations' context, in which the stakeholders are inserted; so, accounting and tax systems should be analysed and interpreted under the received influence from the institutional context. This theory tries to explain this phenomenon (Campbell, 2000) considering the organizations as elements from which habits, thinking and actions are produced and replicated (Burns & Scapens, 2000).

Nowadays, there are a lot of institutionalized rules, and the behaviour of the legal entities and individuals is influenced by the laws, educational system, public opinion, and by the influence of all inserted in the same organizational environment (Meyer & Rowan, 1991). So, the organizations that belong to the same sector feel the same pressures and tend to follow similar accounting and tax practices and, being influenced by their context, they also influence the institutional rules, as the case of the fiscal policy. It is why this theory has been adopted to explain the influence of the different involved actors in the processes of change (Guerreiro et al., 2006; Major et al., 2018).

Besides, different stakeholders make external pressures in the regulation of tax law, in the same way, that the entire stakeholders' chain is involved in the business' results. The cooperation between the involved stakeholders in a business area is fundamental for the success in the reach of results. Freeman (1994) considers that stakeholders have legitimate claims that are established through the existence of exchange relationships, between the provision of critical resources and the satisfaction of the interests. Hill and Jones (1992) explain that investors provide capital and, in return, they expect that it be maximized. Creditors provide financing resources and, in return, they expect the capital reimbursement and the interest receipt. In the same way, the State supplies infrastructures and wait for jobs and tax payments.

The success of the tourism development depends on the role/participation of the interested parts (Amerta, 2017). Presenza and Cipollina (2010) concluded that sustainability in tourism require collaborative and inclusive practices in building consensus among the various stakeholders. The same authors consider the State the preferred stakeholder to support investors on the management and marketing activities of the tourism sector. Although the sustainability in the tourism is the fundamental condition to maintain the tourism growth, Cetin et al. (2017) verified that the tourists are available to pay to have access to better tourism services, though they are resistant to assume responsibilities with the sustainability of the tourist destinations. In this way, the State reinforces its role in tax collection taxes and in applying them to the sustainability of tourism.

METHODOLOGY

In this conceptual framework, this research aims to address three objectives: 1) to study how it has been occurring the taxation of this sector in Portugal, (2) to know if Portuguese legislator adopted tax measures specific to this economic area, and (3) to understand what post-COVID-19 perspectives are in the sector of the fiscal domain. For this purpose, the paper uses the empirical legal research method. This method researches the role of the legal policies and legislation roles (Aldeia, 2021; Epstein & King, 2002). An analysis of European Union legal dispositions was presented, in particular, the case study of Portugal. Several Portuguese data sources were included: the Accounting Law, the Corporate Income Tax Law, Personnel Income Tax Law, Value Added Tax Law, and other juridical structures.

Tourism Companies in the Portuguese Accounting and Tax Law

Accounting Law

Article 3 of the Decree-Law n. ° 158/2009, of 13 July, determines the mandatory application principle of the Portuguese Accounting Standardization System - *Sistema de Normalização Contabilística* (ASS) to the legal and individual companies[3]. ASS is based on eight instruments, and they are framed by the Conceptual Framework of the Accounting Standardization System - *Estrutura Conceptual do Sistema de Normalização Contabilística* (CFASS) (Rodrigues, 2016). The instruments are 1) bases for the presentation of financial statements, 2) models of financial statements, 3) account codes, 4) accounting and financial reporting standards, 5) accounting and financial reporting standard for small entities, 6) accounting and financial reporting standards for entities in the non-profit sector, 7) accounting standard for micro-entities, 8) interpretative norms.

The reference frame is the normative emanating from the International Accounting Standards Board (IASB), based on International Accounting Standards (IAS) and International Financial Reporting Standards (IFRS), and adopted in the European Union (Borges et al., 2010). Although it is mandatory for Portuguese companies to apply the ASS dispositions, in particular situations that are identified in article 4 of the Decree-Law 158/2009, companies must or can prepare financial statements, consolidated or individual accounts, respecting IAS and IFRS. Portuguese accounting legislator allows the existence of different levels of normalization, the grades are defined considering the different requirements of the financial report associated with the entities' categorization (Gonçalves et al., 2017). This happens in both domains, in the context of ASS application, and in the adoption of the IAS and IFRS dispositions. The application of different accounting norms to the Portuguese companies impairs the comparability between entities in a heavy way.

As in other economic areas, in the accommodation and catering sector companies can develop activity as a juridical entity or individual one. Article 9 of the Decree-Law 158/2009, of 13 July, determines categories of entities according to three criteria: (1) annual total of the balance sheet, (2) total of net turnover, (3) average number of employees during the period. Portuguese accounting law foresees four categories of entities: micro, small, medium size and large entities:

1. Micro-entities - not exceeding 2 of the 3 criteria - balance sheet total equal to € 350,000.00; net turnover equal to €700,000.00; average number of employees during the period equal to 10.

2. Small entities - not exceeding 2 of the 3 criteria - balance sheet total equal to € 4,000,000.00; net turnover equal to € 8,000,000.00; average number of employees during the period equal to 50.
3. Medium-sized entities - not exceeding 2 of the 3 criteria - balance sheet total equal to € 20,000,000.00; net turnover equal to €40,000,000.00; average number of employees during the period equal to 250.
4. Large entities - exceeding 2 of the 3 criteria - balance sheet total equal to € 20,000,000.00; net turnover equal to €40,000,000.00; average number of employees during the period equal to 250.

It will be considered a micro-entity if it fulfils the criteria of the topic i), a small entity if respects the limits of the point ii), a medium-sized entity if respects the criterion of item iii), and finally, it will be recognized as a large entity if exceed 2 of the 3 criteria.

Each category has particular requirements to comply with the accountability obligations. The micro-entities are obligate to present the financial position statement and the income statement, they are dismissed to present the statement of changes in equity, cash flow statement and notes[4]. The small entities need to prepare the balance sheet, the income statement, and the notes, it is not required to present the statement of changes in the equity and cash flows statement[5]. The medium and large companies have as mandatory disclosure the preparation of the financial position statement, the comprehensive income statement, the statement of changes in equity, the cash flows statement and notes. Small companies have less disclosure requirements, big entities face more obligations in the information of the financial reporting. Undoubtedly, that the balance sheet and income statement assume preponderance between all the documents of the financial reporting. They show the true and fair view of the business development, so the elements of these financial statements assume particular relevance (Aldeia, 2019).

The financial position statement is composed by three elements: Assets, Liabilities and Equity (ICAEW, 2020). The assets represent the entities resources (investments and rights before debtors, for example). Liabilities are the business' obligations (obligations to financiers and to the State, for example). Equity represents the difference between assets and liabilities, and it is under classified according to the Portuguese commercial societies' law (for example, share capital and reserves). This document allows to determine the composition of the companies' financial statement. Concerning the income statement, incomes and expenses are the elements that compose this financial statement (Melville, 2019). The incomes represent changes in the asset and/or liability with a positive effect on the equity (sales, provided services, interest received, for example). Expenses are changes in the asset and/or liability with a negative impact on the equity (the resources consumed in the provided services, for example). The comprehensive income statement allows determining the entities' net profit, and the result assumes a relevant role in the determination of the corporate income tax base.

The activity development of the accommodation and catering requires several available resources as (1) real states, furniture and machines to produce the meals, in accounting law, they are recognized in the item of building, plant and equipment, and its use implies the recognition of the depreciation spending. Human resources are indispensable to hospitality activity, they allow guaranteeing customer satisfaction, and they imply the accounting recording of personnel expenses. Finally, the goods and services integrated into the manufacturing of the meals and accommodation services, representing the consumption of (1) raw materials by the way of the cost of goods sold and materials consumed (2) and supplies and external services, respectively.

The net profit is determined under two main assumptions, the accruals accounting basis[6] and the balancing between revenue and expenses[7] (Chaudhry et al., 2016). In the tax domain, the profit determined under the accounting standards assumes an important role in the determination of the business income tax (Lapatza, 2006; Nabais, 2018). It is the starting point to calculate the taxable base in general regime of the Portuguese Income Tax Law (Sarmento et al., 2019).

Tax Law, Income Tax

Overall Framework

The Portuguese conceptual framework of income taxation has as base principles and different codes, according to the income titular. The income taxation obtained by the development of activity by the individual entrepreneur has based on personal income tax - *Imposto sobre o Rendimento das Pessoas Singulares* (PIT), and it is underpinned, between others, in the principle of the individual ability to pay (Faustino, 2017). Considering the need to tax the use of public goods and services and to anticipate the taxation of the partners profits, the taxation of the juridical companies' incomes has as base the corporate income tax - *Imposto sobre o Rendimento das Pessoas Coletivas* (CIT). This tax law has underlying the principle of income theory, through the difference between the net equity at the end and the beginning of the fiscal year (Martins, 2017). The Portuguese income tax codes (CIT and PIT Codes) allow several options to the taxation in the hospitality and restaurant activity.

When these businesses are developed by the individual entrepreneur, the incomes related to the business is tax under the PIT, and they can be tax under the simplified or general regime (Pereira, 2019). In the case, entrepreneur has an invoicing lower to €200 000 per fiscal year, the mandatory regime will be the simplified and incomes will be framed in the business and professional income category (B category). Nevertheless, the entity is free to choose its income taxation under the general regime. When business has a turnover higher than €200 000, it is mandatory to apply the general regime, Portuguese tax law doctrine commonly call it as organized accounting regime because it uses the accounting profit as the starting point to fix the taxable income of activities (Courinha, 2019). In the simplified regime the expenses are not real, they are presumed using coefficients (Faustino, 2015). In the accommodation and restaurant activities, the incomes related are subjects to a coefficient of 15%, mainly. It means that in each €100, the net income subject to tax is €15. In the general regime, they are applied the rules of legal companies foresee in the CIT Code, nevertheless it is needed to pay attention to some of specific adjustments fixed in the PIT Code (Faustino, 2017). Independently of the regime used to determine the net taxable income of the activities of that sector; simplified or general regime, the determination of the total taxable income, in the individual domain, results of the accumulation of the net income of the different categories (activity, salaries, pensions, gains, rents, and others). Depending on the income value, it will be taxed under the progressive tax rates from 14.5% to 48% (Morais, 2016)

In the case of the activity development under the legal company, the income taxation occurs under the CIT Code rules, and generally it recurs the general regime (Lapatza, 2007; Pascual & Zamora, 1998). IT assumes the accounting profit as the starting point to determining the taxable income, but it is subject to the tax corrections foresee in the CIT Code, and it is a mandatory regime to the companies that have an annual invoicing of more than €200 000 (Nabais, 2019). In this economic area, the main income comes from the restaurant and accommodation services, that are recognized when they are provided to costumers. Concerning expenses, the main are the foods (raw materials) and labour (personnel expenses), these are recorded as accounting expenses, and they are accepted in the CIT sphere, so there are no adjustments in tax domain. The property, plant and equipment has a different regime, their accounting expenses may not be accepted in the tax domain (Vasconcelos & Pinto, 2011). In the hotel activity buildings, the purchase cost, excluding the land value, is depreciated according to a depreciation rate that can vary from 2.5% to 5%, it represents an useful life between 20 and 40 years (Martins et al., 2020). If in the accounting sphere consider a useful life outside these limit, CIT Code imposes tax corrections, in the settlement of the taxable income (Costa, 2011).

The CIT Code also foresees a simplified regime, it is applied to the small entities that have an annual net turnover that does not exceed the €200 000, and that have a total asset lower than €500 000. Companies with these characteristics can adopt by the CIT general regime. This regime is independent of the accounting, and to determination of the tax profit, it is applied a coefficient of 4% over the value of the accommodation and restaurant services. Both regimes allow to identify the taxable income subject to the tax rate, so after that determination, will be applied the proportional tax rates of 17% and 21% to calculate the due tax to pay from the activity development.

Concerning the profits' distribution to the hospitality companies' individual partners, PIT Code assumes a taxation regime heavier. Regardless the percentage of participation that partner has in the society, partners have to pay PIT over the distributed profits, these incomes were already also taxed in the CIT domain (Pereira, 2019).

Additionally, to the tax over the activities' incomes, companies need to support the autonomous taxation (AT), this tax figure assumes that some expenses are subject to the AT, and it is added to the income tax (Martins et al., 2018). Stand out the spending with the passengers' vehicles, that depending on the purchase value, it can assume rates from 10% to 20% in PIT domain, and from 10% to 35% in the CIT sphere.

Evolution of Taxation in the Sector in the Last Ten Years and Fiscal Measures Introduced in Response to the COVID-19 Pandemic Crisis

In the last decade tax legislator made to differentiate separately the taxation of the accommodation services provided in the house/flat, and bedroom. If companies are taxed by the general regime, this differentiation is not relevant. Nevertheless, companies are framed in the simplified regime, it is needed to take into consideration the coefficients that allow the determining of the taxable net income.

The accommodation establishments exploration in the housing or flat modality had significant changes in 2017. It offers a high increase in the coefficient, if the entrepreneur is taxed under PIT law, there was an increase of 20% (from 15% to 35%). The CIT law presented a higher change, it changes the coefficient from 4% to 35%. The increase happened to promote the permanent lease of houses and flats at the expense of the accommodation services. In 2020, accommodation services face the inclusion of other differentiation in the Portuguese tax law. If the house/flat is located in containment zones and historic areas the coefficient of 35% has an aggravation of 50%. In the bedrooms' modality, the legislator considered included the breakfast service, the coefficient remained 15% in PIT domain, and 4% in the CIT dimension.

Over the last ten years, some tax measures were introduced in the tourism area. In the moment of that creation, tax legislator had different goals, as to incentive companies' investment and to stimulate the increase of tourism companies billing.

To promote the companies' investment, it was created a specific system of tax benefits designed by Tax Regime to the Investment Support - *Regime Fiscal de Apoio ao Investimento* (TRIS). This tax measure is applied to companies that develop tourism activity; however, it is not restricted to this business area. This fiscal incentive allows a tax saving of 25% of realized investment in equipment, for example. The company itself is the beneficiary of the tax saving (Silva et al., 2019).

Considering that the customers of tourism services are mainly individual persons and that they usually pay it with cash, it is known the high level of tax evasion in that economic activity. Thus, in order to increase the company billing, tax lawmaker foresees a tax benefit to incentive costumers to ask the service invoice. It was created a mechanism that allows to Portuguese tourists to deduce to the income tax 15% of VAT supported at the activities the of accommodation and meals. This tax benefit applied also to other expenses as the hair stylist services (Morais, 2016).

During the pandemic situation, this business area was one of the Portuguese economic sector with more restrictions in the economic operations' development, considering the circulation ban as strategies to combat the crises. Portuguese legislator did not pay the due attention to the high losses that this sector faced, comparing with other economic areas (Almeida & Santos, 2020). So, some entrepreneurs tried to transform the threat Covid-19 in an opportunity of value increase and modernization of its establishments and brand through the renovation of spaces. Nevertheless, there was no kind of incentive to compensate these exceptional expenses, the income tax domain (Laranjeiro, 2020).

Consumption Taxation: Value-Added Tax

Overall Framework

The entry into force in Portugal of the Value Added Tax - *Imposto sobre o Valor Acrescentado* (VAT) in 1986, constituted a real reform to the Portuguese Tax System. This change produced a high impact in the companies lives, individual entrepreneurs, consumers and professionals (Basto, 1991). VAT is a community matrix tax, it is mandatory its application in all the European Union (EU) States, adopting a common basis of taxation supported by EU legislation.

This tax is general, ordinarily, it taxes all the transmissions of goods and services, including all phases of the economic cycle (Travanca, 2009). As it settles in the tax credit method, it allows the taxpayer to deduce the tax supported in the acquisition of goods and services, and it does not have cumulative effects (Lima, 2003). The neutrality principle constitutes a basilar principle, both internally and internationally (Palma, 2015). The fundamental idea of the VAT system is that it achieves the highest degree of simplicity and neutrality, and it does not interfere in the decisions of economic agents (Palma, 2015; Vasques, 2015). At the international level, the neutrality of VAT is assured by the adoption of the taxation in the destination country's principle.

There is no a particular regime to specific activities, namely, for the accommodation and restaurant activities. The hospitality sector is subject to a general regime of VAT, where the economic operators that develop an activity assess tax in the provided services and deliveries made (Sanches, 2007). Nevertheless, economic operators have the right to deduce the support VAT of the inputs needed to realize activity (Vasques, 2015).

Generally, they are taxed all transmissions of goods and services, and it is relevant to clearly identify operations that are considered transmissions, and what must be recognized as services, in specific, concerning to the hospitality economic operations. The accommodation and catering servicing fulfil the criteria to be considered as services, to VAT effects.

Respecting to localization of these services, in the accommodation domain, they are localized where the real state is located; and in the case of restaurant operations, it occurs the place where it is realized to the costumer. The service's localization is relevant because it will determine the fiscal jurisdiction and the tax rate applicable.

In Portugal, this tax settles in a structure of three different taxes rates, according to the goods and services to tax, presently, the catering and drinks services are subject to the intermediate VAT tax rate (13%). Nevertheless, it does not contemplate all the goods that can be included in the supply, for example, it does not include alcoholic beverages, soft drinks, juices, nectars, and carbonated water. The accommodation is subject to a reduced tax rate (6%) in the hotel services, although this rate is applicable to the overnight and breakfast when both are billed together.

The right to deduct the VAT paid in the acquisition of goods and services constitutes a nuclear element of the VAT functioning. It is a conditional right that foresees some exceptions, that although they are generic, in particular situations they can affect the activity development in this economic area.

The supported tax in the inputs purchase by accommodation and restaurant sector are subject to the different VAT rates (reduced - 6%, intermediate - 13% and normal - 23% rate). Several goods essential to the activity development are subject to the lower rate, as most of the food items needed by restaurants. The equipment buying impose to support a rate of 23%, and the expenses with salaried personnel are not relevant to effects of the VAT law. Nevertheless, this activity it's a seasonal activity, so companies face this characteristic making contracts for a specific period of time with self-employed, and in that case the labour cost will be subject to the normal rate of VAT.

Furthermore, it is important to refer that there are some restrictions to the deduction right in some purchase of goods, as some vehicles, ships, aircraft. Depending on the company, this equipment can be essential to the activity, so this limitation can represent consequences for the sector. For example, the tourism companies that realizes activities using vehicles and/or ships can be impeded to deduce the supported VAT, when it is the case of passengers vehicles, pleasure boats, helicopters, motorcycles.

In the consumer of services perspective of food and accommodation, if the consumer will be the final consumer and not a company, the VAT paid in the acquisition of that kind of services will not be deductible, and it represents a cost associated to the consumption. Considering the characteristics of that sector, that it is common for the costumer to pay the services in cash, the fraud and tax evasion is sometimes a reality. As way to avoid it Portuguese legislator incorporated in the PIT law a tax benefit applied to individual persons. It allows the deduction directly to the due tax the percentage of the 15% of the supported VAT in the expenses of hospitality and restaurant.

Regarding for business customers, when they acquire accommodation and meals services, the VAT paid is not deductible, generally. Nevertheless, the companies that participate or organize in meetings as congresses, fairs, exhibitions, seminars, conferences, and others similar, can deduct 25% or 50% of VAT of that kind of spending. Even so, these cases need to observe a particular condition, the acquisition of that expenses need to make by a celebrating contract with the suppliers, and company need to proof that these costs are related to economic operations that will be taxed.

Evolution of Taxation in the Sector in the Last Ten Years and Fiscal Measures Introduced in Response to the COVID-19 Pandemic Crisis

The applicable rates to the servicing in the accommodation and restaurant constitutes, in the VAT sphere, the main changes verified in the last years. An analysis of the evolution of the rates in the restaurant sector reveal that until 2012, the service of food and drinks were subject to an intermediate rate: 12% in the Portuguese Continent and 8% in the Autonomous Regions until June 30, 2010; and 13% and 9%

until 31 December 2011, respectively. The 2012 State Budget Law revoked the topic 3.1 of the List II that determined the application of the intermediate rate of VAT to this kind of services. The entrance of this law represented a change in tax rates from 1 January 2012, and the new tax rates were 23% to Continent, 18% and 22% to Autonomous Regions of Açores e Madeira, respectively.

The 2016 State Budget Law introduced new changes at the VAT rates level, that they entered into force on 1 July 2016. This change made a new adjustment to the foods and drinks services, imposing an intermediate rate of 13%, this alteration excludes alcoholics drinks, soft drinks and carbonated waters.

In the VAT rates matter, there is no harmonization in the European Union, for example, the higher rate is 27% in Hungary. Nevertheless, the lower tax rate can not be fixed below 5%, there are several EU States that apply a normal VAT rate in the sector of restaurant[8]. Between the countries that apply the intermediate/ reduced VAT rate, some of them exclude the alcoholic drinks. In the accommodation activity, there are Member States that adopt, as Portugal, the reduced rate and other the maximum one[9].

It is a sector with high relevance to national economy and employment, so some associations representing the sector in the European Union have been complaining to a general decrease of VAT rate in services promoted by this economic area (AHRESP, 2021b). It is claimed the decrease to the lowest VAT rate, as way to guarantee the survival of the sector' companies (AHRESP, 2021a). Nevertheless, Portuguese legislator did not anything about it, Government promoted other measures to stand up for the sector, as lay off and support lines of financing.

One recent temporary measure adopted from the Portuguese lawmaker was the creation of the IVAucher Program. It is a tax incentive system that allows to the taxpayer to accumulate, between 1 June to 31 August, the VAT paid in the sectors of hospitality, restaurant and culture. The value accumulated can after being discounted in buying at the same economic area in the adherents' economic operators.

POST-COVID-19 PERSPECTIVES IN THE TAX DOMAIN

The activity in accommodation and restaurants registered a very high drop (around 40%) due to the pandemic crises, it is one of the sectors that has had the most difficulty in recovering. Since the beginning of the pandemic, the State has made efforts to support the economy and employment, essentially through direct support to companies and not so much with fiscal measures.

The pandemic increased the importance of the State and the need for tax revenues to its finance, decreasing the arguments in favour of the minimum State (Soares & Kupski, 2020). This potential increase of the State must be accompanied of the changes in the tax system, in the sense of the State revenues does not continue to be generated only by the traditional sectors as accommodation and food but also, by other sectors as digital (Macedo, 2020). The last one has been assuming relevance, nevertheless, it only depends on highly qualified workers, so it tends to create fewer workstations, so it represents fewer collaboration to personal income tax collection (Macedo, 2020). The tourism area also has a relevant role in the green economy, because it is a big local employees recruiter and there is a worry about the maintenance and development of environmental preservation (Igamberdievna, 2021). For incentive of sector growth, the income tax needs to decrease as post-COVID economics allows.

In the area of consumption taxation, in specific, in the VAT domain, tourism associations have been asking for the reduction of the VAT rate, as a way to help in companies' treasury (AHRESP, 2021b). This suggestion was adopted by other countries, as Germany, Austria and the United Kingdom, with a positive impact on the reduction in the number of insolvencies and redundancies (AHRESP, 2021a).

Fiscal measures as IVAucher have relevance, but it does not have a significant impact considering the difficulties that the sector is going through.

DISCUSSION AND CONCLUSIONS

This paper has three main goals, as (1) to research how it has been happening the taxation of the accommodation and restaurant sector in Portugal, (2) to study if Portuguese lawmaker approved specific tax measures to this business area, and (3) to comprehend what post-COVID-19 perceptions are in that sector of the tax sphere.

The results show that companies need to attend to the Value Added Tax (VAT). In general, the economic operations involve VAT. This tax has as its main goal to tax consumption, in specific goods and services, and it covers all phases of the economic circuit. The simplicity and neutrality are characteristics of the Value Added Tax, it means that it cannot influence operators' economic decisions and it is expected to be a simplified process. The Value Added Tax Law does not foresee a specific regime to the accommodation and restaurant industry, this sector is included in the general regime of this tax. The framework on this regime represents that the touristic economic operators make a tax assessment on the costumers in the sales and services provided and can deduce the supported tax paid in the acquisition of goods and services. In Portugal, the VAT includes taxation over three different taxes rates, presently they are 6%, 13%, 23%. In general, the food and beverage services are taxed to 13%, however there are some exceptions. The lower VAT rate taxes the accommodation in hotel establishments, and this tax also applies to the breakfast when it is billed with the lodging. The right to deduce the tax paid in the acquisitions is a nuclear element to the VAT functioning. Nevertheless, this possibility is conditioned because it contemplates some exceptions. These specificities can affect lodging and restaurant activity because the VAT paid in some expenses is not deductible. It means that the tax will be considered as spending, to the company. Also, there are some restrictions on the right to deduct tax in specific activity expenses that can have direct consequences to the company, and this is the case of vehicles, ships and aircraft.

The results also show that the corporate income tax law has allowed two ways to tax the accommodation and restaurant industry's income. Small companies can calculate the taxable income under the simplified tax regime. This process is independent from accounting because expenses are assumed, and they are based on coefficients. Depending on the legal form of the activity's development, as an individual company or legal person, it has some differences in the framework and the taxable base's coefficients. Medium and large companies need to determine net income subject to corporate income tax, and in that case, it is done resorting to the accounting records. The activity enhancement presents other differences if it is developed as an individual or a legal company. The differences occur in the tax rates, individual entities are taxed under the progressive tax rates from 14.5% to 48%, and in the legal ones, they are taxed from the proportional tax rates of 17% and 21%. Although corporate profits pay tax, the profits' distributions in juridical entities implies also another tax payment when the beneficiary is an individual person.

The Portuguese legislator did not include in fiscal law specific changes to help the sector to face the pandemic crises of Covid-19. In the last years, there were some relevant changes as it was the case of the alteration in the catering services rate. It was reduced from 23% to 13%, but none of these changes arose from the pandemic situation, except VAT voucher. The resulting consequences from the pandemic crisis has been seriously affecting the lodging and restaurant industry. Considering its relevance in the

Portuguese economy, for its economic contribution and employment, it is considered crucial to decrease the VAT rate for food and beverage services, as a way to guarantee the sector's survival.

As tax issues are part of the economic sphere, managers should pay attention to fiscal questions to make better economic decisions. This paper allows to appreciate the progress and trends of taxation in Portugal's accommodating and lodging economic activity. As well, it identifies the primary desires to be observed to help the sector face the pandemic crisis challenges.

The study explores the Portugal' case study, nonetheless, it could be remarkable to compare the case of Portugal with other countries that are also economically dependent on tourism area, as Italy and Greece.

REFERENCES

AHRESP. (2021a). *Federações europeias da restauração e do alojamento defendem aplicação generalizada da taxa reduzida de IVA em toda a Europa* [European food and accommodation federations advocate widespread application of the reduced RATE of VAT across Europe]. https://ahresp.com/2020/10/descida-do-iva-na-europa/

AHRESP. (2021b). *IVA reduzido constará no próximo orçamento de Estado?* [Will reduced VAT be included in the next State budget?]. https://ahresp.com/2021/08/iva-reduzido/

Aldeia, S. C. R. (2019). The "true and fair view" concept in business and tax law. The Portuguese and Spanish case studies. *Academy of Accounting and Financial Studies Journal, 23*(2), 1-11. https://www.abacademies.org/articles/the-true-and-fair-view-concept-in-business-and-tax-law-the-portuguese-and-spanish-case-studies-8130.html

Aldeia, S. C. R. (2021). Application of the constitutional principle of generality in Spanish companies' taxation. A compared study to Portugal. International Journal of Law and Management. doi:10.1108/IJLMA-01-2021-0008

Almeida, F., & Santos, J. D. (2020). The effects of COVID-19 on job security and unemployment in Portugal. *The International Journal of Sociology and Social Policy, 40*(9/10), 995–1003. doi:10.1108/IJSSP-07-2020-0291

Almeida, F., & Silva, O. (2020). The impact of COVID-19 on tourism sustainability: Evidence from Portugal. *Advances in Hospitality and Tourism Research-Ahtr, 8*(2), 440–446. doi:10.30519/ahtr.775340

Amerta, I. M. S. (2017). The role of tourism stakeholders at Jasri tourism village development, Karangasem Regency. *The International Journal of Social Sciences (Islamabad), 1*, 20–28. doi:10.29332/ijssh.v1n2.32

Basto, J. G. X. d. (1991). A tributação do consumo e a sua coordenação internacional: lições sobre harmonização fiscal na comunidade económica europeia [Taxation of consumption and its international coordination: lessons on tax harmonisation in the European economic community]. *Cadernos de Ciencia e técnica fiscal, 164*.

Borges, A., Rodrigues, A., & Rodrigues, R. (2010). *Elementos de contabilidade geral, SNC - Sistema de Normalização Contabilística* [General accounting elements, CnS - Accounting Standardization System] (25ª Edição). Áreas Editora.

Buffon, M., & Bassani, M. (2013). Benefícios fiscais: uma abordagem à luz da cidadania fiscal e da legitimação constitucional da extrafiscalidade [Tax benefits: an approach in the light of fiscal citizenship and the constitutional legitimation of extrafiscality]. *Revista da AJURIS, 40*(130), 245-280. http://ajuris.kinghost.net/OJS2/index.php/REVAJURIS/article/view/298

Burns, J., & Scapens, R. (2000). Conceptualizing management accounting change: An institutional framework. *Management Accounting Research, 11*, 3–25. doi:10.1006/mare.1999.0119

Callén, Y. F., & Gadea, A. L. (2004). Deficiencias y alternativas de la información contable actual [Deficiencies and alternatives of current accounting information]. *Partida Doble, 152*, 62–73.

Campbell, D. (2000). Legitimacy theory or managerial reality construction? corporate social disclosure in marks and spencer plc corporate reports, 1969-1997. *Accounting Forum, 24*(1), 80–100. doi:10.1111/1467-6303.00030

Cetin, G., Alrawadieh, Z., Dincer, M. Z., Dincer, F. I., & Ioannides, D. (2017). Willingness to pay for tourist tax in destinations: Empirical evidence from Istanbul. *Economies, 5*(2), 15. doi:10.3390/economies5020021

Chaudhry, A., Fuller, C., Coetsee, D., Rands, E., Bakker, E., Merwe, M., ... Balasubramanian, T. (2016). Interpretation and application of international financial reporting standards. Wiley.

Costa, A. P. (2011). *Depreciações e amortizações no SNC: alterações contabilísticas e impacto fiscal* [Depreciation and amortization in the CNS: accounting changes and fiscal impact]. Almedina.

Courinha, G. L. (2019). *Manual do Imposto sobre o rendimento das pessoas coletivas* [Corporate Income Tax Manual]. Almedina.

Epstein, L., & King, G. (2002). The rules of inference. *The University of Chicago Law Review. University of Chicago. Law School, 69*(1), 1–133. doi:10.2307/1600349

Fachin, R. C., & Mendonça, J. R. C. (2003). O conceito de profissionalização e da teoria institucional [The concept of professionalization and institutional theory]. In M. F. Vieira & C. A. Carvalho (Eds.), *Organizações, instituições e poder no Brasil* (pp. 19–41). FGV.

Faustino, M. (2015). Sobre a reforma do IRS (I) [About the reform of the IRS (I)]. *Revista de finanças públicas e direito fiscal, 4*(7).

Faustino, M. (2017). Imposto sobre o rendimento das pessoas singulares [Personal income tax]. In *Lições de Fiscalidade. Princípios gerais e fiscalidade interna*. Almedina.

Fernández, J. I. P., & Rivero, M. S. (2010). Competitiveness versus growth in tourist destinations. An analysis using multivariate techniques. *Cuadernos de Economía (Santiago, Chile), 33*(91), 159–182. doi:10.1016/S0210-0266(10)70061-3

Freeman, R. E. (1994). The politics of stakeholder theory: Some future directions. *Business Ethics Quarterly, 4*(4), 409–421. doi:10.2307/3857340

Gonçalves, C., Santos, D., Rodrigo, J., & Fernandes, S. A. (2017). *Financial accounting explained - Practical Manual*. Vida Económica.

Guerreiro, R., Pereira, C., & Rezende, A. (2006). In search of understanding the formation of habits and routines of management accounting. *Revista de Administração Mackenzie*, *7*(2), 78–101. doi:10.1590/1678-69712006/administracao.v7n2p78-101

Hill, C. W. L., & Jones, T. M. (1992). Stakeholder-agency theory. *Journal of Management Studies*, *29*(2), 131–154. doi:10.1111/j.1467-6486.1992.tb00657.x

ICAEW. (2020). *Financial Accounting and Reporting (IFRS) Q&A 2020. ACA Simplified*. Great Britain Amazon.

Igamberdievna, A. D. (2021). Promising measures for the development of sustainable tourism in the context of the COVID-19 pandemic. *Academic Journal of Digital Economics and Stability*, *4*, 1–6.

Lapatza, J. J. F. (2006). Simplificación; base imponible y renta empresarial. *Quincena fiscal: Revista de actualidad fiscal, 21,* 13-159.

Lapatza, J. J. F. (2007). Resultado contable y base imponible: Deslegalización, autodeterminación y delito fiscal [Accounting result and tax base: Delegalization, self-determination and tax crime]. *Quincena fiscal: Revista de actualidad fiscal, 8,* 9-12.

Laranjeiro, A. (2020). Hotéis continuam a investir apesar da pandemia. *Dinheiro Vivo*. https://www.dinheirovivo.pt/economia/hoteis-continuam-a-investir-apesar-da-pandemia-12895854.html

Lima, E. V. (2003). *VAT: Value added tax, commented and annotated* (9th ed.). Porto Editora.

Lopes, A. S., Sargento, A., & Carreira, P. (2021). Vulnerability to COVID-19 unemployment in the Portuguese tourism and hospitality industry. *International Journal of Contemporary Hospitality Management*, *33*(5), 1850–1869. doi:10.1108/IJCHM-11-2020-1345

Macedo, V. (2020). The opportunity for a new tax paradigm. In Os Impostos e o Combate às Desigualdades (Na era do Covid-19). Rei dos livros.

Machado, A. F., Sousa, B. B., Rocha, A. M. A., & de Abreu, Z. H. L. (2020, Apr 01-02). *Communication Strategies to Combat the Negative Effects of Covid-19 On Tourism* [Conference session]. International Business Information Management Association Conference (IBIMA).

Major, M., Conceição, A., & Clegg, S. (2018). When institutional entrepreneurship failed: The case of a responsibility centre in a Portuguese hospital. *Accounting, Auditing & Accountability Journal*, *31*(4), 1199–1229.

Marques, C. P., Guedes, A., & Bento, R. (2021). Rural tourism recovery between two COVID-19 waves: The case of Portugal. *Current Issues in Tourism*. Advance online publication. doi:10.1080/13683500.2021.1910216

Martins, A., Dinis, A., & Lopes, C. (2018). Interpretative complexity and tax law changes: The taxation of corporate expenses in Portugal. *Journal of Applied Accounting Research*, *19*(4), 537–551. doi:10.1108/JAAR-04-2017-0050

Martins, A., Sá, C., & Taborda, D. (2020). *A dedutibilidade de gasto no IRC. Uma análise económica-fiscal* [The deductibility of expenditure in irc. An economic-fiscal analysis.]. Almedina.

Martins, H. P. (2017). Corporation tax. In *Lições de Fiscalidade. Princípios gerais e fiscalidade interna* (5th ed., Vol. 1). Almedina.

Melville, A. (2019). *International Financial Reporting: A Practical Guide* (7th ed.). Person.

Meyer, J. W., & Rowan, B. (1991). Institutionalized organizations: formal structure as myth and ceremony. In W. W. P. P. J. DiMaggio (Ed.), *The new institutionalism in organizational analysis*. University of Chicago Press.

Morais, R. D. (2016). *About the IRS*. Almedina.

Nabais, J. C. (2018). *Introduction to corporate tax law* (3rd ed.). Almedina.

Nabais, J. C. (2019). *Tax Law* (11th ed.). Almedina.

North, D. C. (1990). *Institutions, institutional change and economic performance*. Cambridge University Press.

Oliveira, N. M. P. (2020). COViD-19, credit, lease and tourism-connected contracts in Portugal. *Revista De Derecho Civil*, *7*(2), 89–102.

Palma, C. (2015). *Public authorities and value added tax: A break in the principle of neutrality*. Almedina.

Pascual, L. A. M., & Zamora, M. P. M. (1998). *El Corporate income tax, general regime*. Centro de Estudios Financieros.

Peci, A. (2006). *The new institutional theory in organizational studies: a critical approach* (Vol. 4). Cadernos EBAPE.

Pereira, F. A. M. (2012). The evolution of institutional theory in organizational studies: a field of research to be explored. *Revista Organizações em Contexto (ROC)*, 275-295. https://www.metodista.br/revistas/revistas-metodista/index.php/OC/article/view/3396

Pereira, P. R. (2019). *Manual de IRS* (2nd ed.). Edições Almedina.

Presenza, A., & Cipollina, M. (2010). Analysing tourism stakeholders networks. *Tourism Review*, *65*, 17–30. doi:10.1108/16605371011093845

Riker, D. G. J., & Pozzetti, V. C., & B., C. (2018). The extratax nature of export and import taxes in environmental preservation. *Percurso*, *2*(25), 36–42. doi:10.21902/RevPercurso.2316-7521.v2i25.3095

Rodrigues, J. (2016). Accounting Standardization System. In SNC explicado (6ª ed.). Porto editora.

Sanches, J. L. S. (2007). *Tax law manual* (3rd ed.). Coimbra Editora.

Sarmento, J. M., Nunes, R., & Pinto, M. M. (2019). *Theoretical-Practical Manual de IRC* (2nd ed.). Almedina.

Silva, E. S., Pereira, A., & Andrade, S. (2019). *Tax Benefits in irc*. Vida Económica.

Soares, J., & Kupski, R. (2020). Covid-19 calls into question defense of the minimum state In *Os Impostos e o Combate às Desigualdades (Na era do Covid-19)*. Rei dos livros.

Sofronov, B. (2017). The economic impact on global tourism. *Annals of Spiru Haret University. Economic Series, 17*(2), 127-139. http://anale.spiruharet.ro/index.php/economics/article/view/1728/pdf

Travanca, D. (2009). *VAT on construction and real estate*. Vida Económica.

Turismo de Portugal. (2017). *Estratégia turismo 2027, liderar o turismo do futuro*. https://www.turismodeportugal.pt/SiteCollectionDocuments/estrategia/estrategia-turismo-2027.pdf

Vasconcelos, A. A., & Pinto, C. (2011). *Tax regime for depreciation and amortization*. Areal Editores.

Vasques, S. (2015). *Value added tax*. Almedina.

ENDNOTES

1 https://gee.gov.pt/pt/publicacoes/estatisticas-tematicas/estatisticas-setoriais

2 https://europa.eu/next-generation-eu/index_pt

3 Nevertheless, individual entities that have a net turnover until 200 000€ are dismissed to apply these accounting rules.

4 N. 3 and n. 4 of article 11, DL 158/2009, of 13 July.

5 N. 3 of article 11, DL 158/2009, of 13 July.

6 According to paragraph 22 of the CSASS

7 Paragraph 93 of the CSASS

8 https://ec.europa.eu/taxation_customs/vat-rates_en

9 Idem

Chapter 11
Mass Masstige Index:
Application in Wine Brands and the Importance in Restaurant Communication

Paula Rodrigues
https://orcid.org/0000-0003-2967-2583
Universidade Lusíada Porto, Portugal & Research Centre in Organizations, Markets and Industrial Management, Porutgal & BRU, Instituto Universitário de Lisboa, Portugal

Ana Pinto Borges
https://orcid.org/0000-0002-4942-079X
ISAG - European Business School, Portugal & Research Center in Business Sciences and Tourism (CICET-FCVC), Portugal & COMEGI, Portugal

Paulo Ramos
https://orcid.org/0000-0001-7534-1246
Universidade Lusíada Porto, Portugal & Universidade Fernando Pessoa, Portugal

Elvira Vieira
European Business School, Portugal & Research Center in Business Sciences and Tourism (CICET-FCVC), Portugal & Polytechnic Institute of Viana Do Castelo, Portugal & UNIAG - Applied Management Research Unit, Portugal

Catarina Alexandra Correia
https://orcid.org/0000-0001-7700-3769
Universidade Lusíada Porto, Portugal & COMEGI, Portugal

ABSTRACT

This work intends to obtain for the first time one mass masstige index for wine brands and analyze the importance of that index in the restaurants' communications. The authors used quantitative research, and information was obtained through a convenience sample of 196 masstige wine brands. The mass masstige index was obtained for a set of wine brands of several countries. The results show that the masstige wine brand index can benefit gastronomic restaurants by having masstige wine brands recognized as top-of-mind brands by tourists. This research fulfills a gap in the tourism literature applying for the first time the masstige theory in wine brands and obtaining a mass masstige index for a set of wine masstige brands. The provider of masstige wine brands in gastronomic restaurants must continue to bet on communication strategies with masstige wine brands as a way to create strong consumer-brand relationships.

DOI: 10.4018/978-1-7998-9148-2.ch011

INTRODUCTION

The Tourism sector and several businesses have begun to explore the unconventional way of marketing luxury goods and services which promise status and superior experiences for the tourists (Correia et al., 2020; Goyal, 2021). Gastronomy is an important factor in the travel experience and its importance is increasing, it is something consumers look for when they choose a destination (Antón, Camarero, Laguna, & Buhalis, 2019). On the other hand, there is a consensus among tourism research that consumption of tourism products has a predominantly hedonic component (Antón et al., 2019; Carlsen & Boksberger, 2015; Kim, 2014). In this way, the consumption of masstige wine brands can translate this hedonic motivation of consumption experience into gastronomy and restaurants choices (Dhar & Wertenbroch, 2000; Kim, 2018).

Wine tourism provides an extraordinary experience to tourists (Santos et al., 2019). This extraordinary experience can be explained by several factors. One of those factors is the communication of the destination's gastronomy and more specifically the communication of the wine brands (Brochado et al., 2021). According to Brochado et al. (2021, p. 597) "wine tourism thus refers to activities motivated by tourists' desire to taste new wines and learn more about the traditions and cultures of wine-producing areas". Then, we can posit that the wine tourism activities need to be focused on sensorial, experiential, symbolic and hedonic attributes (Crespi-Vallbona & Mascarilla-Miró, 2020).

According to Keller (2010), marketing communication is a helpful tool to build and manage brands – directly or indirectly, – thus informing and reminding consumers about functional, symbolic and hedonic attributes of the brand's products. Masstige wine brands can be a source of differentiation in tourist destinations and gastronomic restaurants. Paul (2018b, p. 723) defines masstige marketing as "a phenomenon in which regular products with moderately high prices are marketed to the maximum number of consumers by creating mass prestige without lowering prices or without offering discounts". This implies a strategy of wine brand positioning, aiming at creating mass prestige value for wine brands and for tourism destinations and gastronomic restaurants.

In the wine sector, Portugal is the 11th wine producer in the world and holds the 9th position regarding the global wine exports in value, and the 10th position in volume, with a market share of almost 3% (IVV/Viniportugal, 2021). In 2021, according to data released by ViniPortugal, exports increased by 8,11% in export value compared to the same period in 2020, having exceeded 925 million EUR. The impact of the COVID-19 pandemic has had a huge impact on the tourism sector, both for non-resident visitor tourism and domestic tourism, significantly reducing the global volume of wine being sold through this channel. However, with Portuguese consumers behaviour, we have witnessed the market's recovery during 2021. One of the key factors was the premiumisation (Portugal Wine Landscapes, 2021). Portuguese regular wine drinkers continue to drink more expensive wines, this behaviour hasn't changed with the pandemic. The lockdowns also provided more moments to drink wine, a reality we have observed in Portugal and in several countries, as this drink is a small luxury for those who want to indulge themselves when other indulgent or ambitious options, such as going out or traveling, have been severely limited (Portugal Wine Landscapes, 2021). In this moment, Portugal is the 7th country in the scope of global wine consumption, with a per capita wine consumption (the highest in the world) of almost 51,9 litres per year, which is equivalent to approximately 69 bottles of 750 ml (OIV, 2021). The Portugal Wine Landscapes (2021) report highlights that the biggest growth in wine consumption came from younger consumers – those in the 18-34 and 35-49 year old age brackets.

This chapter intents to make a literature review of masstige brands and the consumers' choice of gastronomic restaurants and the relevance of masstige wine brands. Then, based on the data collection from final consumers, we intend to better understand the perception of wine consumers about masstige wine brands, in the Portuguese case, obtaining a masstige mean score scale index for the chose brands.

BACKGROUND

Masstige Brand

According to Kumar et al., (2021a) masstige means mass prestige associated with brands. The masstige theory arises from an acknowledgment that marketing theory is full of recycled and repeated analyses of different elements of the marketing mix, but it it is not well grounded in how and why certain companies build brand equity (Paul, 2018a).

The author has conducted several studies related to masstige and has focused on the success of masstige marketing strategies based on the 4Ps strategy, rather than exploring or considering the meaning of masstige in the brand's framework of values (Paul, 2015, 2018a, 2019). Following the same thinking, Kumar et al., (2020) designed a review study presenting and understanding the future directions for "masstige marketing", considering the given meaning of masstige to be prestigious and accessible to middle-class mass consumers. They suggested developing a masstige brand through the masstige marketing strategy based on the 4 Ps. They indicated that masstige is different from luxury, but they also concluded that it is important to register the relative newness of masstige as a concept with its potential limitations.

According to the masstige theory, Goyal (2021) states that masstige brands are expressed as a new luxury, different from traditional luxury. It is identified that masstige brands can be distinguished from luxury brands by their functional, financial, social, and individual values. Brands are considered prestigious as luxury brands, but they differ from luxury brands and are purchased in masse (Goyal, 2021). Brands that are able to build mass prestige and brand value can charge high prices from consumers, and consumers are also willing to pay relatively higher prices for these brands compared to mass brands (Kumar et al., 2020).

Consumers looking for masstige brand seek means of self-expression (Chernev, 2011), but middle-class consumers seeking identity through brands may not be willing to spend a lot on them (Zhan & He, 2012). For this type of consumer, masstige brands are good and affordable options (Kumar et al., 2020).

Masstige brands are important means of expression and formidable alternatives to traditional luxury items. Truong et al. (2009, p. 376) posit that masstige brands "are often mass-oriented and less expensive than traditional luxury goods, which have a very confined exclusivity in terms of accessibility and price". This shift from traditional luxury to masstige came to be known as the shift from old luxury to new luxury (Granot et al., 2013).

A consumer of a masstige brand seeks to achieve an image or a desirable ideal through the choice of that brand (Kumar et al., 2020). On the contrary, a consumer who prefers true luxury seeks heritage, family name and country of origin (Kapferer, 2014).

Masstige brands are bringing a new luxury, affordable luxury, luxury consumed in droves. In the existing literature it was found that the definition of this type of brand is for the masses, prestigious, not rare, of high quality and not as expensive as luxury (Goyal, 2021).

Wine as a Product and as a Brand

Wine is a complex product (Hall & Mitchell, 2007). It is an experience product as consumers cannot feel the taste of a wine before buying it (Lockshin et al., 2006). There are few product categories that have so many SKUs as wine in most retail outlets, were it seems too have too many cues for consumers, each one competing with the other (Goodman et al., 2005), making it difficult to decide which wine to choose on the shelf. Wine is also a combination of both intrinsic and extrinsic attributes. The extrinsic attributes (i.e., alcohol, sugar, acidity) will shape its taste, but are not constant as these will vary on every vintage according to non-controllable factors such as climate (Ramos et al., 2011). However, some extrinsic attributes like price, geographical location and branding also play a key role on the perception of a masstige wine brand.

Although there's evidence of growth in the demand and consumption of luxury wine brands (Wolf et al., 2016), according to Nápoles et al. (2020) the luxury wine sector is still underrepresented in Portugal. The exception has been for the fortified wines, like Vintage Porto's and some Madeira. However, in recent years some of the new vintages of iconic wines like Barca Velha by Ferreira/Sogrape, can now reach values of approximately $500-650 in some USA outlets. Thach and Olsen (2019) found that the motivations of the luxury wine buyer are different from the non-luxury wine buyer, and reasons for purchasing luxury wine go beyond mere collecting. Luxury wines must meet some criteria such has the place of origin, high price, scarcity, heritage, and ability to age (Yeung & Thach, 2019). In a previous research, Beverland (2004) indicated that most of the luxury wine buyers would buy them mainly for collection motives. However, more recently, Thach and Olsen (2019) indicated that only 12% of the luxury buyers purchase with the purpose of keeping the bottles in their cellars. Most of the buyers of "Luxury wines" were more motivated for hedonic reasons, like simply enjoying the wine in a social context and comparing with friends. There's is still a critical difficulty in defining what is the price point were a wine could be considered "Luxury". Jarrett and Jarvis (2016) classified them in Ultra-Premium wines $70 - $99 and the Luxury / Icon wines at $73 USD. This, adding to the fact that a bottle of a so called "luxury wine" can be bought and/or shared with a few friends, makes it a potential masstige product that some more average consumers can reach for a unique luxury experience.

Importance of Masstige Wine Brands on Gastronomic Restaurants

Product knowledge and involvement are important aspects when it comes to wine brands, gastronomy and restaurants (Kiatkawsin & Han, 2019). According to several authors, when consumers purchase any product, they look for different sources of information and knowledge to better tailor their purchase to their needs (Bruwer et al., 2017; Ellis & Caruana, 2018; Gustafson et al., 2016; Mason & Bequette, 1998).

Velikova et al. (2019, p. 1751) posit that "wine enhances a restaurant's image and reputation. It enriches guests' dining experiences and maximizes their satisfaction". Thus, restaurant owners continue to encourage their activity, enriching them through the appreciation of several wine options, and gastronomy with specialized wine-oriented events (Velikova et al., 2019). For Ruiz-Molina et al. (2010, p. 98), "good food accompanied by suitable wine can greatly enhance the gastronomical experience". Also, according to Velikova et al. (2019, p. 1754) "diners now ask for more information on which wines complement their selected menu items". By recommending masstige wine brands to complement a gastronomic menu, restaurants can increase their average dinner and provide greater gastronomic satisfaction to their customers (Oliveira-Brochado & Silva, 2014). Recommendations from masstige wine brands that are part of the food menu offered by restaurants, as opposed to being an independent wine list, are associated with sales of the prestige wine brands (Lockshin et al., 2011; Ruiz-Molina et al., 2010). For Ruiz-Molina et al. (2010, p. 98), "the existence of a supply of fine wines has been an excellent parameter for judging restaurant quality". Consumers know that wine in restaurants cost more. But they are willing to pay more for enriching dining experiences, and for learning how to better combine wine with food (Kiatkawsin & Han, 2019; Velikova et al., 2019).

METHODOLOGY

Method

A literature review was carried out on masstige brands, regarding the importance that consumers attach to this type of brands in their choices, and on the impacts of masstige wine brands on gastronomic restaurants.

Afterwards, a structured questionnaire was developed for data collection, using the Masstige Mean Score Scale Index of Paul (2019), applied to wine brands of Portuguese consumers. We used a Likert scale of five points to collect this information.

The questionnaires was divided into three parts: the first contained questions related to the knowledge and behaviour of the respondents' about the wine brands, the second part used the Paul (2019) scale to obtain the MMSSI and, finally, the third part included the questions related to the characterization of the respondents.

Brand Selection and Data Collection

For data collection, we disseminated an online questionnaire available in the period from March 2021 to April 2021. A convenience sample method allowed us to obtain 196 responses.

An initial poll was made amongst 30 wine consumers and wine bloggers, asking them to list the wines they most wished they could buy someday or the ones they had already bought as a treat to themselves and that they could use this as a social flagship (both national and international brands). Some of the brands they suggested can be considered as luxury wines (although most of them can have a less expensive alternative in the same company). However, many of these consumers bought them for a special occasion or to share them with friends, and as a status symbol, in line with what Thach and Olsen (2019) found. Some even replied that they bought these wines splitting the cost with other friends, just to be able to taste them at least once in their lifetime.

RESULTS AND DISCUSSION

Sample Characterization

From the total of 196 respondents, 89.8% said they were wine consumers. Among the wine brands presented, the ones that respondents knew the most and tasted were Ferreira – Barca Velha (60.2%), Cartuxa – Pera Manca (58.2%) and Quinta do Castro (57.7%). These wine brands are all Portuguese, and

Table 1. Sample characterization (n = 196)

Are you a wine consumer?	Nº	%
No Yes	20 176	10,2 89,8
Tell us what brands do you know and have tasted?	**Nº**	**%**
Ferreira - Barca Velha Cartuxa – Pera Manca Mouchão Quinta do Castro Romanée Conti Château Petrus Château Lafitte Veja Sicilia Pingus Sassicaia	118 114 78 113 42 52 59 46 25 29	60,2 58,2 39,8 57,7 21,4 26,5 30,1 23,5 12,8 14,8
Age	**Nº**	**%**
18 to 25 26 to 40 41 to 65 More than 65 years old	12 46 132 6	6,1 23,5 67,3 3,1
Gender	**Nº**	**%**
Male Female	124 72	63,3 36,7

these results can be explained by the fact that the respondents are also Portuguese. 67.3% of respondents are aged between 41 and 65 years old and 63.3% are male (Table 1).

Masstige Mean Score Scale Index (MMSSI)

The masstige mean score scale from Paul (2015) was applied and the statistic results are shown in table 2. The most valued items were M6, M4, M7 and M5, respectively, recognizing that quality and awareness are the main characteristics' of a masstige wine brand. Consumers indicated that they choose the top-of-mind brands, and this is related to the quality that they associate with the wine and the levels of satisfaction, after all, the consumer only remembers what was very negative or positive for him. In addition, standardized brands that meet international standards can provide an association with a status symbol. Lastly, we also observe that the consumers are predisposed to communicate through word-of-mouth to other consumers (friends and relatives).

We analysed the data normality with Skewness and Kurtosis results. No problems were identified with data normality because their values were between the cut fit.

To calculate the brands' masstige scores the values of all 10 items on the MMSSI (Paul, 2019) were aggregated. Paul (2015), who used a 7-point Likert-type scale, argued that a brand needs to have a minimum score of 50 out of 70 on the MMI to be considered masstige (Table 3). As we used a 5-point Likert-type scale we had to convert the original range (50 – 70), which resulted in a range of 35 out of 50 to be considered a masstige brand.

According to the results obtained, it appears that foreign brands are considered by Portuguese consumers as masstige brands – Château Lafitte, Château Petrus, Romanée Conti, Sassicaia and Vega

Table 2 - Scale statistic

Scale	Mean	SD	Skewness	Kurtosis
M1: I like this wine brand because of its masstige	2,81	1,108	-0,194	-0,946
M2: I would buy this wine brand because of its mass prestige	2,93	1,163	-0,027	-0,892
M3: I would pay a higher price for this wine brand for status quo	2,56	1,156	0,212	-0,982
M4: I consider this wine brand, a top-of-mind brand in my country	4,01	0,954	-1,234	1,662
M5: I would recommend this wine brand to friends and relatives	3,96	0,909	-1,437	2,740
M6: I believe this wine brand is known for its high quality	4,05	0,917	-1,342	2,316
M7: I believe this wine brand meets international standards	3,99	0,954	-1,017	0,953
M8: I love to buy this wine brand regardless of price	2,69	1,162	0,053	-0,991
M9: Nothing is more exciting than this wine brand	2,25	1,035	0,256	-0,949
M10: I believe that individuals in my country, perceive this wine brand as prestigious	3,75	0,952	-1,002	1,078

Sicília – unlike Portuguese brands. These results are interesting because consumers have better access to Portuguese brands and can establish more associations with the respective tastes. (Table 3).

DISCUSSION AND CONCLUSION

The multiplicity of wines that consumers have at their disposal poses some complications when making a choice. The growing competitiveness represents a challenge for wine producers and restaurants. In order to respond to consumer needs in an adequate way and with unforgettable experiences and thus achieve a high level of satisfaction, companies must know the perceptions of their customers. However, in the context of wines and gastronomy, there has been no perception of the consumers' motivations that lead them to choose one wine over the other and their experiences during consumption. In this sense, with this chapter we have presented for the first time one masstige index for wine brands. For that, we applied a questionnaire to Portuguese tourists and we presented wine brands not only from Portugal but also

Table 3. Masstige Mean Score Scale Index for Wine Brands

Brand	MMSSI
Cartuxa - Pera Manca	32,36
Château Lafitte	41,5
Château Petrus	35
Ferreira - Barca Velha	34,04
Mouchão	28,5
Quinta do Crasto	31,41
Romanée Conti	37
Sassicaia	37
Vega Sicília	35

from other countries. This research can give a significant contribution for different stakeholders of the wine sector. One of the most important results More specifically, concerns the consumers' perceptions on Portuguese luxury brands. We observed that the foreign wine brands are considered by Portuguese consumers as masstige brands, as opposed to brands from the country itself. We have confirmed an appreciation of the national product (wine), contrary to what it could be expected.

In this, post-COVID-19 moment, we observe a reality which is characterised by new demands and new stakeholders, or the latter with a different positioning. The supply will necessarily have to adapt to face the challenges placed by the recovery period after the pandemic.

The results indicate that a masstige wine brand index can promote gastronomic restaurants through the offer of masstige wine brands identified as top-of-mind brands by tourists. In this case, the restaurants must include the available masstige wine brands in their communication strategies. However, the decline of masstige wine brands should be avoided. To this purpose, it is essential that they adequately define a price and distribution policy strategies. As for the price policy, a premium price must be applied and as for the distribution, the masstige wine brands must only be available in reference restaurants and aimed at middle/high income consumers. If not, widespread brand access tends to occur when purchases by middle-class consumers become frequent or common. The more available a product is, the less prestigious its perception. Additionally, a prestigious environment must be provided around the brand, and both communication and presence must be related to luxury and prestigious places for the brand to appeal to consumers as an "aspirational" brand. Certainly, if the consumer/tourist is involved in a strategy of co-creative wine and food experiences, it will provide greater engagement, stimulating learning and promoting to meaningful and immersive experiences that lead to growing place additions and loyalty (Carvalho et al., 2021). It should be noted that Portugal, in 2020, similar to previous years, continues to be the country in the world with the highest consumption of wine per inhabitant, despite a slight drop of 0,6% in the volume of wine consumed, according to the 2021 report by the International Organization of Vine and Wine (OIV). In terms of global consumption, there was a 3% drop last year, to 234 million hectolitres, the lowest value since 2002. More precisely, in 2020, each Portuguese aged over 15 consumed, on average, 51,9 litters of wine, surpassing the Italians, who consumed 46,6 litters per capita, and the French with 46 litters per capita. Furthermore, it is observed that regular wine consumers continue to prefer to drink more expensive wines, and this behaviour has not changed with the pandemic and that consumption has increased in the 18-34 and 35-49 year old age groups (Portugal Wine Landscapes, 2021). This is an important information for the wine brand managers because it is this target that should be aware of brand knowledge, perceived quality, excitement and status - the masstige dimensions.

This research has some limitations that can be overcome in future research. One of the limitations is the use of a convenience sample. Other limitation is that we consider all the consumers that know the masstige wine brands or have tasted them, and we did not consider their income. The consumer income is a decisive variable for this type of consumption and also the intention to purchase. Furthermore, we do not consider behavioural or emotional variables of brand (e.g., brand loyalty, WOM, e-WOM, happiness and wellbeing). These variables could be contemplated in future research. Lastly, it would be interesting to evaluate the masstige index on another consumer segmentation, namely different nationalities, different types of generations, frequency of visits to restaurants, cultures, and so on.

REFERENCES

Antón, C., Camarero, C., Laguna, M., & Buhalis, D. (2019). Impacts of authenticity, degree of adaptation and cultural contrast on travellers' memorable gastronomy experiences. *Journal of Hospitality Marketing & Management, 28*(7), 743–764. doi:10.1080/19368623.2019.1564106

Beverland, M. B. (2004). An exploration of the luxury wine trade. *International Journal of Wine Marketing, 16*(3), 14–28.

Brochado, A., Stoleriu, O., & Lupu, C. (2021). Wine tourism: A multisensory experience. *Current Issues in Tourism, 24*(5), 597–615. doi:10.1080/13683500.2019.1649373

Bruwer, J., Chrysochou, P., & Lesschaeve, I. (2017). Consumer involvement and knowledge influence on wine choice cue utilisation. *British Food Journal, 119*(4), 830–844. doi:10.1108/BFJ-08-2016-0360

Carvalho, M., Kastenholz, E., & Carneiro, M. J. (2021). Interaction as a Central Element of Co-Creative Wine Tourism Experiences—Evidence from Bairrada, a Portuguese Wine-Producing Region. *Sustainability, 13*, 9374. doi:10.3390u13169374

Carlsen, J., & Boksberger, P. (2015). Enhancing Consumer Value in Wine Tourism. *Journal of Hospitality & Tourism Research (Washington, D.C.), 39*(1), 132–144. https://journals.sagepub.com/doi/abs/10.1177/1096348012471379

Chernev, A. H. R., & Gal, D. (2011). Competing for consumer identity: Limits to self-expression and the perils of lifestyle branding. *Journal of Marketing, 75*, 66–82.

Correia, A., Kozak, M., & Del Chiappa, G. (2020). Examining the meaning of luxury in tourism: A mixed-method approach. *Current Issues in Tourism, 23*(8), 952–970. doi:10.1080/13683500.2019.1574290

Crespi-Vallbona, M., & Mascarilla-Miró, O. (2020). Wine lovers: Their interests in tourist experiences. *International Journal of Culture, Tourism and Hospitality Research, 14*(2), 239–258. doi:10.1108/IJCTHR-05-2019-0095

Dhar, R., & Wertenbroch, K. (2000). Consumer Choice between Hedonic and Utilitarian Goods. *JMR, Journal of Marketing Research, 37*(1), 60–71. https://journals.sagepub.com/doi/abs/10.1509/jmkr.37.1.60.18718

Ellis, D., & Caruana, A. (2018). Consumer wine knowledge: Components and segments. *International Journal of Wine Business Research, 30*(3), 277–291. doi:10.1108/IJWBR-03-2017-0016

Goodman, S., Lockshin, L., & Cohen, E. (2005). Best-worst scaling: a simple method to determine drinks and wine style preferences. *International Wine Symposium*, 1-16.

Goyal, A. (2021). Consumer Perceived Associations of Masstige Brands: An Indian Perspective. *Journal of Promotion Management, 27*(3), 399–416. doi:10.1080/10496491.2020.1838027

Granot, E., Russell, L. T. M., & Brashear-Alejandro, T. G. (2013). Populence: Exploring Luxury for the Masses. *Journal of Marketing Theory and Practice, 21*(1), 31–44. doi:10.2753/MTP1069-6679210102

Gustafson, C. R., Lybbert, T. J., & Sumner, D. A. (2016). Consumer knowledge affects valuation of product attributes: Experimental results for wine. *Journal of Behavioral and Experimental Economics*, *65*, 85–94. doi:10.1016/j.socec.2016.08.004

Hall, M., & Mitchell, R. (2007). *Wine Marketing*. Routledge., doi:10.4324/9780080553283

IVV/Viniportugal. (2021). *Statistics. Portugal in the world.* https://www.viniportugal.pt/Statistics

Jarrett, S., & Jarvis, W. (2016). *Defining the luxury wine category via an assessment of price-tier perceptions* [Conference session]. 9th Academy of Wine Business Research Conference, Adelaide.

Kapferer, J.-N. (2014). The future of luxury: Challenges and opportunities. *Journal of Brand Management*, *21*(9), 716–726. doi:10.1057/bm.2014.32

Keller, K. L. (2010). Brand Equity Management in a Multichannel, Multimedia Retail Environment. *Journal of Interactive Marketing*, *24*(2), 58–70. doi:10.1016/j.intmar.2010.03.001

Kiatkawsin, K., & Han, H. (2019). What drives customers' willingness to pay price premiums for luxury gastronomic experiences at michelin-starred restaurants? *International Journal of Hospitality Management*, *82*, 209–219. doi:10.1016/j.ijhm.2019.04.024

Kim, J.-H. (2014). The antecedents of memorable tourism experiences: The development of a scale to measure the destination attributes associated with memorable experiences. *Tourism Management*, *44*, 34–45. doi:10.1016/j.tourman.2014.02.007

Kim, Y. (2018). Power moderates the impact of desire for exclusivity on luxury experiential consumption. *Psychology and Marketing*, *35*(4), 283–293. doi:10.1002/mar.21086

Kumar, A., Paul, J., & Starčević, S. (2021a). Do brands make consumers happy?- A masstige theory perspective. *Journal of Retailing and Consumer Services*, *58*, 102318. doi:10.1016/j.jretconser.2020.102318

Kumar, A., Paul, J., & Unnithan, A. B. (2020). 'Masstige' marketing: A review, synthesis and research agenda. *Journal of Business Research*, *113*, 384–398. doi:10.1016/j.jbusres.2019.09.030

Lockshin, L., Jarvis, W., Hauteville, F., & Perrouty, J.-P. (2006). Using simulations from discrete choice experiments to measure consumer sensitivity to brand, region, price and awards in wine choice. *Food Quality and Preference*, *17*(3/4), 166–178.

Lockshin, L., Cohen, E., & Zhou, X. (2011). What Influences Five-star Beijing Restaurants in Making Wine Lists? *Journal of Wine Research*, *22*(3), 227–243. doi:10.1080/09571264.2011.596200

Mason, K., & Bequette, J. (1998). Product experience and consumer product attribute inference accuracy. *Journal of Consumer Marketing*, *15*(4), 343–357. doi:10.1108/07363769810225984

Nápoles, N., Nobre, H., & Meneses, R. (2020). The Portuguese Luxury Wine Market: A Study on the Perceptions of Wine Brand Managers. In V. Chkoniya, A. Madsen, & P. Bukhrashvili (Eds.), *Anthropological Approaches to Understanding Consumption Patterns and Consumer Behavior* (pp. 303–324). IGI Global. doi:10.4018/978-1-7998-3115-0.ch017

OIV. (2021). *State of the Vitivinicultural World In 2020.* https://www.oiv.int/public/medias/7880/oiv-state-of-the-vitivinicultural-world-2020-ppt.pdf

Oliveira-Brochado, A., & Silva, R. (2014). The Wine List Design by Upscale Restaurants. World Academy of Science, Engineering and Technology, Open Science Index 88. *International Journal of Humanities and Social Science*, *8*(4), 1052–1056.

Paul, J. (2015). Masstige marketing redefined and mapped. *Marketing Intelligence & Planning*, *33*(5), 691–706. doi:10.1108/MIP-02-2014-0028

Paul, J. (2018a). Toward a'masstige'theory and strategy for marketing. *European Journal of International Management*, *12*, 722–745.

Paul, J. (2018b). Toward a 'masstige' theory and strategy for marketing. *European Journal of International Management*, *12*(5-6), 722–745. https://www.inderscienceonline.com/doi/abs/10.1504/EJIM.2018.094466

Paul, J. (2019). Masstige model and measure for brand management. *European Management Journal*, *37*(3), 299–312. doi:10.1016/j.emj.2018.07.003

Portugal Wine Landscapes. (2021). *Encouraging Signs for Portugal's wine market in 2021*. https://www.wineintelligence.com/encouraging-signs-for-portugals-wine-market-in-2021/

Ramos, P. M. G., Martins, F. V., & Barandas, H. G. (2011). Differences in the perception of wine attributes: A comparative view between consumers, producers and intermediaries. *Bulletin de l'OIV*, *84*(1), 271–306.

Ruiz-Molina, M. E., Gil-Saura, I., & Berenguer-Contrí, G. (2010). Instruments for Wine Promotion in Upscale Restaurants. *Journal of Foodservice Business Research*, *13*(2), 98–113. doi:10.1080/15378021003784483

Santos, V. R., Ramos, P., Almeida, N., & Santos-Pavón, E. (2019). Wine and wine tourism experience: A theoretical and conceptual review. *Worldwide Hospitality and Tourism Themes*, *11*(6), 718–730. doi:10.1108/WHATT-09-2019-0053

Thach, L., & Olsen, J. (2019). Luxury wine: Analyzing motivations of luxury wine buyers in the US market. *Applied Studies in Agribusiness and Commerce*, *13*(3-4), 51–58. doi:10.19041/APSTRACT/2019/3-4/6

Truong, Y., McColl, R., & Kitchen, P. J. (2009). New luxury brand positioning and the emergence of Masstige brands. *Journal of Brand Management*, *16*(5), 375–382. doi:10.1057/bm.2009.1

Velikova, N., Canziani, B., & Williams, H. (2019). Small winery-restaurant relationship building: Challenges and opportunities. *International Journal of Wine Business Research*, *31*(1), 5–11. doi:10.1108/IJWBR-07-2018-0038

Wolf, H. L., Morrish, S. C., & Fountain, J. (2016). A conceptualization of the perceptions and motivators that drive luxury wine consumption. *International Journal of Wine Business Research*, *28*(2), 120–133. doi:10.1108/IJWBR-09-2015-0038

Yeung, P., & Thach, L. (2019). *Luxury Wine Marketing: The art and science of luxury wine branding*. Infinite Ideas.

Zhan, L., & He, Y. (2012). Understanding luxury consumption in China: Consumer perceptions of best-known brands. *Journal of Business Research*, *65*(10), 1452–1460. doi:10.1016/j.jbusres.2011.10.011

KEY TERMS AND DEFINITIONS

Luxury Wine: Wines that meet criteria such as the place of origin, high price, scarcity, heritage, and ability to age.

Masstige: Can be consider as mass prestige associated with brands.

Masstige Mean Score Scale Index (MMSSI): Is an instrument of performance of masstige brands.

Chapter 12
Restaurant Industry in the Tourist City of Eastern Europe: Post-COVID-19 Perspectives – The Case of Lviv, Ukraine

Mykhailo Rutynskyi
https://orcid.org/0000-0003-3142-8723
Ivan Franko National University of Lviv, Ukraine

Halyna Kushniruk
https://orcid.org/0000-0002-0242-9746
Ivan Franko National University of Lviv, Ukraine

ABSTRACT

This chapter is devoted to the issue of post-COVID-19 revitalization of the restaurant industry in Lviv – one of the most representative tourist cities in Eastern Europe. Lviv is a fashionable and inexpensive destination of international gastronomic tourism. The modern restaurant industry developed in Lviv until 2020. Statistical trends of losses of the city's restaurant industry, in 2020 to the the first half of 2021, are unprecedentedly: 27% of establishments were closed; more than 55% of staff lost their jobs. However, the status of the cultural and tourist capital saved the restaurant industry of Lviv from collapse. Thanks to the acquired image and promotional campaigns in 2020–2021, domestic tourists actively visited Lviv during periods of easing of quarantine restrictions and enjoyed original local gastronomic products. The strategic goal in the post-COVID-19 era is to complete the transformation of the traditional business partnership into an updated smart hospitality cluster in Lviv.

INTRODUCTION AND BACKGROUND

Due to the COVID-19 pandemic, which had led to social distancing, remote working, and even the closure of the restaurant business, many restaurants have faced enormous operational challenges. However, in 2021, the restaurant business is expected to grow, mainly due to companies rearranging their operations

DOI: 10.4018/978-1-7998-9148-2.ch012

and recovering from the COVID-19 impact. The global restaurants market is expected to grow from \$2193.94 billion in 2020 to \$2619.27 billion in 2021 at a compound annual growth rate (CAGR) of 19.4% (Restaurants and Mobile Food Services Global Market Report, 2021).

The COVID-19 pandemic caused massive restrictions on the restaurant market in 2020, as governments imposed restrictions on domestic and international travel, limiting the need for services offered by these establishments. In 2020-2021, Ukraine proved to be the weakest country in Europe in terms of the effectiveness of counteracting the COVID-19 pandemic and supporting business entities. In particular, the tourism sector and the hospitality industry have been the most vulnerable to restrictions. As a result, tourist centers in Ukraine have suffered large-scale destructive losses (Danylyshyn, 2020; European Bank Report, 2020; etc.). The hospitality industry of Ukraine consists of almost 100 thousand commercial facilities – restaurants and cafes, hotels, tour operators and travel agencies. From March 2020, all these facilities were completely closed until the summer for the period of total lockdown. Further functioning of the hospitality sector graphically resembles a sinusoid, where situational restarts chronologically coincide with the easing of quarantine restrictions introduced by the government, and alternate with periods of stagnation in the relevant periods of total lockdown in Ukraine.

According to official data, the hospitality business in Ukraine by the end of 2020 lost at least \$ 1.5 billion. More than 400,000 HoReCa employees have lost their jobs. The hospitality industry of major tourist cities of Ukraine has suffered huge financial and infrastructural losses. In particular, in 2020 the restaurant industry of Ukraine lost almost 4 thousand restaurants (20.7% of the total). As of January 1, 2021, there were 14,786 restaurants, cafes and bars (vs 18,636 establishments in 2020). The capacity of the restaurant market decreased in 2020 by almost 30% and amounted to \$ 0.5 billion (State Statistics Service of Ukraine, 2021).

At the beginning of 2020, about 550,000 people worked in the restaurant sector. Now there are a third less of them. Almost 150,000 employees in the restaurant industry have lost their jobs due to lockdowns and quarantine restrictions in Ukraine. These are the people who worked in cafes and restaurants that closed due to the pandemic COVID-19 and those who were fired due to falling restaurant attendance. Most of these people will not return to work in the restaurant industry. Most of them have found work in other fields. The professional labor market, which has been formed over the years, has suffered huge losses. In 2021, this caused a big problem for the restaurant industry, many restaurant establishments simply cannot find enough professionals.

The biggest financial and infrastructural losses were suffered by the hospitality industry of large tourist cities of Ukraine, which was market-oriented to serve tourist flows, which grew steadily until 2020 (Rutynskyi & Kushniruk, 2020a). In Kyiv, Lviv and Odessa, every seventh restaurant closed after the spring lockdown (14% of the total). The number of other food establishments decreased by 12%.

Lviv is one of the largest tourist cities in Eastern Europe. It should be noted that the historical center of Lviv is included in the list of the UNESCO World Heritage Sites (L'viv – the Ensemble of the Historic Centre, 2021). The city is located near the eastern border of the EU and thanks to popular low-cost flights is tightly integrated into the pan-European international travel industry. Lviv is closely connected with more than a hundred large cities of Great Britain, the EU, Turkey and other countries of the Eastern Mediterranean, the Caucasus, and the Arabian Peninsula through a network of low-cost routes.

The Telegraph, influential British editorial, named Lviv as one of 12 most exciting new routes in 2018. Lviv was on the 5th place in the list. Fiona Duncan, after visiting for Telegraph Travel, said: "Though Lvivians of today are known for both their fervent nationalism and for their churchgoing, their city has an easy-going, almost frivolous air, filled with university students, embellished by its frothy confection of Renaissance, Baroque, Belle Epoque and Art Nouveau buildings and scented with aromas from its hundreds of Viennese style coffee houses" (Morris, 2017). Today, coffee tourism is one of the leading specializations of Lviv in the international tourism market (Rutynskyi & Kushniruk, 2020b). The British newspaper The Independent called Lviv an ideal place for winter holidays in the 2018-2019 season, noting that "there's never a bad time to go to this beautiful city" (Henderson, 2019).

The chapter is devoted to the post-COVID-19 perspectives of the restaurant industry in Lviv, the historic center of which is included in the UNESCO World Heritage List. First of all, the growing potential of Lviv's restaurant industry in the Pre-Covid-19 Era is substantiated. Then the statistical trends of losses of the city's restaurant industry, which were complicated by pandemic challenges in 2020 – the first half of 2021, are highlighted. Finally, noteworthy post-COVID-19 trends in the recovery of Lviv's restaurant industry are analyzed.

The study of the dynamics of restaurant industry development, changes in the market structure of food establishments and other sectoral transformations involves the use of a system of methods for assessing current processes in them and determining their impact on the development of relevant destinations. General and special scientific methods were used in the research. To correctly analyze the data sets from various agencies and departments in Lviv, the authors used statistical methods described in the works on quantitative methods in tourism and hospitality (Altinay et al., 2015).

LITERATURE REVIEW

The restaurant industry is one of the most dynamic sectors of the economy and leisure of society, and therefore is the subject of interdisciplinary research in a number of sciences. Dozens of scientific works of leading scientists, business analysts and experts-practitioners are annually devoted to the study of the functioning of the restaurant business. The positive dynamics of these publications is reflected in the specialized Internet platforms Scopus, Web of Science, etc. (Rodríguez-López et al., 2020).

In the 21st century, the world's leading tourist destinations have entered a fierce competition for tourist flows and tourist loyalty. And one of the main components of such competition is gastronomic identity (Bittman et al., 2012; Bellini et al., 2018). Today, the success of the tourist center, along with the architectural heritage, is determined by the colorfulness of its gastronomic heritage, the ability of restaurants to reveal the ethnographic specifics of the cuisine of the local population (Sim, 2009; Bertan, 2020; Wen et al., 2020; Kim et al., 2020a). Some scholars directly consider the potential of the restaurant business of a tourist destination as a source of knowledge "Ethnographies of Where we Eat" (Beriss & Sutton, 2007; Tsai & Lu, 2012; Činjarević & Pestek, 2014; Etcheverria, 2020).

Scientific research in recent years reveals the trend of intensification of competition of international and local restaurant brands for the loyalty of tourists. At one pole of this struggle – global fast food chains (McDonald's, KFC, etc.), at the other – the brands of local cuisines of individual tourist centers and ethnic communities (Rolim de Albuquerque Meneguel et al., 2019; Hernandez-Rojas et al., 2021).

Achieving profitability and service excellence in the guest experience in such realities is based on new principles of hospitality management, taking into account society's demands for greening of technological processes, healthy low-calorie food, ethno-animation, customer involvement in choice, cooking and food culture, etc. (Erkuş-Öztürk & Terhorst, 2016; Lee et al., 2019; Lee et al., 2020; Ford & Sturman, 2020).

The COVID-19 crisis has become another challenge for the restaurant industry. At the same time, it stimulated the development of new technologies and the trend of digitalization of the restaurant business.

A recent study by Yiqi & Jewoo (2021) details the interconnectedness between online review valence, brand, and restaurant performance. Thus, in the digital era, the commercial success of a restaurant becomes derived from the success of its representation on the Internet (Zhai et al., 2015; Brewera & Sebby, 2021; Park et al., 2021; Dirsehan & Cankat, 2021).

Ukraine is an integral part of European tourist and gastronomic diversity (Kiptenko & Doan, 2015; Nistoreanu, 2020). Ethnographic cuisine of Ukraine is one of the richest and most colorful in Central and Eastern Europe. Today the state is in the process of international legal consolidation of the top ten of its food geographical indications. And such restaurant dishes of Ukrainian cuisine as borshch, galushky, varenyky with cherries or cheese, potato pancakes (deruny), Carpathian dishes with white mushrooms and others are recognized tourist cards of the state.

In the neighboring EU countries, restaurants and gastronomic culture are one of the representative elements of the cultural heritage of cities (Ganzaroli et al., 2017; Hernandez-Rojas et al., 2021). In the cities-centers of mass tourism of Ukraine the restaurant industry is also closely connected with tourist business, and its economic indicators directly correlate with incomes from tourism. In turn, Ukraine's tourism and restaurant sectors in the 21st century depend on external geopolitical, economic, pandemic challenges such as the financial crisis of 2008, the military invasion of the Russian Federation into Ukraine in spring 2014, COVID-19 total lockdown in spring 2020 (Ivanov et al., 2016; Wisła & Nowosad, 2020; Tomczewska-Popowycz & Quirini-Popławski, 2021). Each crisis has a destructive effect on the consumer behavior of locals and tourist (Senbeto & Hon, 2020). However, despite the crisis challenges, the national restaurant industry demonstrates high adaptive and evolutionary potential (State Statistics Service of Ukraine, 2021).

In 2020, the world faced an unprecedented COVID-19 crisis. Coordinated counteraction to the spread of the pandemic has led to the systemic transformation of all spheres of human life. Researchers uncover the global, regional and sectoral scales of these transformations, particularly in the food industry, food supply, tourism, and hotel and restaurant business (Abbas et al., 2021; Grunert et al., 2021; Panzone et al., 2021; Rivera-Ferre et al., 2021; Song et al., 2021; etc.)

The COVID-19 crisis has been a challenge for the restaurant industry. At the same time, it stimulated the development of new technologies and digitalization of the restaurant business. A recent study by Gavilan et al. (2021) reveals the current vectors of innovation in online food delivery. Yiqi & Jewoo (2021) details the interconnectedness between online review valence, brand, and restaurant performance. Thus, in the digital age, the commercial success of a restaurant depends on the success of its representation on the Internet (Brewer & Sebby, 2021; Park et al., 2021; Dirsehan & Cankat, 2021; Kim et al., 2021; Li et al., 2021; Luo & Xu, 2021; Zhai et al., 2015).

The restaurant industry in all regions of the world simultaneously overcame lockdown stagnation, transformed and developed mechanisms for quarantine and market adaptation in the realities of the devastating COVID-19 crisis (Gössling et al., 2020; Neise et al., 2021; Nhamo et al., 2020; Rutynskyi & Kushniruk, 2020a; Song et al., 2021). Nevertheless, catering establishments (fast food outlets, roadside cafes, "dark kitchen", coffee houses with summer terraces, bars, nightclubs) in 2020-2021 demonstrate effective business strategies and better recovery trends compared to other HoReCa sectors (Kostynets et al., 2021; Li et al., 2021; Min et al., 2021; Neise et al., 2021; Singh et al., 2021; Yost et al., 2021). In particular, in large cities of Ukraine, and first of all in Lviv, the restaurant business plays a significant role in overcoming the negative economic and socio-psychological consequences of the COVID-19 crisis and demonstrates the statistical trend of active recovery (State Statistics Service of Ukraine, 2021).

PRESENTATION OF DATA AND RESULTS

The Growth Potential of Lviv's Restaurant Industry in the Pre-Covid-19 Era

More than 760 thousand people live in Lviv (30% of the population of Lviv region). The average annual foreign direct investment per capita in Lviv for 2014-2019 ranged from $ 800-900.

As of January 1, 2020, there were more than 1,012 establishments of the restaurant industry in Lviv (Figure 1). Only in the central part of the city there are 805 fast food outlets, cafes, bars, restaurants, including 92 restaurants at hotels (Tourism Office of Lviv City Council, 2021; TripAdvisor Restaurants near Old Town, Lviv, Ukraine, 2021).

Since the end of the twentieth century, Lviv has been showing the highest growth rates of tourists among the cities of Ukraine. Before the beginning of the COVID-19 crisis, Lviv annually served up to 2.5 million tourists and excursionists. According to the growth rate of the number of overnight stays of foreign tourists among the TOP 100 tourist cities in Europe, Lviv in 2017 topped the ranking table with

Figure 1. Comparative ratio of the number of catering establishments per 100 thousand inhabitants of some European cities in 2020
Source: calculated by the authors according to: Tourism Office of Lviv City Council, 2021

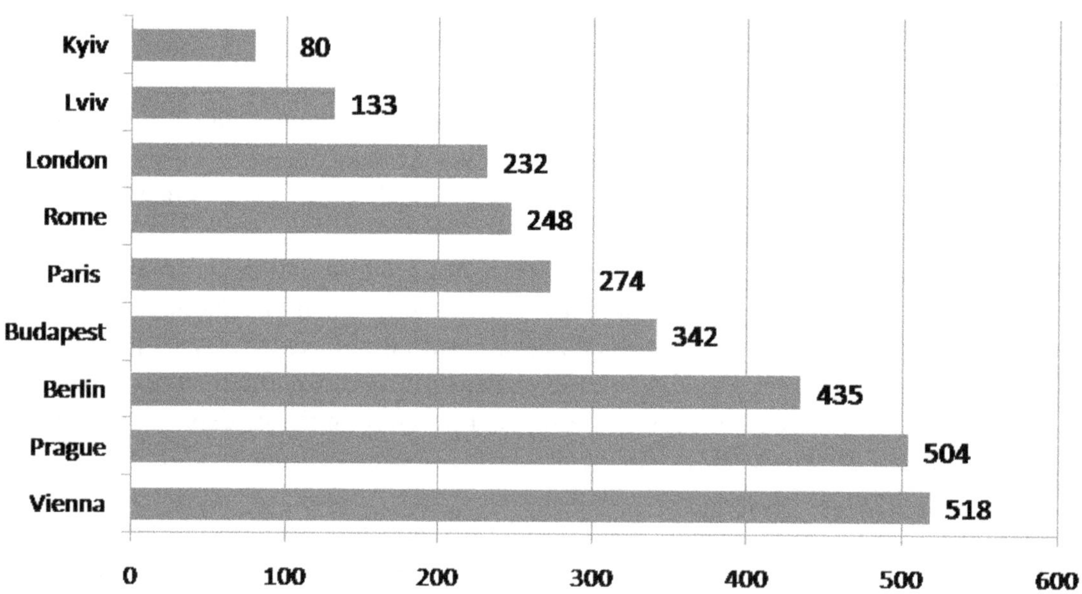

a relative value of 73.9%.

Every year Danylo Halytskyi International Airport "Lviv" expands its flight geography. In 2018, passenger traffic to Lviv increased by 46.3% compared to last year and reached over 1.5 million passengers. Changes in the structure of foreign tourists correlate with new flights to the EU. For example, in 2018, the opening of direct flights from London operated by low cost airlines Wizzair and Ryanair, according to expert estimates, has led to an increase: by 86% of tourists from India, by 71% of tourists from Australia, by 62% of tourists from the United Kingdom, by 21% of tourists from the USA and Canada (Tourism Office of Lviv City Council, 2021).

The tourism and hospitality infrastructure of Lviv has been developing dynamically over the last two decades, focusing on high standards of service and requests from tourists from Central and Western Europe, Canada and the USA.

The following factors have stimulated the development of the hospitality sector and the economic profile of Lviv as the "tourist capital of Ukraine":

1. The process of successful market transformations in Ukraine;
2. Active integration of the economy and society of the state into the global system of world order;
3. Upholding European democratic values by the Ukrainian nation (Orange Revolution in 2004, Dignity Revolution in 2013-2014);
4. Ukraine's introduction of a visa-free regime for EU tourists in 2005, further simplification of visa formalities for tourists from other regions of the world;
5. The geographical position of Lviv near the EU border;
6. Development of the international Lviv airport and land transport network with major EU cities;
7. Active involvement of local authorities and businesses in EU grant projects for the development of tourism infrastructure;
8. Realization of promotional measures on the tourist promotion of Lviv in the domestic and international markets;
9. Development of event tourism and corresponding brand "Lviv – city of 100+1 festivals";
10. Active representation of the city's guest potential on the Internet;
11. Concentration in the city of the largest number of higher education establishments in Ukraine, which train specialists in the tourism, hotel and restaurant business;
12. Providing the best qualified staff of the hospitality sector of Lviv, as a rule, able to kindly communicate with foreign tourists in English and often in one of the languages of other countries (Polish, German, Italian, etc.).

In the rating of the travel company Momondo, published by the American publication Business Insider, Lviv with a rate of £ 25.46 ($ 31.27) took first place among the cheapest tourist cities in the world. In particular, a tourist pays about $ 25 for accommodation in Lviv, $ 5 for dinner for two, and only 20 cents for a 1 km taxi ride (Hancock, 2016).

In 2017 in the ranking of 26 ideal European cities where you can live comfortably for less than £ 600 a month, published by the British newspaper The Independent, Lviv took second place (£ 405.51 / € 458.56) (Edsor, 2017).

In 2019, Lviv not only became the only city in Ukraine to be included in the tourist rating by Airbnb and Post Office Travel Money, but also topped this rating (Table 1). The barometer cost in Lviv showed only £ 137 against, for example, £ 246 in Antalya or £ 378 in Lisbon (Post Office Celebration Break Report, 2019).

The rating showed how much a short holiday weekend for two in one of the popular cities costs on average. It includes two nights of accommodation booked through Airbnb, transfer from the airport round trip, travel ticket per day by public transport, bicycle rental per day, the cost of visiting tours and museums, the cost of a bottle of sparkling wine, dinner for two with a bottle of wine, snacks and cock-

Table 1. Lviv in the rating of affordability for tourists conducted by Post Office Travel Money and Airbnb, in 2019

ITEMS	1	2	3	4	5
	Ukraine LVIV	Albania TIRANA	Poland KATOWICE	Slovakia KOSICE	Lithuania KAUNAS
TOTAL BAROMETER COST	£137.42	£153.43	£191.72	£195.52	£236.79

Source: Post Office Celebration Break Report, 2019

tails for two. Before the COVID-19 crisis, a couple of tourists spent an average of £137.42 on a whole package of services in Lviv.

During the period 2010–2021, an array of data on monitoring tourist flows was accumulated (Lviv Tourist Barometer, 2021). The offices of the Tourism Development Centre of Lviv and the City Institute regularly conduct sociological surveys of incoming tourists at the request of the Lviv City Council.

The results of sociology correlate with the periodization of the development of the restaurant industry in Lviv proposed below. The following results are generalized by sociological methods; the geography of foreign tourists for the last pre-crisis 2016-2019 outlined four dominant vectors of arrivals: 1) neighboring countries, 2) countries of the West Europe, 3) the United States and Canada, 4) Turkey and the Middle East. In the most successful 2017, the top ranking of foreign guests was headed by tourists from Poland (22.1%), Belarus (7.9%), Turkey (7.3%), Germany (3.6%), USA (2%), Lithuania (1.6%), Great Britain (1.5%), Czech Republic (1.2%), Canada (0.7%). The average length of stay in Lviv is 3.6 days; 82% of tourists book accommodation through Booking.com; 38% live in hotels, 28% – in hostels. For comparison, guests spend in Lviv: Americans – €131, the British – €112, Turks – €111, Ukrainians – €35 per person per night.

On average, 77-81% of tourists come to Lviv for recreation and entertainment. For 48.2% of respondents Lviv is associated with coffee and coffee tourism, for 47.9% – with historical architecture, for 27.6% – with cultural, gastronomic and event tourism. Accordingly, during 2015-2021, Lviv City Council and business focus their efforts on promoting these guest brands and tourism products in Lviv. Targeted promotional campaigns are aimed at promoting, in particular, the brand "Lviv – a center of gastronomic tourism" (growth rate was 6.4% until 2020, after zeroing due to the COVID-19 pandemic, according to data for December 2020-January 2021 – 7.1%). Interesting results were obtained on the question "How do tourists spend their leisure time in Lviv?" In 2015-2019, an average of 85% of respondents chose city tours on their own, 82% – gatherings in cafes and restaurants, 58% – visits to museums, galleries and theaters (summarized by the authors according to: Tourism Office of Lviv City Council, 2021).

The city of Lviv from year to year strengthens its attractive image as a fashionable and inexpensive destination of international gastronomic tourism. The development of tourism until 2020 stimulated the development of a modern restaurant industry in Lviv with the traditions of local cuisines and high standards of service.

In the development of the restaurant industry in Lviv, taking into account the pace of infrastructure development, growth of the range of gastronomic products and the volume of visitor service, the authors recommend to highlight the following periods:

- 1991-1999 – period of post-socialist transformation and spontaneous market,

- 2000-2008 – period of formation of a civilized market and approximation to EU service standards,
- 2009-2013 – period of intensive market development, implementation of large-scale infrastructure projects, capitalization of Lviv's PR in the domestic and international markets,
- 2014-2019 – period of maintaining the achieved positions and coordinated efforts to stimulate the market of domestic and inbound gastronomic tourism in the conditions of a protracted military-geopolitical crisis in Ukraine caused by the military and information aggression of the Russian Federation,
- 2020-2021 – period of COVID-19-depression of the city's restaurant industry,
- from 2022 – period of post-COVID-19 restoration of capacity of the restaurant industry, active promotion of inbound gastronomic tourism and competition for the attention of the global consumer.

The rapid development of entertainment and gastronomic tourism in Lviv in 2010-2019 was based on the extensive, growing and diversified infrastructure of restaurant business establishments of different types and price categories (Figure 2). Over the decade, as of January 1, 2020, the network of gastronomic locations has tripled due to the entry into the market of powerful gastronomic holdings, as well as the

Figure 2. The growth rate of the number of establishments in the restaurant industry in Lviv, %
Source: calculated by the authors according to: Lviv Tourist Barometer, 2021; Tourism Office of Lviv City Council, 2021

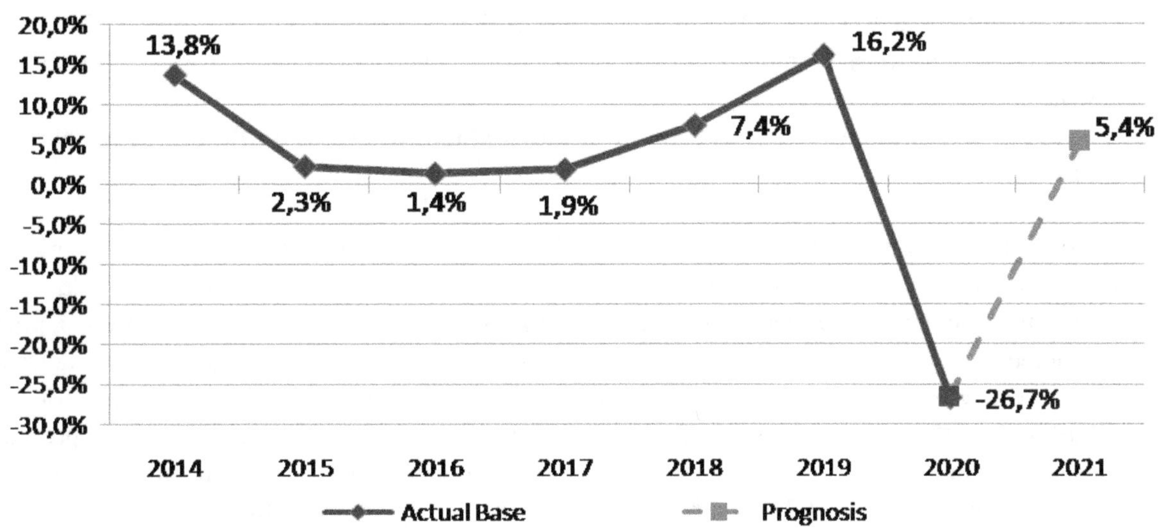

opening of small street catering establishments by representatives of small family businesses.

From 6 to 14 new establishments were opened in the city every year, primarily focused on servicing the steadily growing tourist flow. On average, each such new location of gastronomic tourism employs from 8-30 to more than 100 people. For example, as of January 1, 2020, the staff of the "Viennese Coffee House" was 31 employees, the restaurant "Park. Art of Rest" – 19, the themed restaurant "Shakespeare" – 17, the cozy restaurant "Tante Sophie" – 7 employees; in contrast – the staff of the restaurant "Stargorod" reached 122 employees, the restaurant-brewery "Kumpel" – 148 employees.

In total, the number of full-time employees of the restaurant industry within the tourist center of the city increased from 1589 people in 2016 to 2157 people in 2019. For example, in 2019, the Holding of Emotions "!FEST", only one industry leader, employed 1443 people (mostly communicative youth with fluent English) in 22 hotel and restaurant establishments.

The development of restaurant infrastructure in Lviv has a number of special features. Firstly, until recently, it was difficult for top transnational restaurant brands (except fast food) to enter the market due to the objective and subjective difficulties of investing and doing business in Ukraine. Foreign investors are still in no hurry to enter the Lviv market with new facilities. Restaurants of the elite segment operate at 4-5* hotels. Instead, the greatest financial return is provided by restaurants of the middle price category, conceptually aimed at serving the mass tourist flow.

Over the last decade, restaurant businesses within the tourism cluster in partnership with the City Council have purposefully invested millions in the promotion of Lviv as a top destination for gastronomic tourism in Eastern Europe. It gave a new impetus to the development of mass tourism. Thus, before the COVID-19 crisis, Lviv earned the following titles:

- Top destination for gastronomic tourism in Central and Eastern Europe (the main location, where tourists have a unique opportunity to taste more than half a thousand dishes and drinks of ethnic cuisine of Ukrainians, as well as a range of traditional ethnic dishes of Jewish, Polish and other cuisines),
- The capital of coffee tourism of Ukraine and Eastern Europe (Rutynskyi & Kushniruk, 2020b),
- The capital of beer tourism (pubs offer more than 120 varieties of craft beer), as well as the capital of wine & cheese tourism in Ukraine,
- The capital of gastronomic and romantic slow tourism (TripAdvisor Lviv Romantic Restaurants, 2021).

The city of Lviv has formed a historic urban environment, which is quite difficult to fit new restaurants with a capacity of more than 50 seats. Instead, tourist and gastronomic holdings and small businesses offer Lviv guests a wider range of cozy small restaurants every year – coffee houses, pubs, pastries, confectioneries, noodle houses, kebab houses, BBQ restaurants, conceptual mini-restaurants with author's kitchens and drinks cards. Among them, about 50% are tiny coffee houses in historic 19th century buildings with summer tables on narrow streets that operate in a speak easy format and promote the values of slow tourism for gourmets (TripAdvisor Cafés in Lviv, 2021). The opposite market niche (11%) is occupied by spacious BBQ Restaurants for large groups (TripAdvisor BBQ Restaurants for Large Groups in Lviv, 2021). It should be noted that 23% of Lviv's restaurants attract with live music – that is, not only delicious food, but also a concert atmosphere (jazz, pop hits of the last century, playing the piano, saxophone, etc.). Another 17% of catering establishments are stylized as thematic living museums.

During the period 2010-2019 in Lviv, the number of small tourist-oriented gastronomic establishments increased 4.5 times. The growth of the tourist flow guarantees a quick payback and high profitability of the gastronomic industry (Erkuş-Öztürk & Terhorst, 2016).

Sociological surveys of coffee house owners in different years allow using the extrapolation method to obtain the following approximate data:

1. Tourists make up an average of 65-75% of coffee house visitors within the tourist center of the city;

2. During the year, tourists left in a network of coffee houses within the historic city center from €28.5 million (in 2010) to €74.5 million (in 2019).

The global Internet platform TripAdvisor has recorded a very high growth rate for Ukraine in the popularity of Lviv gastronomic establishments (TripAdvisor Restaurants near Old Town, Lviv, Ukraine, 2021), for example, "Kryjivka", "Masoch Café", "House of Legends", "Lviv Coffee Mining Manufacture", "Baczewski Restaurant", pub-museum "Gas Lamp", "The Most Expensive Galician Restaurant (Masonic Lodge Restaurant)", the first restaurant-brewery "Kumpel", five-storey complex of concept restaurants on Rynok Square "36Po", "Champagneria X&X", coffee house "At the Golden Rose", Italian cuisine restaurant "Capriccio", pizzeria-pasteuria "Tisto", wine bar "Cannoli Wine Club", gourmet restaurant "Grand SoLie" (as part of the "Grand Hotel Lviv Luxury & Spa"), etc.

Over the last decade, the stereotype of Lviv restaurant hospitality has become stable: original food + positive emotions. Travel websites and numerous bloggers today primarily focus on such a competitive advantage of Lviv's gastronomic and entertainment business. Restaurateurs and consumers jointly postulate the thesis: during the visit to the city, people visit as many of its restaurants and cafes as possible not only to eat, but also to experience unforgettable emotions. Thus, some restaurants in Lviv offer their visitors the opportunity to go back two or three centuries and discover the forgotten dishes of European retro recipes, or sit in the atmospheric dungeons, or taste exclusive ancient dishes and drinks of Galician locavore cuisine, or taste dishes of Jewish and Asian ethnic cuisines, or become participants in original gastronomic shows. Here are some examples:

- Coffee shop-museum "Under the Blue Bottle" immerses in the atmosphere of the 18th-19th centuries, offers exclusive special dishes, such as coffee on fire, or a variety of fondue: cheese, meat, chocolate. There are more than 200 lamps and chandeliers of the 19th – early 20th centuries in the interior. The waiters arrange a pyrotechnic show to order.
- The First Lviv Grill Restaurant of Meat and Justice invites you to descend into the semi-dark casemates of the defensive wall of the Bernardine monastery, built in the first half of the 17th century in the style of Italian-Dutch Mannerism. The theme of the restaurant is medieval Lviv cuisine using dark animation with a city torturer and tools of medieval justice, including a guillotine, which cuts the thickest sausage in Ukraine.
- Masoch Café immerses in the atmosphere of the era of literary work Leopold von Sacher-Masoch, a famous writer who was born in Lviv. The cafe is decorated with the appropriate attributes of the founder of the literary and erotic direction of masochism.
- Restaurant "Trout, Bread and Wine" specializes in baked trout and dishes with home-made bread, prepared according to ancient recipes of the inhabitants of the mountain villages of the Eastern Carpathians. Before entering the pub, the musician-animator acquaints visitors with millennial traditions, song and musical heritage of the Ukrainian mountaineers of the Eastern Carpathians.
- Galician Jewish restaurant "At the Golden Rose" is located near the historic Golden Rose synagogue and recreates the atmosphere of ancient Jewish life of the 18th-19th centuries. Before eating, guests traditionally wash their hands and get acquainted with the history of the Jewish community of Lviv. A feature of the menu is the lack of prices, which allows visitors and waiters to bargain for the cost of ordered dishes.

Taking into account the requests of tourists, a separate niche in 2012-2019 was formed by panoramic restaurants and cafes in Lviv. These are locations where you can combine gourmet drinks and meals with photo shoots and selfies and with an unforgettable overview of the historic landscape of the city, which is included in the UNESCO World Heritage List. Today in the TripAdvisor rating high positions are occupied by at least five such panoramic locations: "Cloud no.7 Panorama", "36Po", "Pretty High Kitchen", "Lviv Handmade Chocolate" and "Gas Lamp".

During the period 2011-2021, the infrastructure of night gastronomic and entertainment establishments in Lviv increased 9 times. The influx of foreign tourists stimulated the formation of another specialization of the city as a center for night parties of various art formats and price categories. Minimal (compared to EU and US nightclubs) COVID-19 formalities for guests of this type of establishments during periods of easing of quarantine restrictions in 2020 have further increased their popularity among foreign guests. In TripAdvisor's 2019 and 2020 rankings (despite the COVID-19 crisis), the top 10 nightclubs were led by the following: Split Lviv Club, Malevich Night Club, EROS Show Bar, White Rabbit – Show Bar, Cazanova Show Bar, !FESTrepublic.club, Cloud no.7 Panorama, Faces and others (TripAdvisor Lviv Late Night Restaurants, 2021).

In 2019, 50 restaurants located in the tourist center of the city paid €0.45 million of personal income tax to the city budget, of which: 73.6% of the amount of payments was received from 22 restaurants of the Holding of Emotions "!FEST", 5.2% – establishments of Kumpel' Group, 2.9% – restaurants-breweries "36Po" and "Royal Brewery", 18.3% – other small restaurants, pubs and cafes. In order to reduce tax deductions to the budget, more than 50% of institutions document minimum wages (without bonuses and other allowances). In contrast, thanks to the honest attitude to taxation, in 2019 four MacDonald's establishments transferred almost €0.23 million to the budget of Lviv (Tourism Office of Lviv City Council, 2021).

On September 7, 2019, the IV Festival of Hospitality and Hotel and Restaurant Business "HoReCa Show Lviv" took place in Lviv – the largest hospitality forum in Ukraine. The mission of this international forum is to make Ukraine's hotels and restaurants attractive, Ukrainian cuisine recognizable, and the service that will attract tourists from around the world. The HoReCa sector is recognized by the City Council as one of the priorities of the city's economic recovery in 2021-2022 in the post-COVID-19 era. At the beginning of November 2021, the anniversary HoReCa Show Lviv-2021(platform for post-COVID-19 restart of the restaurant business of Ukraine in cooperation with leading foreign partners and investors) is planned to be held in compliance with the COVID-19 restrictions.

The Restaurant Industry in the COVID-19 Crisis

In Ukraine, the coronavirus causes the greatest economic losses to both large companies (airlines, tour operators) and centers of mass city tourism. Since the beginning of March 2020, the hospitality industry (including financially vulnerable private small tourist business) of Kyiv, Lviv, Odessa, Chernivtsi and other popular centers of city tourism has died down.

Since March 16, 2020, Lviv city authorities had to close all mass leisure, tourism, and trade establishments (except grocery stores, pharmacies, and gas stations). A few days before the border closes, the cancellation of flights and the railway connection, practically all tourists left Lviv hotels.

Since March 17, for two weeks, the network of gastronomic tourism establishments of Lviv was closed for quarantine – one of the most profitable sub-branches of the city, its tourist business card, which promotes new fashion trends of city gastronomic tourism in Ukraine. Famous Lviv restaurateur Vardkes

Arzumanyan, the owner of the elite top restaurants popular with foreign tourists, was the first who announced the closure of his restaurants: Champaneria X&X, Para Janov & Port wine, Taron, Mons Pius, Centaur and Libraria. Next, a similar decision was made by the owners of the concept restaurant chain Holding of Emotions "!FEST", an integral part of the tours to Lviv, the top leader of the sympathies of hundreds of thousands of tourists in global ratings. Finally, in accordance with the decision of the City Council, the doors of small business establishments, which traditionally associated with leisure tourists in Lviv, were forced to close – stylish coffee shops (element of the architectural and cultural environment of the historical center of Lviv, included in the UNESCO Heritage).

Chronicle of national counteraction measures COVID-19:

- On March 11, 2020, the World Health Organization (WHO) announced a coronavirus pandemic. On March 11, the Cabinet of Ministers has decided to introduce quarantine in Ukraine from March 12 to April 3. In particular, the educational institutions were closed for three weeks; all mass events in the country were canceled;
- On March 15, Poland and Ukraine have closed borders for foreigners;
- On March 16, the Cabinet of Ministers has decided to suspend the operation of air, rail and bus services between cities and regions of Ukraine;
- On March 17, the Verkhovna Rada of Ukraine has adopted a law on preventing the spread of the coronavirus disease COVID-19;
- Since March 18, Ukrzaliznytsya has canceled all passenger traffic in Ukraine;
- From May 22 to December 31, 2020, the "adaptive quarantine" was introduced in Ukraine. In Lviv, most restrictions were lifted; the activity of summer terraces at restaurants and the sale of takeaway food were allowed;
- On November 11, 2020, the Government introduced a "weekend lockdown";
- From January 8 to 24, 2021, the Cabinet of Ministers introduced a "total lockdown" in Ukraine;
- From January 25 to April 30, 2021, the Cabinet of Ministers extended the quarantine restrictions.

As of May 31, 2020, only 70% of catering establishments in Ukraine have resumed their activities. The average revenue after the opening of summer terraces at restaurants reached 50%, in August – 72% of pre-crisis figures.

The consequence of the year of quarantine restrictions for Lviv is as follows: from April 2020 to April 2021, 27% of restaurant establishments went bankrupt and closed due to quarantine. If the restrictions continue until the end of the year, in 2021 we will forecast further industry losses of up to 34-44% in general in the city (at the same time in the sleeping areas of the city – to more than 47%).

In 2020, the staff of the restaurant industry in Lviv has more than halved. Prolonged coronavirus crisis has pushed a significant part of Lviv's hotel and restaurant establishments to return to the "tax shadow". Following the results of 2020, financial analysts of the Department of Economic Development of the Lviv City Council have discovered that most institutions do not indicate the real official salary of their employees in order not to pay income tax in full, which had led to additional large-scale losses in the city budget.

In 2020, business and the City Council coordinated the information campaign about Lviv through TV and Internet communication channels. Journalists of leading TV channels and travel bloggers were invited to the city, promotional videos were shot, and Lviv's charms were actively advertised on social networks.

Due to the lack of tourists, interesting promotional campaigns (#I'm – Tourist in my own city, #For their own) were launched to activate the residents to discover the tourist attractions of Lviv. Some coffee houses stimulated demand with promotions "second cup of coffee – free". More than 40% of the city's hotels supported the global campaign "Work from hotel" and offered favorable conditions for living & working in hotel rooms (mostly for IT workers and workers whose families were infected infected with COVID-19 and were in self-isolation).

Lviv City Council has introduced a number of benefits under the business support program in the context of the COVID-19 pandemic. In particular, from March 2020, entrepreneurs who leased communal property were exempted from rent (this is a shortfall in the budget of UAH 40 million). In addition, by the end of 2020, more than 850 entrepreneurs in the field of tourism, hotel and restaurant business were exempted from paying the tourist fee and for six months from paying the single tax (this is a shortfall in the budget of UAH 60 million).

Entrepreneurs were forced to comply with all quarantine restrictions imposed by the government. However, during periods of easing of quarantine restrictions, they were actively involved in the competition for potential customers. Since May 2020, summer terraces of restaurants and cafes have started working, hotels and apartments have resumed work according to new quarantine standards.

The next challenges for the city's hospitality industry were government restrictions in the fall of 2020 and winter of 2021. According to the extrapolation forecast, Lviv's restaurant business has lost up to 1.4 million customers due to the introduction of weekend quarantine in autumn 2020, and then total lockdowns (8.01-24.01.2021 and 19.03-4.04. 2021), and thus the extension of quarantine restrictions for February-May 2021.

Post-COVID-19 Perspectives

The global COVID-19 crisis has prompted the restaurant industry to comprehensively reorganize the subsystems of interpersonal communication, sanitation, hygiene and disinfection in the organization of food supply, cooking, customer service in the hall and takeaway, waste disposal. The goal is to ensure safety, reduce the risk of disease transmission and the spread of coronavirus in accordance with new guidelines and standards developed internationally (World Health Organization, 2020b; Food and Drug Administration, 2020; Food and Agriculture Organization, 2020; Food and Agriculture Organization of the United Nations and World Health Organization, 2020).

The human factor is the main in the restaurant business, and therefore the set of sanitary and hygienic renovations is a new evolutionary stage of the restaurant industry. New standards for the restaurant industry will never be the same as in the pre-COVID-19 era. They guarantee the effectiveness of sanitary control at all technological levels, cleaning and ventilation of premises, maintaining the social distance between producers and consumers of food in restaurants, the spread of alternative channels of contactless delivery of restaurant food home (Brizek et al., 2021; Byrd et al., 2021; Chang et al., 2021; Cronin & Evans, 2021; Gavilan et al., 2021; Hu et al., 2021; Li et al., 2021; Singh et al., 2021; Sung et al., 2021). The trust and motivational choice of consumers of a particular restaurant today is determined by a combination of these new guarantees of customer safety (Dedeoğlu & Boğan, 2021; Grunert et al., 2021; Hakim et al., 2021; Kim et al., 2021; Yost & Cheng, 2021).

From mid-2020 the restaurant industry of Ukraine is guided by the new international standards and methodical principles developed by WHO, FAO, FDA. Thus, Ukraine simultaneously coordinates its national strategy with the neighboring countries of the European Union, where public strategies for

Figure 3. Age structure of visitors to Lviv restaurant business establishments in the first half of 2021
Source: calculated by the authors according to: Tourism Office of Lviv City Council, 2021; Lviv Tourist Barometer, 2021.

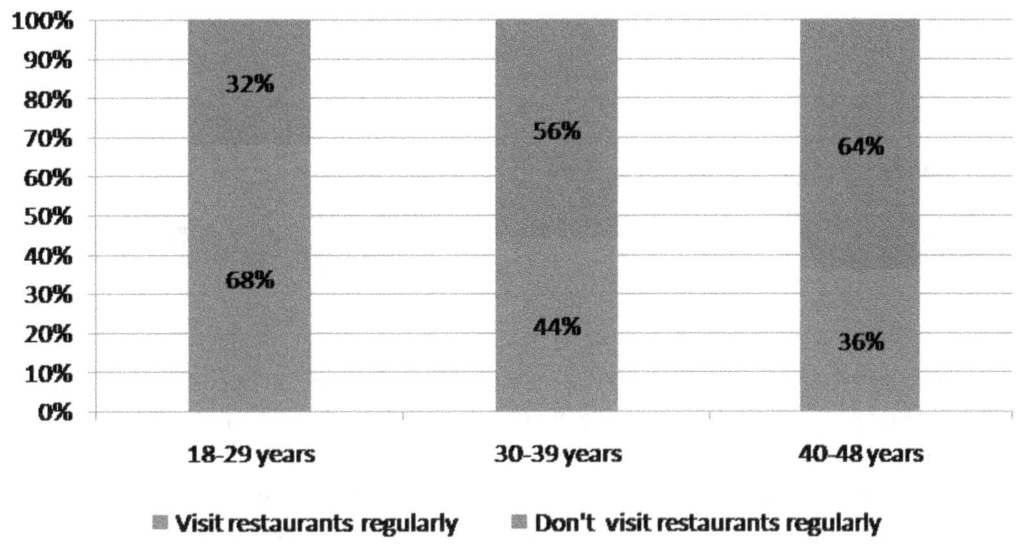

rescuing the hospitality industry after the impact of COVID-19 are being systematically implemented (Sanabria-Díaz et al., 2021).

Prolonged COVID-19 crisis did not nullify the tourist flow to the city of Lviv. The status of a fashionable destination on the map of gastronomic tourism in Eastern Europe, as well as the image of the cultural and tourist capital of Ukraine "worked" as motivators for thousands of Ukrainians: not to be locked in their homes for a whole year and dare to travel positively and emotionally inside their country.

Despite all the restrictions during 2020, tourists from all over Ukraine amateurly visited Lviv, although not as massively as in the pre-COVID-19 era. Summarizing a number of posts on social networks, it is worth noting the emergence of a new socio-therapeutic-tourist trend: for many visitors a weekend in Lviv has become an effective form of relieving psycho-emotional stress, positive psychological relief, restoring thirst for life, tone and stress resistance.

Sociological monitoring "portrait of a guest of the city" revealed a number of new features of tourism during the COVID-19 crisis. Among them: the average length of stay of a foreign tourist in Lviv increased from 2.2 nights (in 2015) to 4-6 nights (in 2020). The number of domestic tourists staying in the city for more than 3 nights increased 6.8 times (the reason is international quarantine isolation) (Tourism Office of Lviv City Council, 2021).

Sociological monitoring of customers in the summer of 2021 revealed a post-COVID-19 trend of increasing the share of regular and irregular visitors to restaurants of the age group of young people (18-30 years) (Figure 3). Obviously, the reasons for this are, firstly, the desire of young people to communicate with friends due to long-term COVID-19 isolation, and secondly, the democratization of the pricing policy of Lviv restaurants in the COVID-19 crisis to stimulate customer turnover.

On August 30, 2021, the first Lviv ethno-animation restaurant "Kryjivka" (it was opened in 2007) served its 1,200,000[th] visitor, which was recorded in the Book of Records of Ukraine (Holding of Emotions !FEST, 2021).

Figures 4. Typological structure of Lviv restaurant industry establishments in 2019 and 2021,%
Source: calculated by the authors according to: Lviv Tourist Barometer 2021; Tourism Office of Lviv City Council 2021

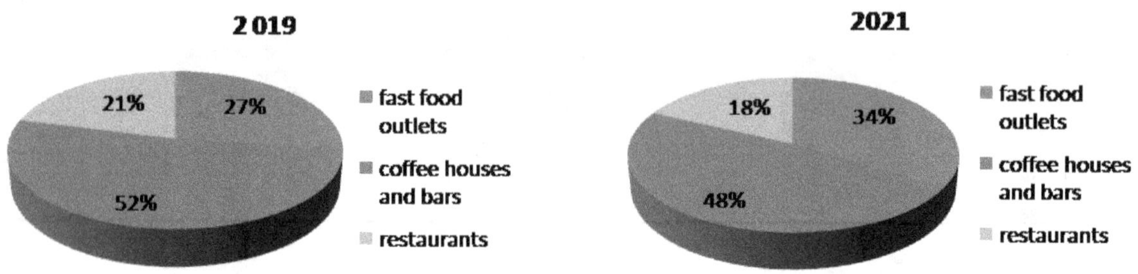

The most dynamic trend of post-COVID-19 recovery was demonstrated in the first half of 2021 by fast food outlets (Figure 4). Their share in the structure of the city's restaurant business has grown to 34% for a year and a half (vs 27% in pre- COVID-19 crisis).

Another noteworthy post-COVID-19 trend: lounge cafes with an anti-stress atmosphere became especially popular due to the COVID-19 crisis among Lviv residents and tourists. Such new-fashioned establishments of the city as "Trava lounge café", "Purple Lounge", "Hashtag Lounge Bar", "Hariacha Chasha Lounge Bar", "M'yata" and "Hubble Bubble" by 2021 gained the customer's commitment to brands and increased the number of serviced customers by 1.6-2.4 times.

Another post-COVID-19 trend: rapid growth (by 270% vs 2019) of small businesses in the "dark kitchen" format (online-only restaurant). They specialize exclusively in the delivery of semi-finished and ready meals, communicate with customers through chats in messengers (Viber, etc.) and actively promote virtual showcases and food art menus on the Internet.

Business and Lviv City Council are developing a strategy for rapid post-COVID-19 recovery of the potential of the city's restaurant, hotel and tourism industry by the end of 2022. In particular, measures are planned to comprehensively promote the Lviv brand as an inexpensive center of cultural and gastronomic tourism, a smart city on the map of Europe.

In 2021, popular among tourists restaurants, cafes and street food establishments in Lviv intensified work in the following directions:

- Large-scale digital transformation and expansion of the functionality of smart technologies in the city's restaurant service;
- Development of the presence of institutions of the urban hospitality cluster in the global Internet space;
- Development of augmented and virtual reality for co-creation of visitor's experience;
- Restoration of the range of gastronomic events: street food festivals, ethnic cuisine festivals, festivals of favorite dishes and drinks (including coffee, craft beer, cheese & wine, etc.), festivals-competitions of baristas and haute cuisine professionals, as well as fairs of ecological and local farm products;
- Revival of cooperation with fashionable food bloggers (social media sites: Instagram, YouTube, Facebook, Tik Tok, Twitter, etc.);
- Expanding the locations of fast and slow open-air cuisine;

- Rebranding and expansion of the menu of dishes of ethno-authentic cuisines of Ukrainian, Jewish, Armenian and other nations;
- Development of original products of gastronomic and entertaining tourism "made in Lviv"; the goal is to create memorable experiences.

Let's illustrate with some examples:

1. The innovative fast food restaurant "AppFood" was opened in December 2019 by an IT businessman. The feature of the restaurant is the lack of waiters. All a guest needs is his smartphone. The visitor downloads the application, chooses what he likes, pays with his smartphone and receives his order in 3-4 minutes.
2. Craft & Kumpel' – Futura Hub Restaurant was opened in 2019. In 2020, the restaurant entered the top 10 in popularity due to its creative concept of service. It is located in the Futura Hub in Lviv – Event and Art space, which implements the concept of "Eat & work", because for the birth and implementation of creative ideas need these components. In compliance with the quarantine requirements, it is offered to combine gourmet food with cultural and business events and Coworking Platforma.

The start of a number of new investment projects for the development of restaurant infrastructure in 2022 has been announced. Among them, for example, the Norwegian restaurant "Egersund" with live fish in aquariums, which is caught to order in the presence of the client and cooked according to Scandinavian recipes.

In 2021, the City Council and businessmen of the hospitality cluster restarted the tradition of holding colorful gastronomic festivals in Lviv, in particular:

- On May 14-16, 2021 Lviv effectively hosted the 8th Craft Beer & Vinyl Music Festival with the participation of 46 breweries and 7 more home brewers from all regions of Ukraine. Thanks to tourist announcements, more than 5,000 tourists arrived in Lviv for this holiday. In total, more than 17,000 people took part in tastings and music shows.
- On September 1-3, 2021, the IV Street Food So Good Festival took place in the Khmelnytsky Landscape Park of Culture and Recreation in compliance with the COVID-19 restrictions. For 3 days the festival was visited by about 35 thousand gourmets.

To successfully overcome the destructive consequences of the COVID-19 crisis, the city and business have declared joint actions to promote an information campaign in 2021-2022 through TV and Internet communication channels about Lviv as a destination where, in compliance with the quarantine requirements, they are waiting for tourists and have something to offer them.

Today, Lviv restaurateurs invest heavily in online advertising; they work closely with the Association of Guides and private guides, tour operators and sites for booking tourist services in Lviv. In the post-COVID-19 era, the norm for more than 80% of gastronomic tourism establishments in Lviv by the end of 2022 will be the introduction of QR-codes with access via smartphone to the English-language menu with a description of ingredients and local traditions of cooking and serving. In 2022 it is planned to include the restaurant infrastructure of Lviv in the free application RSRVIT (on the Google Play platform) for online booking of tables and dishes with the ability to search for places by geolocation or cuisine.

CONCLUSION AND RECOMMENDATIONS

Since the middle of spring 2021, the lost positions of the restaurant business in Ukraine have been actively restored. There is a rebranding, design and service re-equipment of existing restaurants. Premises of most closed restaurants in large cities are bought by investors and reorganized into fundamentally new restaurants & leisure. Foreign investors are no less active in the Ukrainian market. For example, on April 27, 2021, McDonald's opened the first restaurant in Ivano-Frankivsk; on May 21, the 100th restaurant of the chain in Ukraine was opened in Kharkiv – it operates in the format of a food court, has seven self-service terminals. It will be recalled that the first McDonald's in Ukraine was opened on May 24, 1997 in Kyiv. The press service on the official website of McDonald's announced for 2021 to invest in Ukraine in the opening of new facilities and renovation of existing nearly $ 42 million.

Such activity of the market leader encourages other restaurant brands to develop their own business in Ukraine more actively. For example, the press service on the official website of the Lviv Holding of Emotions "!FEST" reported that in 2020 the company successfully overcame the crisis not only without closing any restaurants in the chain, but in 2020 opened 12 new restaurants and bars in other cities of Ukraine (mostly franchising) (Holding of Emotions !FEST, 2021).

The status of the cultural and tourist capital saved the restaurant industry of Lviv from complete collapse. Thanks to the acquired image and targeted promotional campaigns in 2020-2021, tourists from other cities and regions of Ukraine during periods of easing quarantine restrictions actively visited Lviv and enjoyed a range of original local entertainment and gastronomic products, drinks and dishes of ethnic cuisines, locations of slow gastronomic tourism for gourmets.

From the middle of 2021, the following trends are expressed, which will be inherent in the post-COVID-19 prospects for the development of the restaurant industry in Lviv until 2030:

- Proactive marketing,
- Deepening cooperation with global Internet platforms (Google, TripAdvisor, Booking, etc.),
- Support for food bloggers,
- Development of Internet services and mobile applications,
- Intensification of communication with potential consumers through social networks and messengers,
- Digitalization, QR-menus and contactless services in catering establishments,
- Development of loyalty programs,
- Providing delivery services in restaurants (TripAdvisor Lviv Delivery Restaurants, 2021),
- Launching new projects and hybrid formats, primarily in the casual dining segment,
- Development of new emotional, animated and local-kitchen brands,
- Global popularization of Lviv gastronomic and tourist Customer Experience,
- Development of event and gastronomic tourism, participation in city festivals with a gastronomic component,
- Development of collaboration and partnership within the urban & suburban hospitality cluster and production of local products,
- Strengthening competition at the level of concepts, locations and services,
- Promotion of Gastro Getaway (Gastro & Relax Getaway) products,
- Development of a "dark kitchen" network,
- Staff development and improvement of company standards, etc.

The analysis showed that, despite the pandemic, the instability of the military-political and economic situation in the country, the forecast for the future development of the restaurant industry in Lviv – a representative tourist city in Eastern Europe – is optimistic. The market of gastronomic service in Lviv has favorable conditions for its recovery and successful development. First of all, it is due to the significant growth of the city's tourist popularity in the world, the post-COVID-19 resumption of intensive passenger flights by low-cost, global fashion for gastronomic tourism and knowledge of authentic culinary traditions, global digitalization of information about the range of original gastronomic products in Lviv.

Lviv restaurateurs have been uniting to implement the strategy of post-COVID-19 recovery of the restaurant industry. The strategic goal in the post-COVID-19 era is to transform the traditional business partnership into a smart hospitality cluster in Lviv.

REFERENCES

Abbas, J., Mubeen, R., Iorember, P. T., Raza, S., & Mamirkulova, G. (2021). Exploring the impact of COVID-19 on tourism: Transformational potential and implications for a sustainable recovery of the travel and leisure industry. *Current Research in Behavioral Sciences*, *2*, 100033. doi:10.1016/j.crbeha.2021.100033

Altinay, L., Paraskevas, A., & Jang, S. (2015). *Planning Research in Hospitality and Tourism* (2nd ed.). Routledge. doi:10.4324/9781315723662

Bellini, N., Clergeau, C., & Etcheverria, O. (2018). *Gastronomy and Local Development: The Quality of Products, Places and Experiences*. Routledge. doi:10.4324/9781315188713

Beriss, D., & Sutton, D. (2007). *The Restaurants Book: Ethnographies of Where we Eat*. Berg Publishers. doi:10.5040/9781350044913

Bertan, S. (2020). Impact of restaurants in the development of gastronomic tourism. *International Journal of Gastronomy and Food Science*, *21*, 100232. doi:10.1016/j.ijgfs.2020.100232

Bittman, M., Oseland, J., & Bush, A. (2012). *Food Lover's Guide to the World: Experience the Great Global Cuisines*. Lonely Planet.

Brewer, P., & Sebby, A. G. (2021). The effect of online restaurant menus on consumers' purchase intentions during the COVID-19 pandemic. *International Journal of Hospitality Management*, *94*, 102777. doi:10.1016/j.ijhm.2020.102777 PMID:34785837

Brizek, M. G., Frash, R. E., McLeod, B. M., & Patience, M. O. (2021). Independent restaurant operator perspectives in the wake of the COVID-19 pandemic. *International Journal of Hospitality Management*, *93*, 102766. doi:10.1016/j.ijhm.2020.102766 PMID:33230361

Byrd, K., Her, E., Fan, A., Almanza, B., Liu, Y., & Leitch, S. (2021). Restaurants and COVID-19: What are consumers' risk perceptions about restaurant food and its packaging during the pandemic? *International Journal of Hospitality Management*, *94*, 102821. doi:10.1016/j.ijhm.2020.102821 PMID:34866742

Chang, H. S., Capuozzo, B., Okumus, B., & Cho, M. (2021). Why cleaning the invisible in restaurants is important during COVID-19: A case study of indoor air quality of an open-kitchen restaurant. *International Journal of Hospitality Management, 94*, 102854. doi:10.1016/j.ijhm.2020.102854 PMID:33776189

Činjarević, M., & Pestek, A. (2014). Tourist perceived image of local cuisine: The case of Bosnian food culture. *British Food Journal, 116*(11), 1821–1838. doi:10.1108/BFJ-01-2014-0046

Cronin, C. J., & Evans, W. N. (2021). Total shutdowns, targeted restrictions, or individual responsibility: How to promote social distancing in the COVID-19 Era? *Journal of Health Economics, 79*, 102497. doi:10.1016/j.jhealeco.2021.102497 PMID:34325285

Danylyshyn, B. (2020). The peculiarities of economic crisis due to COVID-19 pandemic in a developing country: Case of Ukraine. *Problems and Perspectives in Management, 18*(2), 13–22. doi:10.21511/ppm.18(2).2020.02

Dedeoğlu, B. B., & Boğan, E. (2021). The motivations of visiting upscale restaurants during the COVID-19 pandemic: The role of risk perception and trust in government. *International Journal of Hospitality Management, 95*, 102905. doi:10.1016/j.ijhm.2021.102905

Dirsehan, T., & Cankat, E. (2021). Role of mobile food-ordering applications in developing restaurants' brand satisfaction and loyalty in the pandemic period. *Journal of Retailing and Consumer Services, 62*, 102608. doi:10.1016/j.jretconser.2021.102608

Edsor, B. (2017, November 30). 26 European cities where you can live on less than £600 a month. *The Independent.* https://www.independent.co.uk/travel/news-and-advice/european-cities-live-less-ps600-month-dnipro-lviv-kharkiv-pristina-craiova-a8084766.html

Erkuş-Öztürk, H., & Terhorst, P. (2016). Innovative restaurants in a mass-tourism city: Evidence from Antalya. *Tourism Management, 54*, 477–489. doi:10.1016/j.tourman.2016.01.003

Etcheverria, O. (2020). *The Restaurant, a Geographical Approach: from invention to gourmet tourist destinations*. John Wiley & Sons.

European Bank Report. (2020). *The COVID-19 pandemic and implications in the tourism sector in Ukraine*. https://www.ntoukraine.org/assets/files/EBRD-COVID19-Report-UKR.pdf

Food and Agriculture Organization. (2020). *Q&A: COVID-19 pandemic–impact on food and agriculture.* https://www.fao.org/2019-ncov/q-and-a/impact-on-food-and-agriculture/en/

Food and Agriculture Organization of the United Nations & World Health Organization (2020). *COVID-19, and food safety. Guidance for food businesses*. Author.

Food and Drug Administration (2020). *Shopping for food during the COVID-19 pandemic - information for consumers*. FDA.

Ford, R. C., & Sturman, M. C. (2020). *Managing Hospitality Organizations: Achieving Excellence in the Guest Experience*. Sage Publications Inc.

Ganzaroli, A., De Noni, I., & van Baalen, P. (2017). Vicious advice: Analyzing the impact of TripAdvisor on the quality of restaurants as part of the cultural heritage of Venice. *Tourism Management, 61*, 501–510. doi:10.1016/j.tourman.2017.03.019

Gavilan, D., Balderas-Cejudo, A., Fernández-Lores, S., & Martinez-Navarro, G. (2021). Innovation in online food delivery: Learnings from COVID-19. *International Journal of Gastronomy and Food Science, 24*, 100330. doi:10.1016/j.ijgfs.2021.100330 PMID:34745390

Gössling, D., Scott, D., & Hall, M. (2020). Pandemics, tourism and global change: A rapid assessment of COVID-19. *Journal of Sustainable Tourism, 29*(1), 1–20. doi:10.1080/09669582.2020.1758708

Grunert, K. G., De Bauw, M., Dean, M., Lähteenmäki, L., Maison, D., Pennanen, K., Sandell, M. A., Stasiuk, K., Stickel, L., Tarrega, A., Vainio, A., & Vranken, L. (2021). No lockdown in the kitchen: How the COVID-19 pandemic has affected food-related behaviours. *Food Research International, 150*, 110752. doi:10.1016/j.foodres.2021.110752 PMID:34865770

Hakim, M. P., Zanetta, L. D., & da Cunha, D. T. (2021). Should I stay, or should I go? Consumers' perceived risk and intention to visit restaurants during the COVID-19 pandemic in Brazil. *Food Research International, 141*, 110152. doi:10.1016/j.foodres.2021.110152 PMID:33642018

Hancock, E. (2016, December 23). The 12 cheapest holiday destinations in the world. *Business Insider*. https://www.businessinsider.com/cheapest-holiday-destinations-to-visit-in-2017-2016-12

Henderson, E. (2019, February). Why Lviv makes the best weekend break this winter. *The Independent*. https://www.independent.co.uk/travel/48-hours-in/iviv-what-do-travel-guide-winter-weekend-break-away-ukraine-cheap-holiday-a7955531.html

Hernandez-Rojas, R., Folgado-Fernandez, J., & Palos-Sanchez, P. (2021). Influence of the restaurant brand and gastronomy on tourist loyalty. A study in Córdoba (Spain). *International Journal of Gastronomy and Food Science, 23*, 100305. doi:10.1016/j.ijgfs.2021.100305

Holding of Emotions! (2021). *FEST*. https://www.fest.lviv.ua

Hu, X., Yan, H., Casey, T., & Wu, C.-H. (2021). Creating a safe haven during the crisis: How organizations can achieve deep compliance with COVID-19 safety measures in the hospitality industry. *International Journal of Hospitality Management, 92*, 102662. doi:10.1016/j.ijhm.2020.102662 PMID:32904503

Ivanov, S., Gavrilina, M., Webster, C., & Ralko, V. (2016). Impacts of political instability on the tourism industry in Ukraine. *Journal of Policy Research in Tourism, Leisure & Events*. Advance online publication. doi:10.1080/19407963.2016.1209677

Kim, K., Bonn, M. A., & Cho, M. (2021). Clean safety message framing as survival strategies for small independent restaurants during the COVID-19 pandemic. *Journal of Hospitality and Tourism Management, 46*, 423–431. doi:10.1016/j.jhtm.2021.01.016

Kim, S., Park, E., & Xu, M. (2020). Beyond the authentic taste: The tourist experience at a food museum restaurant. *Tourism Management Perspectives, 36*, 100749. doi:10.1016/j.tmp.2020.100749

Kiptenko, V., & Doan, P. (2015). Destination Ukraine: Tourism litmus of the transition to a green economy. In M. Reddy & K. Wilkes (Eds.), *Tourism in the Green Economy*. Routledge. doi:10.4324/9781315885681

Kostynets, V., Kostynets, I., & Olshanska, O. (2021). Pent-up Demand's Realization in the Hospitality Sector in the Context of COVID-19. *Journal of International Students*, *14*(1), 89–102. doi:10.14254/2071-8330.2021/14-1/6

L'viv – the Ensemble of the Historic Centre. (2021). *UNESCO: Culture: World Heritage Centrest: World Heritage List*. https://whc.unesco.org/en/list/865

Lee, C., Hallak, R., & Sardeshmukh, S. (2019). Creativity and innovation in the restaurant sector: Supply-side processes and barriers to implementation. *Tourism Management Perspectives*, *31*, 54–62. doi:10.1016/j.tmp.2019.03.011

Lee, S., Bee-Lia Chua, & Han, H. (2020). Variety-seeking motivations and customer behaviors for new restaurants: An empirical comparison among full-service, quick-casual, and quick-service restaurants. *Journal of Hospitality and Tourism Management*, *43*, 220–231. doi:10.1016/j.jhtm.2020.04.004

Li, B., Zhong, Y., Zhang, T., & Hua, N. (2021). Transcending the COVID-19 crisis: Business resilience and innovation of the restaurant industry in China. *Journal of Hospitality and Tourism Management*, *49*, 44–53. doi:10.1016/j.jhtm.2021.08.024

Luo, Y., & Xu, X. (2021). Comparative study of deep learning models for analyzing online restaurant reviews in the era of the COVID-19 pandemic. *International Journal of Hospitality Management*, *94*, 102849. doi:10.1016/j.ijhm.2020.102849 PMID:34785843

Lviv Tourist Barometer. (2021). http://www.barometer.lviv.ua

Min, J., Kim, J., & Yang, K. (2021). How generations differ in coping with a pandemic: The case of restaurant industry. *Journal of Hospitality and Tourism Management*, *48*, 280–288. doi:10.1016/j.jhtm.2021.06.017

Morris, H. (2017, December 14). Where you should fly in 2018: The 12 most exciting new routes. *The Telegraph*. https://www.telegraph.co.uk/travel/news/new-flight-routes-2018-holiday-destinations/

Neise, T., Verfürth, P., & Franz, M. (2021). Rapid responding to the COVID-19 crisis: Assessing the resilience in the German restaurant and bar industry. *International Journal of Hospitality Management*, *96*, 102960. doi:10.1016/j.ijhm.2021.102960

Nhamo, G., Dube, K., & Chikodzi, D. (2020). Restaurants and COVID-19: A focus on sustainability and recovery pathways. In G. Nhamo, K. Dube, & D. Chikodzi (Eds.), *Counting the cost of COVID-19 on the global Tourism Industry* (pp. 205–224). Springer Nature Switzerland AG. doi:10.1007/978-3-030-56231-1_4

Nistoreanu, P. (2020). *New Trends and Opportunities for Central and Eastern European Tourism*. IGI Global. doi:10.4018/978-1-7998-1423-8

Panzone, L. A., Larcom, S., & She, P.-W. (2021). Estimating the impact of the first COVID-19 lockdown on UK food retailers and the restaurant sector. *Global Food Security*, *28*, 100495. doi:10.1016/j.gfs.2021.100495

Park, C. W., Sutherland, I., & Seul, K. L. (2021). Effects of online reviews, trust, and picture-superiority on intention to purchase restaurant services. *Journal of Hospitality and Tourism Management, 47*, 228–236. doi:10.1016/j.jhtm.2021.03.007

Post Office Celebration Break Report. (2019). https://www.postoffice.co.uk/travel-money/city-cost-celebration-report

Restaurants And Mobile Food Services Global Market Report. (2021). https://www.thebusinessresearch-company.com/global-market-reports

Rivera-Ferre, M. G., López-i-Gelats, F., Ravera, F., Oteros-Rozas, E., di Masso, M., Binimelis, R., & El Bilali, H. (2021). The two-way relationship between food systems and the COVID19 pandemic: Causes and consequences. *Agricultural Systems, 191*, 103134. doi:10.1016/j.agsy.2021.103134

Rodríguez-López, M.-E., Alcántara-Pilar, J.-M., Del Barrio-García, S., & Muñoz-Leiva, F. (2020). A review of restaurant research in the last two decades: A bibliometric analysis. *International Journal of Hospitality Management, 87*, 102387. doi:10.1016/j.ijhm.2019.102387

Rolim de Albuquerque Meneguel, C., Mundet, L., & Aulet, S. (2019). The role of a high-quality restaurant in stimulating the creation and development of gastronomy tourism. *International Journal of Hospitality Management, 83*, 220–228. doi:10.1016/j.ijhm.2018.10.018

Rutynskyi, M., & Kushniruk, H. (2020a). The impact of quarantine due to COVID-19 pandemic on the tourism industry in Lviv (Ukraine). *Problems and Perspectives in Management, 18*(2), 194–205. doi:10.21511/ppm.18(2).2020.17

Rutynskyi, M., & Kushniruk, H. (2020b). Coffee Tourism in Lviv in the Context of World Coffee Tourism. *Annales Universitatis Mariae Curie-Skłodowska. Sec. B., 75*, 87–113. doi:10.17951/b.2020.75.0.87-113

Sanabria-Díaz, J. M., Aguiar-Quintana, T., & Araujo-Cabrera, Y. (2021). Public strategies to rescue the hospitality industry following the impact of COVID-19: A case study of the European Union. *International Journal of Hospitality Management, 97*, 102988. doi:10.1016/j.ijhm.2021.102988 PMID:34092893

Senbeto, D. L., & Hon, A. H. (2020). The impacts of social and economic crises on tourist behaviour and expenditure: An evolutionary approach. *Current Issues in Tourism, 23*(6), 740–755. doi:10.1080/13683500.2018.1546674

Sim, R. (2009). Food, place and authenticity: Local food and the sustainable tourism experience. *Journal of Sustainable Tourism, 17*(3), 321–336. doi:10.1080/09669580802359293

Singh, P., Brown, D. M., Chelekis, J., Apostolidis, C., & Dey, B. L. (2021). Sustainability in the beer and pub industry during the COVID-19 period: An emerging new normal. *Journal of Business Research.* Advance online publication. doi:10.1016/j.jbusres.2021.11.066

Song, H. J., Yeon, J., & Lee, S. (2021). Impact of the COVID-19 pandemic: Evidence from the U.S. restaurant industry. *International Journal of Hospitality Management, 92*, 102702. doi:10.1016/j.ijhm.2020.102702 PMID:33024347

State Statistics Service of Ukraine. (2021). http://www.ukrstat.gov.ua

Sung, Y.-K., & Hu, H.-H., & King, B. (2021). Restaurant preventive behaviors and the role of media during a pandemic. *International Journal of Hospitality Management*, *95*, 102906. doi:10.1016/j.ijhm.2021.102906

Tomczewska-Popowycz, N., & Quirini-Popławski, Ł. (2021). Political Instability Equals the Collapse of Tourism in Ukraine? *Sustainability*, *13*(8), 4126. doi:10.3390u13084126

Tourism Office of Lviv City Council. (2021). https://city-adm.lviv.ua

TripAdvisor. (2021a). *BBQ Restaurants for Large Groups in Lviv*. https://www.tripadvisor.com/Restaurants-g295377-c6-zfp9-Lviv_Lviv_Oblast.html

TripAdvisor. (2021b). *Cafés in Lviv*. https://www.tripadvisor.com/Restaurants-g295377-c8-Lviv_Lviv_Oblast.html

TripAdvisor. (2021c). *Restaurants near Old Town, Lviv, Ukraine*. https://www.tripadvisor.com/RestaurantsNear-g295377-d598874-Old_Town-Lviv_Lviv_Oblast.html

TripAdvisor: Lviv Delivery Restaurants. (2021). https://www.tripadvisor.com/Restaurants-g295377-zfp19-Lviv_Lviv_Oblast.html

TripAdvisor: Lviv Late Night Restaurants. (2021). https://www.tripadvisor.com/Restaurants-g295377-zfp8-Lviv_Lviv_Oblast.html

TripAdvisor: Lviv Romantic Restaurants. (2021). https://www.tripadvisor.com/Restaurants-g295377-zfp3-Lviv_Lviv_Oblast.html

Tsai, C.-T., & Lu, P.-H. (2012). Authentic dining experiences in ethnic theme restaurants. *International Journal of Hospitality Management*, *31*(1), 304–306. doi:10.1016/j.ijhm.2011.04.010

Wen, H., Leung, X., & Pongtornphurt, Y. (2020). Exploring the impact of background music on customers' perceptions of ethnic restaurants: The moderating role of dining companions. *Journal of Hospitality and Tourism Management*, *43*, 71–79. doi:10.1016/j.jhtm.2020.02.007

Wisła, R., & Nowosad, A. (2020). *Economic Transformation in Poland and Ukraine. National and Regional Perspectives*. Routledge. doi:10.4324/9781003041375

World Health Organization. (2020b). *Water, Sanitation, Hygiene and Waste Management for COVID-19: Technical Brief.* https://www.who.int/publications-detail/watersanitation-hygiene-and-waste-management-for-covid-19

Yiqi, W., & Jewoo, K. (2021). Interconnectedness between online review valence, brand, and restaurant performance. *Journal of Hospitality and Tourism Management*, *48*, 138–145. doi:10.1016/j.jhtm.2021.05.016

Yost, E., & Cheng, Y. (2021). Customers' risk perception and dine-out motivation during a pandemic: Insight for the restaurant industry. *International Journal of Hospitality Management*, *95*, 102889. doi:10.1016/j.ijhm.2021.102889

Yost, E., Kizildag, M., & Ridderstaat, J. (2021). Financial recovery strategies for restaurants during COVID-19: Evidence from the US restaurant industry. *Journal of Hospitality and Tourism Management, 47*, 408–412. doi:10.1016/j.jhtm.2021.04.012

Zhai, S. (2015). Mapping the popularity of urban restaurants using social media data. *Applied Geography (Sevenoaks, England), 63*, 113–120. doi:10.1016/j.apgeog.2015.06.006

KEY TERMS AND DEFINITIONS

COVID-19 Crisis: A global complex crisis caused by the COVID-19 pandemic. The outbreak began in December 2019 in Wuhan (Hubei, China), and was recognized by the WHO as a pandemic on March 11, 2020. Restaurant industry has faced huge financial losses due to international quarantine measures.

Gastronomic Tourism: It is a type of tourism associated with acquaintance with national and local culinary traditions of the world and with the tasting of local cuisine. Gastronomic tourism, in addition to the actual consumption of unusual (exotic) dishes and drinks, includes a number of other events. Among them: culinary master classes; visiting festivals and fairs of food and drink; participation in thematic gastronomic excursions to local cuisine establishments; participation in specialized gastronomic ceremonies; purchase of local food products; visits to farms and plantations (coffee, tea, etc.), food markets and producers of popular local food.

Local Cuisine: It is a cuisine based on the traditions of cooking and serving dishes typical of the culture of different nationalities and local communities. Features of local cuisine: geographical mosaic of ethnocultures of the world; the influence of different natural and agro-climatic living conditions, mentality, religious and ritual and other traditions; geographical diversity of historical traditions and cooking, etc. As a rule, local cuisine has noticeable differences from the national cuisine of the country.

Lviv (Ukraine): The city is located in the west of Ukraine, near the eastern border of the EU. It is one of the most representative tourist cities in Eastern Europe. Lviv is a fashionable and inexpensive destination of international gastronomic tourism. The historical center of Lviv is included in the list of the UNESCO World Heritage Sites.

Pre-COVID-19 Era: A period of stable dynamic development of the restaurant industry in the 21st century (except for the collapse of 2008 due to the financial crisis), which lasted until March 2020, that is, before the global COVID-19 crisis.

Post-COVID-19 Perspectives: A period of recovery of the potential of the global and regional markets of the restaurant industry and gastronomic tourism after the lifting of quarantine travel restrictions imposed by governments around the world in 2020-2021 to counter the spread of the COVID-19 pandemic.

Restaurant Industry: It is an industry of restaurants, cafes, bars, fast food and other food service providers that allow customers to enter, order food, and eat on the premises, as well as order takeaway food. The restaurant industry is part of the hospitality industry.

Ukrainian Ethnic Cuisine: It is one of the richest and most colorful national cuisines in Central and Eastern Europe. Today, Ukraine is at the stage of international legal consolidation of the top ten of its food geographical indications. And such restaurant dishes of Ukrainian cuisine as borshch, galushky, varenyky with cherries or cheese, potato pancakes (deruny), Carpathian dishes with white mushrooms and others are recognized tourist cards of the state.

Chapter 13
Co-Creation of Destination Image and Food Tourism of the Emerging Economies

Rahmat Hashim
https://orcid.org/0000-0003-0082-4248
Taylor's University, Malaysia

Mohd Raziff Jamaluddin
Universiti Teknologi MARA, Malaysia

ABSTRACT

The changing foodscape in society has redefined the role of food beyond the biological needs for survival. Food has always played an essential role in travel experiences; nevertheless, the concept of visiting a place just for its food is claimed to be a recent phenomenon. With the break-out of the COVID-19 pandemic, food tourism and branding have been affected worldwide, including Malaysia. Based on the literature review related to food tourism during the pandemic period, this chapter attempts to contribute to the food tourism literature by combining the concept of co-creation and image building. The Malaysian tourism industry needs to be transformational to continue playing a vital role in the economy in the endemic period. Food tourism that combines cultural and heritage, experience, and sustainable aspects co-created between relevant parties will become the critical success factor to educate and offer a meaningful experience to tourists in terms of heritage and traditional food.

INTRODUCTION

Tourism is a dynamic industry that touches on many different elements of human existence. As a result, every human activity may be converted and promoted into tourism goods and services that also made the industry the single largest industry globally. However, studying tourism's impact on humans and the environment has become fascinating but challenging. In the early phases of tourism development, food is taken for granted, and the subject of food and tourism is not a critical research area among tourism

DOI: 10.4018/978-1-7998-9148-2.ch013

and social scientists (Cohen & Avieli, 2004). The changing foodscape in society has redefined the role of food beyond the biological needs for survival.

As the world progressed after WWII, the advancement of diplomacy between nations produced stiff competition among tourism sites, leading many nations to adopt a differentiation strategy in order to carve their own identity. Food has become a symbol of national identity and pride in most countries. According to Anderson et al., (2017), "Food and tourism are, and have always been, closely integrated." Notably, food is viewed as a strong reflection mirror of society. Like enotourism, gastronomy tourism is moving towards becoming a popular tourism product and research topic because of its propensity to become a niche market that capitalizes on the economic, heritage, and ecotourism market (Lee & Scott, 2015).

Meal experience can become a powerful marketing communication tool (even though, sometimes, food-related activities are not the main travel motive) in describing the broader experiential benefits of a destination of its socio-cultural impressions. Gastronomy or food tourism, particularly heritage food, promotes a nation's identity through differentiation and sustainable destination development. As Andersson et al. (2017) mention, the strong relationship between food and local culture is a symbol and a sense of quality and authenticity for the tourists to engage (Borghini, 2014). Since the diverse food consumption behavior on holidays reflects the tourist's unique experiences, understanding food consumption and demands is a highly complex issue. Destination marketers and government agencies must understand tourists' food-related activities' demand and supply aspects to create a memorable experience to cherish (Lee & Scott, 2015).

Food tourism provides an opportunity for a developing country. Malaysia, for example, is ethnically diverse, with ancestral traditions, culture, native language, and culinary history still maintained today. Over time, ethnic mixing led to distinct Malaysian food culture. This chapter aims to revisit the concept of food tourism in reviving Malaysian food tourism post-pandemic period. This will be achieved through looking at food tourism from the co-creation lens to reposition Malaysian brand image as a destination of choice. The chapter starts by looking at the concept of food tourism. Next, it will be preceded by discussing the issues and challenges in the co-creation concept in food tourism. The chapter then moves on to discuss food tourism and sustainability. The chapter concludes with some suggestions and recommendations.

FOOD TOURISM

Omar *et al.* (2020) noted that food, gastronomy, or culinary tourism originated from the same concept and evolved into a significant cultural indicator for a tourism destination and has become an essential part of sustainable destination image development. Specifically, promoting regional food through tourism has the potential for local economic prosperity. Wolf describes:

Tourism inter-linked with local foods helps support the livelihoods of local agricultural producers and promotes the maintenance of high quality and purity in food and drink. When tourists fill otherwise empty restaurant tables, more sales are made. more profit, and more capital to reinvest in the community (Wolf, 2002, p. 11).

Hall and Mitchell (2001,p. 308) define food tourism as the

...visitation to primary and secondary food producers, food festivals, restaurants and specific locations for which food tasting and/or experiencing the attributes of specialist food production regions are the primary motivating factor for travel.

Many countries and regions across the globe recognize the potential provided by food and tourism. According to Lynch (2021), "food is powerful. it can shape your journey and define your memories." on the other hand, Jones and Jenkins say that

Food is now used to develop new niche markets, support regional identities, and develop quality and sustainable tourism. Therefore, food has developed from a basic necessity for tourist consumption to an essential element of regional culture.

(Jones & Jenkins, 2002, p. 115)

Long first introduced the concept of culinary tourism in 1998 as a manifestation of an appreciation of food culture (Wolf, 2002). Long defines culinary tourism as

...adventurous eating, eating out of curiosity, exploring other cultures through food, intentionally participating in the foodways of an "other," and developing food as tourist destination and attraction."

It is also common to see the term gastronomy and culinary, which focuses on the motivational factor of the visitation. For example, Kivela and Crotts (2005, p. 42) use the term 'gastronomy' tourism as

...traveling for the purpose of exploring and enjoying the destination's food and beverage and to savour unique and memorable gastronomy experiences.

This has resulted in many countries pushing the idea for competitive advantage and differentiating their destinations (Sotiriadis, 2015). For example, "gastrodiplomacy" was initially found in The Economist magazine in 2002 to describe the Thai government's promoting Thai food through an ambitious "Global Thai" strategy. Based on Miles (1998) and Bourdieu et al. (2004), the development of tourism consumption is has evolved from serving to the elites, shifted from catering to the elites to the masses (consumption by undifferentiated tourists), and now the post-Fordist model to address the diverse tourists' demands where tourism consumption is becoming increasingly aesthetics (Urry, 1995, 151).

The impact of globalization has negative and positive implications for developing and emerging economies. Globalization is more than just about trade; it also has a cultural component. To some, globalization contributes to the expansion of the Western hegemony. As a result, many ethnic communities' national and cultural identities are undermined (Hodos, 2010). Nevertheless, the era of McDonaldization was considered an encroachment of the local culinary scene. For example, Tomlinson's (1991) cultural imperialism and Ritzer's (1995) McDonaldization are believed to lead to homogeneity of 'global taste and cuisine' (Richards, 2003). On the other side of the coin, globalization has made food available and shared with people all over the world, giving rise to the concept of "cultural diplomacy" or "gastrodiplomacy" (Ichijo & Ranta, 2016, p. 108). Glocalization, like globalization, has kept pace with the evolving of "both universalizing and particularising impulses" to protect the identity of local culture, products, and history (Robertson, 2012).

According to Getz and Robinson (2014), the glocalization of local food dramatically influences the tourists' destination choices and their intention to revisit the destination. Today, tasting local food was not considered an impediment to the tourist experience. Before, tourists wanted to be in a familiar environment, including their meal (Cohen & Avieli, 2004). Apart from the taste, tourists sometimes find food hygienic standards and sustainable practices challenging when traveling. On the other hand, food tourism should be fostered to promote long-term development within a place. According to many food social scientists, food can transfer interests within the domains of socio-cultural and political agendas, fueled by its recent inclusion in heritage and cultural management frameworks. Today, food, culinary, or gastronomy tourism is a significant cultural heritage product, alongside architecture, monuments, events, and many others (Ellis et al., 2018; Grimaldi et al., 2019). According to UNESCO's records, Mexico was the first nation to submit their gastronomic cuisine to be included in the 2005 Proclamation of Masterpieces of the Oral and Intangible Heritage of Humanity (Oliveira et al., 2018).

Nonetheless, encapsulating a visitor experience at a particular destination concerning food is a psychologically complex process. It is challenging to provide a concise description of the idea because it involves many features. First, it is better to grasp the nature of food tourism and its characteristics to realize the interweaving nature. To ensure that the tourist industry performs its critical role in bolstering the nation's cultural heritage asset and economy, the tourism authorities or government must distinguish food tourism as a distinct type of tourism to develop a strong brand formation and identity.

MAIN FOCUS OF THE CHAPTER

Issues, Controversies, Problems

Co-creation and Food Tourism Experience

The concept of co-creation in tourism was introduced by Campos to indicate that

A co-creation tourism experience is the sum of the psychological events a tourist goes through when contributing actively through physical and/or mental participation in activities and interacting with other subjects in the experience environment. (Campos et al., 2016, p. 3)

Tourism co-creation is a construct that provides narrative experiences to promote tourists and tourism providers by stressing the tourist's need as a crucial success factor in tourism. The primary objective is to ensure that the tourist enjoys integrating the host's daily life, resulting in a joyful interaction. According to the United Nations World Travel Organization (2012), the co-creation of tourism is when "tourists and visitors organize their journeys partially or entirely to enjoy the cuisine of the location or to carry out activities related to gastronomy." Etymologically speaking, there are differences when various terms are used to represent this alternative tourism. In 2010, UNESCO acknowledged food as an element of intangible heritage (Miguel-Molina et al., 2016). In this chapter, the word food tourism will be adopted to represent the various terms used previously. This umbrella term (Figure 1) best describes those motivated by a desire and need to learn about a destination's culture and people through its unique food or eating as part of their travel itinerary (Hall et al., 2003).

Figure 1. Food Tourism as a Special Interest Tourism
Source: Hall et al. (2003)

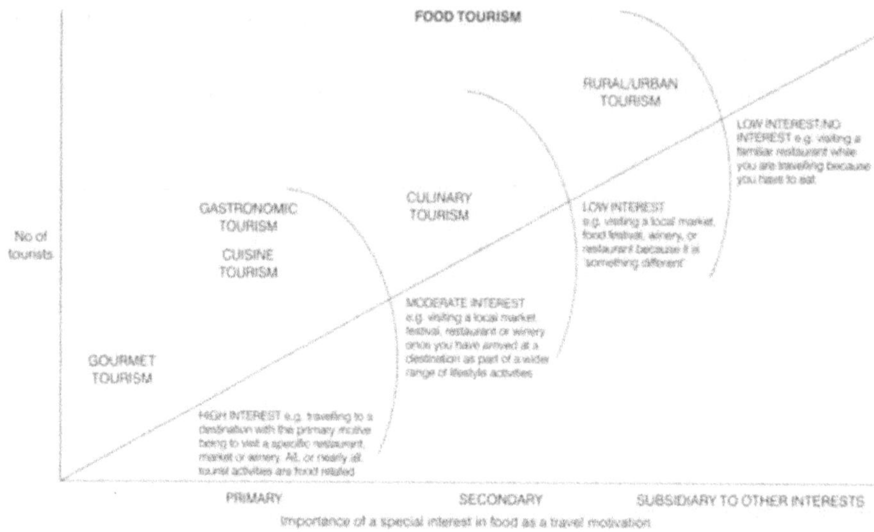

However, critics contend that co-creation can only exist when there are direct interactions on the consumers' level of participation in the co-creation process and the assessment of value that will be generated. According to Chathoth et al. (2013), value creation is mostly production-centric, originating in the manufacturing process rather than the consumption phase. While co-creation participation is occasional, customer-driven customization participation is continuous. Manufacturing and service researchers disputed whether a firm's resources should be directed around the process of value creation.

Food Tourists

The changing foodscape in society has redefined the role of food beyond the biological needs for survival. People now are willing to travel, near or far, just to experience food. Food tourists are the most committed tourists to experiencing local cuisine as part of their travel itinerary. Hall and Mitchell (2001, pp. 74-75) have categorized the food tourist (Figure 1) as to how vital food is to them as a reason to travel. Gourmet/cuisine and gastronomic tourism are the first two subsets of great interest. This is often linked with upscale restaurants, be it a stand-alone or hotel restaurant, which serves French or Western cuisine. An upscale restaurant, in particular, has been associated with esteem and status.

The second subset of food tourism ideals is tourists who do not have food as their primary concerns as a meaningful experience (culinary, rural/urban tourism). These are the group of individuals who love to explore and see the different activities or "because it is something different when food is concerned. Within this subset, another group considers food a biological need and is not interested in learning what the locals eat. The second subgroup of food tourism principles (labeled culinary, rural/urban tourism) caters to visitors who are not interested in food as a personal and meaningful experience. When it comes to food, this group of people enjoys exploring and seeing 'new things' about a place. Another group

Figure 2. The scale of Market Entities
Source: Levitt, 1972

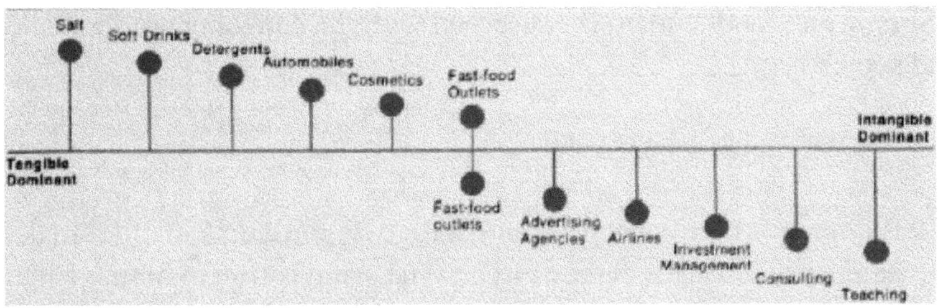

within this subgroup views food as a biological requirement and is uninterested in learning the local's diet and consumption.

Tourism is a service industry; hence, most of its products are intangible. Food tourism is not about dining conspicuously in the upscale restaurant, not mainly about eating. It is instead an experience of dining and learning. Food tourism entails much more than a list of restaurants serving haute cuisine with excellent services. Food tourism needs to also focus on a range of activities, events, and heritage explorations. It refers to any action involving food to bring people, places, and time together (Hall et al., 2003). According to Anton et al. (2019), a location's culinary component attracts tourists and helps them understand its uniqueness and distinguishing characteristics. Hence, authentic cuisines, unique culinary outlets within a location such as street food, night bazaars, and food and beverage festivals, in particular, can boost a destination's appeal (Getz & Robinson, 2014; Henderson, 2009; Sotiriadis, 2015).

Characteristics of Food Tourism

Many have claimed that tourism products are intangible, and the tourism industry is a service industry. As mentioned earlier, tourism is the sole industry that involves any tourist-drawing activities. This may be one of the reasons why the industry is recognized as the single largest. Nevertheless, the tourism industry is a combination of tangible and intangible products. According to Levitt (1972) "There are no such things as service industries, and there are only industries whose service components are greater or less than those of other industries. Everybody is in service."

Kotler and Keller (2016), on the other hand, describe the service industry as

A service is any activity or benefit that one party can offer to another that is essentially intangible and does not result in the ownership of anything. Its production may or may not be tied to a physical product.

The diagram above depicts which industry is physical or intangible dominant. The unique nature of the tourism industry is well made up of both the tangible and intangible aspects of a purchase. This holds true in most tourism products and services. At the other extreme of goods, dominant offerings are pure services, in which the product consists of predominantly intangible services like legal travel advice and consulting. In other situations, a traveler, for example, dines at a restaurant and order her or his favorite food. Because once food (tangible) is consumed, there is no physical evidence that one has eaten the food. The restaurant's service is the only memory the person may take with them when they return home. Food's social and cultural importance in tourism has also been recognized as a critical component in a traveler's experience visiting a place (Long, 1998; Hall & Mitchell, 2001; Haven-Tang & Jones, 2006). Simply put, food tourism is concerned with the enjoyment of food and its associations with locations and people.

Authenticity

Trilling (1972, p.93) introduced the concept of authenticity to museums in order to determine if an art object appeared to be precisely what it claimed to be. The same is true in heritage tourism. If heritage tourism is to be as significant as other forms of tourism, Cohen (1988) believes that authenticity must be one of the essential elements. As a form of cultural heritage (Bessière, 1998), food tourism must be promoted as a genuine (MacCannel, 1973). MacCannel (1973) deserves the credit for bringing the notion of authenticity into the study of tourism (Cohen, 1988; Hughes, 1995; Reisinger & Steiner, 2006). This happens when social scientists criticize the tourist sector for commodifying cultural assets due to modernization and development (MacCannel, 1973). The issue remains whether food tourism as a strategy to preserve tradition or staged authenticity' (MacCannell, 1973). 'Stage authenticity' has no place in today's travelers since they expect to be co-created by the host and tourist to have a personally meaningful experience (Prahalad & Ramaswamy, 2004). It all comes down to a natural desire and striking a balance to promote cultural heritage to people seeking authenticity or existential authenticity (Wong, 2015).

Tourist Co-creation and Experience

The concept of co-creation is becoming more relevant and perceived as an essential element in today's tourism operating; the notion of co-creation is becoming increasingly significant and seen as a vital aspect in tourism functioning in an experience economy (Pine & Gilmore, 2011; Suntikul & Jachna, 2016). Participative experiences allow tourists to participate actively in different environments to develop self-consciousness and connect with fellow tourists and the locals rather than merely gaze at them (Eraqi, 2011). The travelers have to deliver the service actively; otherwise, the service delivery will be thwarted.

Food Tourism Providers in Post-COVID-19 for Emerging Economies

In Malaysia, food tourism branding posts the COVID-19 pandemic need to improve its public image and brand awareness among its prospective international visitors. The ethnic cuisines of Malaysia will be used to brand its destinations post the COVID-19 pandemic through a meaningful relationship between consumption experiences, the community, and food in the area (Lin et al., 2011). Arguably, many tourists are attracted to explore ethnic and regional food as part of their preference to have an outstanding experience. In conclusion, Malaysia's food tourism and branding post the COVID-19 pandemic will be characterized by redefining the concept of quality food and services offered. In addition, the post-COVID-19 study might look into customer return on investment, as seen in destinations like Penang and Bangkok, where food tourism is promoted as a food haven and a good value for money destination. Hence, potential tourists will be attracted to Malaysian destinations that adhere to the redefinition of dining experience and/or relationship with the communities and food heritage and quality.

According to Lai et al. (2019), gastronomic tourism entails traveling to participate in a particular culture associated with certain people and places while eating and drinking. However, with the break-out of the COVID-19 pandemic, food tourism and branding have been affected worldwide, including Malaysia. The food and beverage manufacturing sector deals with a COVID-19 problem that affects operations, safety, supply chain management, training, emergency response, awareness, incident management, business model reinvention, digitalization, and other unintended repercussions (Telukdarie et al., 2020). Furthermore, COVID-19 has influenced customer perceptions concerning food. Gerritsen et al. (2021) noted that adults who are experiencing "cognitive overload" during times of increased stress are even more susceptible to advertising, and it is clear that the COVID-19 pandemic was a time of unprecedented change in daily life, resulting in decreased psychological well-being for many people worldwide.

The pandemic entailed incremental and sometimes unexpected expenditures for food firms, which were obliged to use financial resources to keep items on shelves while simultaneously adapting to a new normal. To ensure a regular supply of components in the face of shortages, COVID-19 outbreaks, and transit delays, certain companies were obliged to overcome difficulties and change suppliers. Other industries, particularly those in the meat industry, have invested hundreds of millions of dollars in manufacturing worker safety and health protection. Customers began ordering food online through grocery stores, third-party shopping services, and direct-to-consumer websites.

This could provide an avenue for planners and authorities to self-reflect and revisit the charting of the future direction of tourism development, particularly food tourism. Hence, beyond the post-COVID-19, food tourism must be redefined to continue to play a crucial role in the total experience of travelers visiting Malaysia (Andersson et al., 2017) and sustainable initiatives. The industry will characterize the period beyond post-COVID-19 by health and safety issues and touchless technology and responsibility. Malaysia's food service businesses or restaurants must now focus on maintaining cleanliness and hygiene because the pandemic has changed tourists' food consumption behavior that demands a clean, safe, and sustainable environment.

Food Tourism and Sustainable Development

According to The Brundt Land Report, sustainable development is defined as "… development that meets the needs of the present without compromising the ability of future generations to meet their own needs" (United Nations, 1987). Tourism occupies a significant place in the global economy, representing a significant success in developing some countries. Tourism has expanded its forms because, over time, the inspirations of tourists have evolved. Food tourism, also referred to as gastronomic tourism, is one of the newest forms of tourism. Gastronomic tourism incorporates cultural, thematic elements, sustainable tourism, and aims to educate and satisfy tourists in terms of traditional and local cuisine and therefore brings multiple benefits to residents and helps the economic sector.

Ştefanovici and Mazilu (2021) explain that as a new form of tourism, food tourism is part of cultural tourism that shows the culture of a particular area and highlights traditional values. Food is a vital visitor attraction in various forms and enhances or is key to the tourist experience. It has taken a significant role in tourist satisfaction and making decisions, tourist products, and place promotion tactics. Therefore, it can be an essential tool of general and destination growth. The tourism industry and tourists across the globe have the same interest in food. At the base of food tourism are traditional discoveries and the values of residential communities and entities, combining traditional resident elements, gastronomic reserves by rediscovering regional and local values (Ştefanovici & Mazilu, 2021).

Food tourism is about much more than food. It mirrors various peoples' traditions, heritage, culture, and sense of community. It is a means of enhancing understanding among various cultures and bringing individuals and traditions closer together. Food tourism has been emerging as a significant protector of cultural heritage, and the sector helps build opportunities, including employment, particularly in rural destinations (Henderson, 2009). Gastronomy is part of people's cultural and social heritage, mirroring a particular lifestyle specific to each geographical area.

Food and tourism have a very tight connection, and food is a vital resource in tourism. It is essential for physical provisions, and all visitors must eat during their travels, but food can be a crucial draw and key influence for some, which gratifies a variety of psychological and other wants and necessities (Florek & Gazda, 2021). Food tourism promotes local and regional sustainable development. Eating places are a significant product of tourism, and drink and food channels may be built primarily for visitors or rely significantly on them. Visitors also take raw and uncooked foods from consistent stores and buy domain souvenir products and cuisine books. Farmer markets are increasing and attracting locals and tourists with their garden-fresh products, some of which are biological. Cooking institutions are another form of tourism food produce, occasionally run by or linked with famous cooks, often selling a blend of accommodation and tuition. Food plants and farms have also been opening their sites to the people, some revelatory building centers, and arts centers (Florek & Gazda, 2021).

Studies have found increasing gratitude of homegrown food by visitors, with a considerable number ready to pay more for a locally recognizable product. Research shows that there has been a surge in food eating and much more comprehensive gratitude for food. Food tourism has also been acting to lengthen the tourist period, with multiple benefits. Gastronomy tourism is a genuine avenue of tourism studies and can add a mislaid viewpoint to the knowledge on tourism, particularly within the cultural sustainability agenda (Florek & Gazda, 2021).

Florek and Gazda (2021) explain that food and food-connected businesses can be vital to establishing local identities, authenticating the hypothetical link between food and identity. Food tourism also gives an alternative means of regional and local development, with the possibility to reinforce identity, improve gratitude of the environment and inspire the restoration of the economy and local culture. Food tourism has led to increasing demand for characteristic, local products and has inspired the restitution of traditional celebrations, encouraged the rise of educational tourist attractions, relinked customers with the local regions, motivated the retaining and growth of agricultural and culinary skills (Florek & Gazda, 2021). Additionally, food tourism has provided the dying traditional businesses and small industries with a lifeline. There is evidence that food tourism can lead to the triple bottom line of social, financial, and environmental sustainability.

Food tourism provides a way of improving and lengthening the tourist without conceding an area's cultural, societal, or ecological fabric. It is advised that an improved comprehension of the food visitors in connection to typologies of visitor inspiration, behavior, and features should allow the growth of more improved decision-making on top of contributing to broader academic dissertations of sustainability and tourism (Ştefanovici & Mazilu, 2021).

Food tourism has much promise, but the tourist infrastructure must create favorable circumstances to thrive for it to be appreciated. A strategy ought to be adopted and executed to ensure the improvement and development of food tourism in tourist destinations. Food tourism has massive significance in rural regions, providing authenticity to regions, diversifying and revitalizing the tourist offer of tourist destinations (Ştefanovici & Mazilu, 2021). It plays a significant role in promoting and protecting community values, portraying local identity. Tourist destinations whose objective is to develop grounded on food tourism must turn their attention to the best possible quality of goods, ensuring that tourists recognize the local products. Food tourism could play an important role, and its characteristics have significantly contributed to local and regional sustainable development.

SOLUTIONS AND RECOMMENDATIONS

Increased sensitivity, complexity, and interdependency characterize world tourism in the 21st century. No industry in the world can function in a vacuum. Limited resources and intense rivalry among tourism destinations have forced tourism authorities to adopt a differentiation strategy to reflect their distinctive character to attract niche markets. As the COVID-19 pandemic is moving toward an endemic, it is apparent that Malaysia's tourism industry would need to reconsider its product and service offerings, as well as its marketing strategy. Malaysia's food tourism and branding post-COVID-19 pandemic will be characterized by redefining the concept of travel experience, quality food, and services offered. The relationship between food and tourism in destination differentiation attempts to introduce uniqueness and contribute to economic development and sustainability. Cuisine embodies the essence of a country's history, customs, heritage, and values, making it an essential component of its culture and identity to promote nation-building and nation-branding. Food has long been an essential component of human travel. However, food tourism has grown in popularity due to these competing concepts of diversity and cultural hegemony. Through research and strategic efforts, developing nations must use every opportunity to manage tourism more sustainably.

CONCLUSION

Food tourism promotes sustainability through environmental and socio-cultural authenticity in host communities, in conjunction with other key stakeholders. This will ensure long-term prosperity as suggested by the triple bottom line strategy (Elkington, 1998). Tourism has taken on new shapes as travelers' motivations have changed throughout time. Unlike other forms of tourism, food tourism is a fantastic experience that emphasizes sustainability, experiential learning, and inclusivity. Despite its negative perception of McDonaldization and other issues in 'stage authenticity' as a threat to local culinary identities, it has provided a chance to reinforce local culinary identities at the same time. Food tourism (or call it by any other terms) combines cultural, thematic, and sustainable aspects of an experience to educate and offer a meaningful experience to tourists in terms of heritage and traditional food, resulting in several benefits for locals and their socioeconomic status well-being. The COVID-19 pandemic has profoundly impacted the travel and tourist business worldwide. Tourists will avoid visiting high-density and crowded places to ensure a safe and healthy experience.

As the Malaysian tourism industry progresses to an endemic stage, it must adopt a strategy that promotes itself, offering a distinctive experience (brand identity) for visitors through co-creation and reality reconstruction.

REFERENCES

Andersson, T. D., Mossberg, L., & Therkelsen, A. (2017). Food and tourism synergies: Perspectives on consumption, production, and destination development. *Scandinavian Journal of Hospitality and Tourism, 17*(1), 1–8. doi:10.1080/15022250.2016.1275290

Anton, C., Camarero, C., Laguna, M., & Buhalis, D. (2019). Impacts of authenticity, degree of adaptation and cultural contrast on travellers' memorable gastronomy experiences. *Journal of Hospitality Marketing & Management, 28*(7), 743–764. doi:10.1080/19368623.2019.1564106

Bessière, J. (1998). Local development and heritage: Traditional food and cuisine as tourist attractions in rural areas. *Sociologia Ruralis, 38*(1), 21–34. doi:10.1111/1467-9523.00061

Borghini, A. (2014). Authenticity in Food. In Encyclopedia of food and agricultural ethics. Springer.

Campos, A. C., Mendes, J., Valle, P. O., & Scott, N. (2016). Co-creation of tourist experiences: A literature review. *Current Issues in Tourism, 21*(4). Advance online publication. doi:10.1080/13683500.2015.1081158

Chathoth, U. G. R., Harrington, R. J., Altinay, L., Okumus, F., & Chan, E. S. W. (2013). Conceptualization of value co-creation in the tourism context. In N. K. Prebensen, J. S. Chen, & M. S. Uysal (Eds.), Creating experience value in tourism. CABI.

Cohen, E. (1988). Authenticity and commoditization in tourism. *Annals of Tourism Research, 15*(3), 371–386. doi:10.1016/0160-7383(88)90028-X

Cohen, E., & Avieli, N. (2004). Food in tourism. *Annals of Tourism Research, 31*(4), 755–778. doi:10.1016/j.annals.2004.02.003

Elkington, J. (1998). *Cannibals with Forks: the triple bottom line of 21st century business*. New Society Publishers.

Eraqi, M. I. (2011). Co-creation and the new marketing mix as an innovative approach for enhancing tourism industry competitiveness in Egypt. *International Journal of Services and Operations Management, 8*(1), 76–91. https://www.inderscienceonline.com/doi/abs/10.1504/IJSOM.2011.037441

Florek, M., & Gazda, J. (2021). Traditional food products— Between place marketing, economic importance and sustainable development. *Sustainability, 13*(3), 1277. doi:10.3390u13031277

Gerritsen, S., Sing, F., Lin, K., Martino, F., Backholer, K., Culpin, A., & Mackay, S. (2021). The timing, nature and extent of social media marketing by unhealthy food and drinks brands during the COVID-19 pandemic in New Zealand. *Frontiers in Nutrition, 8*, 1–13. doi:10.3389/fnut.2021.645349 PMID:33748177

Getz, D., & Robinson, R. N. S. (2014). Foodies and food events. *Scandinavian Journal of Hospitality and Tourism, 14*(3), 315–330. doi:10.1080/15022250.2014.946227

Grimaldi, P., Fassino, G., & Porporato, D. (2019). Introduction. In C. Petrini (Ed.), Culture, heritage, identity and food: a methodological approach. Stampa: Geca Industrie Grafiche.

Hall, C. M., & Mitchell, R. (2001). Wine and food tourism. In N. Douglas, N. Douglas, & R. Derrett (Eds.), *Special interest tourism* (pp. 307–329). John Wiley and Sons.

Hall, C. M., & Sharples, L. (2003). The consumption of experiences or the experience of consumption: an introduction to the tourism of taste. In C. M. Hall, L. Sharples, R. Mitchell, N. Macionis & B. Cambourne (Eds.), Food tourism around the world: Development, management and markets. Butterworth-Heinemann.

Hall, C. M., Sharples, L., Mitchell, R., Macionis, N., & Cambourne, B. (2003). *Food tourism Around the World: Development, management and markets*. Butterworth-Heinemann.

Haven-Tang, C., & Jones, E. (2006). Learning provision, the labour market and skills needs of the tourism and related sectors in Wales. *Journal of Human Resources in Hospitality & Tourism*, *5*(2), 13–35. doi:10.1300/J171v05n02_02

Henderson, J. C. (2009). Food tourism reviewed. *British Food Journal*, *111*(4), 317–326. doi:10.1108/00070700910951470

Hodos, T. (2010). Globalization and colonization: A view from Iron Age Sicily. *Journal of Mediterranean Archaeology*, *23*(1), 81–106. doi:10.1558/jmea.v23i1.81

Hughes, G. (1995). Authenticity in tourism. *Annals of Tourism Research*, *22*(4), 781–803. doi:10.1016/0160-7383(95)00020-X

Ichijo, A., & Ranta, R. (2016). *Food, national identity and nationalism: From everyday to global politics*. Palgrave Macmillan.

Jones, A., & Jenkins, I. (2002). A taste of Wales-Blas Ar Gymru: institutional mailaise in promoting Welsh food tourism product. In A.-M. Hjalager & G. Richards (Eds.), Tourism and gastronomy. Routledge.

Kivela, J., & Crotts, J. C. (2005). Gastronomy tourism: A meaningful travel market segment. *Journal of Culinary Science & Technology*, *4*(2/3), 39–55. doi:10.1300/J385v04n02_03

Kotler, P., & Keller, K. L. (2016). *A framework for marketing management*. Pearson.

Lai, M. Y., Wang, Y., & Khoo-Lattimore, C. (2020). Do food image and food neophobia affect tourist intention to visit a destination? The case of Australia. *Journal of Travel Research*, *59*(5), 928–949. doi:10.1177/0047287519867144

Lee, K. H., & Scott, N. (2015). Food tourism reviewed using the paradigm funnel approach. *Journal of Culinary Science & Technology*, *13*(2), 95–115. doi:10.1080/15428052.2014.952480

Levitt, T. (1972). Production-line approach to service. *Harvard Business Review*, *50*(5), 41–52.

Lin, Y. C., Pearson, T. E., & Cai, L. A. (2011). Food as a form of destination identity: A tourism destination brand perspective. *Tourism and Hospitality Research*, *11*(1), 30–48. doi:10.1057/thr.2010.22

Long, L. M. (1998). Introduction to the special issue on culinary tourism. *Southern Folklore*, *55*(3), 179.

Lynch, L. (2021, April 26). Why I travel for food and so should you. *The Planet D: Adventure Travel Blog*. https://theplanetd.com/culinary-travel-why-i-travel-for-food/

MacCannell, D. (1973). Staged authenticity: Arrangements of social space in tourist settings. *American Journal of Sociology*, *79*(3), 589–603. https://www.jstor.org/stable/2776259

Miguel-Molina, M., Miguel-Molina, B., Santamarina, V., & Segarra-Oña, M. V. (2016). Intangible Heritage and Gastronomy: The Impact of UNESCO Gastronomy Elements. *Journal of Culinary Science & Technology*, *14*(4), 293–310. doi:10.1080/15428052.2015.1129008

Oliveira, B. S., Tricárico, L. T., Sohn, A. P. L., & Pontes, N. (2018). The culinary intangible cultural heritage of UNESCO: A review of journal articles in EBSCO platform. *Journal of Culinary Science & Technology*. Advance online publication. doi:10.1080/15428052.2018.1513882

Omar, S. R., Ab Karim, S., Isa, S. S., & Omar, S. N. (2020). Enlivening the effect of tourists' knowledge and involvement on Malaysian heritage food (MHF) towards food tourism image in Malaysia. *International Journal of Tourism and Hospitality Management in the Digital Age*, 268–290.

Pine, B. J., & Gilmore, J. H. (2011). *The Experience Economy, Uptade Edition*. Harvard Business Review Press.

Prahalad, C. K., & Ramaswamy, V. (2004). Co-creation experiences: The next practice in value creation. *Journal of Interactive Marketing*, *18*(3), 5–14. doi:10.1002/dir.20015

Reisinger, Y., & Steiner, C. J. (2006). Reconceptualizing object authenticity. *Annals of Tourism Research*, *33*(1), 65–86. doi:10.1016/j.annals.2005.04.003

Richards, G. (2003). Gastronomy: an essential ingredient in tourism production and consumption? In *M.-A. Hjalager, & G. Richards, Tourism and gastronomy* (pp. 17–34). Routledge. doi:10.4324/9780203218617

Ritzer, G. (1995). *The McDonaldization of society*. Sage Publication.

Robertson, R. (2012). Globalisation or glocalisation? *Journal of International Communication*, *18*(2), 191–208. doi:10.1080/13216597.2012.709925

Sotiriadis, M. D. (2015). Culinary tourism assets and events: Suggesting a strategic planning tool. *International Journal of Contemporary Hospitality Management*, *27*(6), 1214–1232. doi:10.1108/IJCHM-11-2013-0519

Ştefanovici, M., & Mazilu, M. (2021). The importance of gastronomic tourism in economic development of Romania. *Quaestus*, *18*, 365–376.

Suntikul, W., & Jachna, T. (2016). The co-creation/place attachment nexus. *Tourism Management*, *52*, 276–286. doi:10.1016/j.tourman.2015.06.026

Telukdarie, A., Munsamy, M., & Mohlala, P. (2020). Analysis of the Impact of COVID-19 on the Food and Beverages Manufacturing Sector. *Sustainability*, *12*(22), 9331. doi:10.3390u12229331

The Economist. (2002, February 21). Thailand's gastro-diplomacy. *The Economist*. https://www.economist.com/asia/2002/02/21/thailands-gastro-diplomacy

Tomlinson, J. (1991). *Cultural imperialism: A critical introduction*. Johns Hopkins University Press.

Trilling, L. (1972). *Sincerity and authenticity*. Oxford University Press.

United Nation. (1987). *Report of the World Commission on Environment and Development: Our Common Future.* http://www.un-documents.net/ocf-02.htm#I

United Nations World Travel Organization. (2012). *Global report on food tourism.* http://cf.cdn.unwto.org/sites/all/files/ docpdf/amreports4-foodtourism.pdf

Urry, J. (1995). *Consuming Places.* Routledge.

Wolf, E. (2002). *Culinary tourism: a tasty economic proposition.*

Wong, I. A. (2015). A multimethod multilevel study of heritage transmission: The role of culture on tourist interest and authenticity. *Journal of Travel Research, 54*(5), 672–685. doi:10.1177/0047287514532368

Compilation of References

Abbas, J., Mubeen, R., Iorember, P. T., Raza, S., & Mamirkulova, G. (2021). Exploring the impact of COVID-19 on tourism: Transformational potential and implications for a sustainable recovery of the travel and leisure industry. *Current Research in Behavioral Sciences*, *2*, 100033. doi:10.1016/j.crbeha.2021.100033

Abbaspourrad, A., Padilla-Zakour, O., Wiedmann, M., Moraru, C. I., & Goddard, J. M. (2017). Taking a systems approach to clean label challenges. *Food Technology*, *71*(11), 32–43.

Abubakar, A. M., Rezapouraghdam, H., Behravesh, E., & Megeirhi, H. A. (2021). Burnout or boreout: A meta-analytic review and synthesis of burnout and boreout literature in hospitality and tourism. *Journal of Hospitality Marketing & Management*, 1–46. doi:10.1080/19368623.2022.1996304

Aday, S., & Aday, M. S. (2020). Impact of Covid-19 on the food supply chain. *Food Quality and Safety*, *4*(4), 167–180. doi:10.1093/fqsafe/fyaa024

Adebo, G. M., & Ayelari, T. A. (2011). Climate change and vulnerability of fish farmers in Southwestern Nigeria. *African Journal of Agricultural Research*, *6*(18), 4230–4238. doi:10.5897/AJAR10.948

Advance London. (2020, June). *Food that doesn't cost the earth.* https://www.foodmadegood.org/wp-content/uploads/2020/06/Food_that_doesnt_cost_the_earth_Full_guide-1.pdf

Agir Sustentável. (n.d.). https://agirsustentavel.com.br/

Ahmad, A. (2013). *New age globalization: Meaning and metaphors.* Springer.

AHRESP. (2021a). *Federações europeias da restauração e do alojamento defendem aplicação generalizada da taxa reduzida de IVA em toda a Europa* [European food and accommodation federations advocate widespread application of the reduced RATE of VAT across Europe]. https://ahresp.com/2020/10/descida-do-iva-na-europa/

AHRESP. (2021b). *IVA reduzido constará no próximo orçamento de Estado?* [Will reduced VAT be included in the next State budget?]. https://ahresp.com/2021/08/iva-reduzido/

Akhshik, A., Ozturen, A., & Rezapouraghdam, H. (2021). A passionate travel to mind green turtles-Unpacking the complexity of visitors' green behavior. *International Journal of Tourism Research*, *23*(3), 301–318. doi:10.1002/jtr.2407

Aksoy, M., & Sezgi, G. (2015). Gastronomy Tourism and Southeastern Anatolia Region Gastronomic Elements. *Journal of Tourism and Gastronomy Studies*, *3*(3), 79–89.

Aldeia, S. C. R. (2019). The "true and fair view" concept in business and tax law. The Portuguese and Spanish case studies. *Academy of Accounting and Financial Studies Journal, 23*(2), 1-11. https://www.abacademies.org/articles/the-true-and-fair-view-concept-in-business-and-tax-law-the-portuguese-and-spanish-case-studies-8130.html

Aldeia, S. C. R. (2021). Application of the constitutional principle of generality in Spanish companies' taxation. A compared study to Portugal. International Journal of Law and Management. doi:10.1108/IJLMA-01-2021-0008

Alex Chin, J. C., Perera, M., Hui, K., Yen, H. L., Chan, M., & Poon, L. (2020). Stability of SARS-CoV-2 in different environmental conditions. *The Lancet Microbe*, *1*(1). Advance online publication. doi:10.1101/2020.03.15.20036673

Alipour, H., Olya, H. G., Hassanzadeh, B., & Rezapouraghdam, H. (2017). Second home tourism impact and governance: Evidence from the Caspian Sea region of Iran. *Ocean and Coastal Management*, *136*, 165–176. doi:10.1016/j.ocecoaman.2016.12.006

Alipour, H., Rezapouraghdam, H., & Akhshik, A. (2021). Heritage redemption and the curse of tourism: The case of world's last inhabited troglodyte village. *Tourism Planning & Development*, *18*(1), 68–85. doi:10.1080/21568316.2020.1775114

Alipour, H., Rezapouraghdam, H., & Esmaeili, B. (2020). Public health and tourism: A personalist approach to community well-being: A narrative review. *Iranian Journal of Public Health*, *49*(3), 447–453. doi:10.18502/ijph.v49i3.3127

Almeida, A. M., Gouveia, A. T., Prazeres, C., Ramos, S., & Graça, P. (2016). *Manual "Selo Saudável"* [Manual "Healthy Seal"]. Academic Press.

Almeida, A., & Garrod, B. (2017). Experiences with local food in a mature tourist destination: The importance of consumers' motivations. *Journal of Gastronomy and Tourism*, *2*(3), 173–187. doi:10.3727/216929717X14870140201116

Almeida, F., & Santos, J. D. (2020). The effects of COVID-19 on job security and unemployment in Portugal. *The International Journal of Sociology and Social Policy*, *40*(9/10), 995–1003. doi:10.1108/IJSSP-07-2020-0291

Almeida, F., & Silva, O. (2020). The impact of COVID-19 on tourism sustainability: Evidence from Portugal. *Advances in Hospitality and Tourism Research-Ahtr*, *8*(2), 440–446. doi:10.30519/ahtr.775340

Almeida, I., & Silveira, L. (2021). Olive-oil tourism in Portugal: Contextualization of the current supply. *Journal of Tourism & Development*, *36*(2), 153–165. doi:10.34624/rtd.v36i2.23289

Alonso, A. D., & Liu, Y. (2011). The potential for marrying local gastronomy and wine: The case of the 'fortunate islands'. *International Journal of Hospitality Management*, 974–981. doi:10.1016/j.ijhm.2011.02.005

Alonso-Almeida, M. D. M., Bagur-Femenias, L., Llach, J., & Perramon, J. (2018). Sustainability in small tourist businesses: The link between initiatives and performance. *Current Issues in Tourism*, *21*, 1–20. doi:10.1080/13683500.2015.1066764

Alonso, D., & Gonzáles, J. (2021). Museumization of olive oil mills: Instruments for conservation of the heritage of the olive grove culture. *Ge-conservacíon*, *20*, 173–183. doi:10.37558/gec.v20i1.1036

Alsaawi, A. (2014). A critical review of qualitative interviews. *European Journal of Business and Social Sciences*, *3*(4), 149–156. doi:10.2139srn.2819536

Altinay, L., Paraskevas, A., & Jang, S. (2015). *Planning Research in Hospitality and Tourism* (2nd ed.). Routledge. doi:10.4324/9781315723662

Altomare, R., Cacciabaudo, F., Damiano, G., Palumbo, V. D., Gioviale, M. C., & Bellavia, M. (2013). The mediterranean diet: A history of health. *Iranian Journal of Public Health*, *42*(5), 449–457. https://www.researchgate.net/publication/242018034 PMID:23802101

Aman, F., & Masood, S. (2020). How Nutrition can help to fight against Covıd-19 Pandemic. *Pakistan Journal of Medical Sciences*, *36*(S4), 121. doi:10.12669/pjms.36.COVID19-S4.2776

Amed. (n.d.). *Amed de establecimientos promotores de la Alimentación Mediterránea* [Amed of establishments promoting Mediterranean Food]. http://www.amed.cat/home.php

American Marketing Association. (2017). *What is marketing?* https://www.ama.org/the-definition-of-marketing/

Amerta, I. M. S. (2017). The role of tourism stakeholders at Jasri tourism village development, Karangasem Regency. *The International Journal of Social Sciences (Islamabad)*, *1*, 20–28. doi:10.29332/ijssh.v1n2.32

Ammar, A., Brach, M., Trabelsi, K., Chtourou, H., Boukhris, O., & Masmoudi, L. (2020). Effects of COVID-19 Home Confinement on Eating Behaviour and Physical Activity: Results of the ECLB-COVID19 International Online Survey. *Nutrients*, *12*(6). Advance online publication. doi:10.3390/nu12061583 PMID:32481594

Andersson, T. D., Mossberg, L., & Therkelsen, A. (2017). Food and tourism synergies: Perspectives on consumption, production, and destination development. *Scandinavian Journal of Hospitality and Tourism*, *17*, 1–8. doi:10.1080/15022250.2016.1275290

ANSES. (2020, March 9). *Opinion on an Urgent Request to Assess Certain Risks Associated with COVID-19*. https://www.anses.fr/fr/system/files/SABA2020SA0037-1EN.pdf

Antón, C., Camarero, C., Laguna, M., & Buhalis, D. (2019). Impacts of authenticity, degree of adaptation and cultural contrast on travellers' memorable gastronomy experiences. *Journal of Hospitality Marketing & Management*, *28*(7), 743–764. doi:10.1080/19368623.2019.1564106

Aquino, R. S., & Hyde, K. F. (2018). Unearthing the geotourism experience: Geotourist perspectives at Mount Pinatubo, Philippines. *Tourist Studies*, *18*(1), 41–62. doi:10.1177/1468797617717465

Armah, F. A., Ekumah, B., Yawson, D. O., Odoi, J. O., Afitiri, A. R., & Nyieku, F. E. (2018). Access to improved water and sanitation in sub-Saharan Africa in a quarter century. *Heliyon (London)*, *4*, e00931. doi:10.1016/j.heliyon.2018.e00931

Armitage, C. J., & Conner, M. (1999). Distinguishing perceptions of control from self-efficacy: Predicting consumption of a low-fat diet using the theory of planned behavior. *Journal of Applied Social Psychology*, *29*, 72–90. doi:10.1111/j.1559-1816.1999.tb01375.x

Arslan, G., & Yildirim, M. (2021). Meaning-based coping and spirituality during the COVID-19 pandemic : Mediating effects on subjective well-being. *Frontiers in Psychology*, *12*. Advance online publication. doi:10.3389/fpsyg.2021.646572 PMID:33935906

Astil, G. (2020, March 2). *Food Loss: Why Food Stays on the Farm or off the Market*. Economic Research Service. https://www.ers.usda.gov/amber-waves/2020/march/food-loss-why-food-stays-on-the-farm-or-off-the-market/

Atkinson, J. W. (1957). Motivational determinants of risk-taking behavior. *Psychological Review*, *64*(6), 359–372. doi:10.1037/h0043445 PMID:13505972

Atlas turístico de Mexico. (n.d.). https://atlasturistico.sectur.gob.mx/AtlasTuristico/inicio.do

Avşar, Ö. Y., & Seçim, Y. (2020). The Process of Adaptation of the Culinary Culture of Syrian Immigrants; The Case of Konya. *Journal of Recreation and Tourism Research*, *7*(2), 209–221. doi:10.31771/jrtr.2020.61

Azurmendi Restaurant. (n.d.). https://azurmendi.restaurant

Babio, N., Bullo, M., & Salas-Salvado, J. (2009). Mediterranean diet and metabolic syndrome: The evidence. *Public Health Nutrition*, *12*(9a), 1607–1617. doi:10.1017/S1368980009990449 PMID:19689829

Bach-Faig, A., Berry, E. M., Lairon, D., Reguant, J., Trichopoulou, A., & Dernini, S. (2011). Mediterranean diet pyramid today. Science and cultural updates. *Public Health Nutrition*, *14*(12), 2274–2284. PMID:22166184

Baidu. (2020, June 10). *New travel, new experience: Baidu tourism industry study 2020.* https://yingxiao.baidu.com/new/home/product/details?id=7206&ly=article_next

Bakalis, S., Valdramidis, V. P., Argyropoulos, D., Ahrne, L., Chen, J., Cullen, P., & … . (2020). Perspectives from CO+RE: How COVID-19 changed our food systems and food security paradigms. *Current Research in Food Science, 3*, 166. doi:10.1016/j.crfs.2020.05.003 PMID:32908972

Barbaros, B. (2015). *Yüksek, Orta ve Düşük Akdeniz Diyeti Uyumu Olan Bireylerde Oksidatif Stres ve Total Antioksidan Kapasitenin Belirlenmesi* [Determination of Oxidative Stress and Total Antioxidant Capacity in Individuals with High, Medium and Low Mediterranean Diet Compliance]. Yüksek Lisans Tezi, Doğu Akdeniz Üniversitesi, Beslenme ve Diyetetik Anabilim Dalı. Gazimağusa.

Barbaros, B., & Kabaran, S. (2014). Akdeniz Diyeti ve Sağlığı Koruyucu Etkileri [Mediterranean Diet and Health Protective Effects]. *Beslenme ve Diyet Dergisi, 42*(2), 140–147.

Barış, Z. (2015). *Turizm İşletmelerinde Organik Gıda Kullanımı Algı ve Tutumlarının Araştırılması* [Investigation of Perceptions and Attitudes of Organic Food Use in Tourism Enterprises]. Yüksek Lisans Tezi. İzmir Katip Çelebi Üniversitesi Sosyal Bilimler Enstitüsü Turizm İşletmeciliği Anabilim Dalı.

Barman, A., Das, R., & De, P. K. (2021). Impact of COVID-19 in food supply chain: Disruptions and recovery strategy. *Current Research in Behavioral Sciences, 2*, 100017. doi:10.1016/j.crbeha.2021.100017

Barney, J. B. (1991). Firm resources and sustained competitive advantage. *Journal of Management, 17*(1), 99–120. doi:10.1177/014920639101700108

Başaran, B., & Purut, H. P. (2021). The Impact of the Covıd-19 Pandemic on the Frequency of Food Consumption. *Journal of Tourism and Gastronomy Studies, 9*(1), 47–66. doi:10.21325/jotags.2021.776

Bastenegar, M. (2019). Spiritual understanding and experience in the creative tourism of gastronomy. *International Journal of Tourism and Spirituality, 3*(2), 43–67.

Basto, J. G. X. d. (1991). A tributação do consumo e a sua coordenação internacional: lições sobre harmonização fiscal na comunidade económica europeia [Taxation of consumption and its international coordination: lessons on tax harmonisation in the European economic community]. *Cadernos de Ciencia e técnica fiscal, 164.*

Baysal, A. (1996). Sağlıklı Beslenme ve Akdeniz Diyeti [Healthy Eating and Mediterranean Diet]. *Beslenme ve Diyet Dergisi, 25*(1), 21–29.

Beck, C. (1986). Education for spirituality. *Interchange, 17*(2), 148–156.

Bellini, N., Clergeau, C., & Etcheverria, O. (2018). *Gastronomy and Local Development: The Quality of Products, Places and Experiences.* Routledge. doi:10.4324/9781315188713

Bellisle, F. (2009). Infrequently asked questions about the Mediterranean diet. *Public Health Nutrition, 12*(9a), 1644–1647. doi:10.1017/S1368980009990498 PMID:19689834

Belsinger, S. (n.d.). *Herb harvest fall festival featuring plants of South America.* Fine Gardening. https://www.finegardening.com/article/herb-harvest-fall-festival-featuring-plants-of-south-america

Berbegal, C., Fragasso, M., Russo, P., Bimbo, F., Grieco, F., Spano, G., & Capozzi, V. (2019). Climate changes and food quality: The potential of microbial activities as mitigating strategies in the wine sector. *Fermentation (Basel, Switzerland), 5*(4), 85. doi:10.3390/fermentation5040085

Berbel-Pineda, J., Palacios-Florencio, B., Ramírez-Hurtado, J. M., & Santos-Roldán, L. (2019). Gastronomic experience as a factor of motivation in the tourist movements. *International Journal of Gastronomy and Food Science*, *18*, 100171. doi:10.1016/j.ijgfs.2019.100171

Beriss, D., & Sutton, D. (2007). *The Restaurants Book: Ethnographies of Where we Eat*. Berg Publishers. doi:10.5040/9781350044913

Berry, E. M., Arnoni, Y., & Aviram, M. (2011). The Middle Eastern and biblical origins of the Mediterranean diet. *Public Health Nutrition*, *14*(12a), 2288–2295. doi:10.10171368980011002539 PMID:22166186

Bertan, S. (2020). Impact of restaurants in the development of gastronomic tourism. *International Journal of Gastronomy and Food Science*, *21*, 100232. doi:10.1016/j.ijgfs.2020.100232

Bessière, J. (1998). Local development and heritage: Traditional food and cuisine as tourist attractions in rural areas. *Sociologia Ruralis*, *38*(1), 21–34. doi:10.1111/1467-9523.00061

Beverland, M. B. (2004). An exploration of the luxury wine trade. *International Journal of Wine Marketing*, *16*(3), 14–28.

Bevilacqua, D. (2020). Food Safety, Coronavirus and Risk Prevention: Future Perspectives in Four Proposals. *European Journal of Risk Regulation*, *13*(1), 56–77. doi:10.1017/err.2020.81

Bezerra, R., & Correia, A. (2018). The potential of olive oil for creative tourism experiences in the northern region of Portugal. *Revista Portuguesa de Estudos Regionais*, 56-72.

Bhalla, R., Chowdhary, N., & Ranjan, A. (2021). Spiritual tourism for psychotherapeutic healing post COVID-19. *Journal of Travel & Tourism Marketing*. Advance online publication. doi:10.1080/10548408.2021.1930630

Bittman, M., Oseland, J., & Bush, A. (2012). *Food Lover's Guide to the World: Experience the Great Global Cuisines*. Lonely Planet.

Björk, P., & Kauppinen-Räisänen, H. (2016). Local food: A source for destination attraction. *International Journal of Contemporary Hospitality Management*, *28*(1), 177–194. doi:10.1108/IJCHM-05-2014-0214

Boddy, C. R. (2016). Sample size for qualitative research. *Qualitative Market Research*, *19*(4), 426–432. doi:10.1108/QMR-06-2016-0053

Bode, M., Craven, M., Leopoldseder, M., Rutten, P., & Wilson, M. (2020). *Contact Tracing for COVID-19: New Considerations for Its Practical Application*. https://www.mckinsey.com/industries/public-and-social-sector/our-insights/contact-tracing-for-covid-19-new-considerations-for-its-practical-application

Bonaccio, M., Iacoviello, L., de Gaetano, G., & Moli-Sani, I. (2012). The Mediterranean diet: The reasons for a success. *Thrombosis Research*, *129*(3), 401–404. doi:10.1016/j.thromres.2011.10.018 PMID:22100317

Bone, J. (2005). Breaking bread : Spirituality, food and early childhood education. *International Journal of Children's Spirituality*, *10*(3), 307–317. doi:10.1080/13644360500347607

Borges, A., Rodrigues, A., & Rodrigues, R. (2010). *Elementos de contabilidade geral, SNC - Sistema de Normalização Contabilística* [General accounting elements, CnS - Accounting Standardization System] (25ª Edição). Áreas Editora.

Borghini, A. (2014). Authenticity in Food. In Encyclopedia of food and agricultural ethics. Springer.

Bosch, A., Gkogka, E., Le Guyader, F. S., Loisy-Hamon, F., Lee, A., & van Lieshout, L. (2018). Foodborne viruses: Detection, risk assessment, and control options in food processing. *International Journal of Food Microbiology*, *285*, 110–128. doi:10.1016/j.ijfoodmicro.2018.06.001 PMID:30075465

Boyle, D. (2004). *Authenticity: Brands, fakes, spin and the lust for real life*. Harper Perennial.

Boyne, S., Williams, F., & Hall, D. R. (2002). On the trail of regional success: tourism, food production and the Isle of Arran Taste Trail. In A. M. Hjalager & G. Richards (Eds.), *Tourism and Gastronomy* (pp. 91–114). Routledge.

Bozbulut, R., & Akbulut, G. (2016). Kansere Karşı Koruyucu Yeni Bir Öge: Skualen [A New Anti-Cancer Element: Skualen]. *Beslenme ve Diyet Dergisi, 44*(2), 160–168.

Bracale, R., & Vaccaro, C. M. (2020). Changes in food choice following restrictive measures due to Covıd-19. *Nutrition, Metabolism, and Cardiovascular Diseases, 30*, 1423–1426. doi:10.1016/j.numecd.2020.05.027 PMID:32600957

Brewer, P., & Sebby, A. G. (2021). The effect of online restaurant menus on consumers' purchase intentions during the COVID-19 pandemic. *International Journal of Hospitality Management, 94*, 102777. doi:10.1016/j.ijhm.2020.102777 PMID:34785837

Brizek, M. G., Frash, R. E., McLeod, B. M., & Patience, M. O. (2021). Independent restaurant operator perspectives in the wake of the COVID-19 pandemic. *International Journal of Hospitality Management, 93*, 102766. doi:10.1016/j.ijhm.2020.102766 PMID:33230361

Brochado, A., Stoleriu, O., & Lupu, C. (2021). Wine tourism: A multisensory experience. *Current Issues in Tourism, 24*(5), 597–615. doi:10.1080/13683500.2019.1649373

Bruwer, J., Chrysochou, P., & Lesschaeve, I. (2017). Consumer involvement and knowledge influence on wine choice cue utilisation. *British Food Journal, 119*(4), 830–844. doi:10.1108/BFJ-08-2016-0360

Bucak, T., & Aracı, Ü. E. (2013). Türkiye'de gastronomi turizmi üzerine genel bir değerlendirme [A general assessment of gastronomy tourism in Turkey]. *Balıkesir Üniversitesi Sosyal Bilimler Enstitüsü Dergisi, 16*(30), 203–216.

Bucak, T., & Yiğit, S. (2021). The future of the chef occupation and the food and beverage sector after the Covıd-19 outbreak: Opinions of Turkish chefs. *International Journal of Hospitality Management, 92*, 102682. doi:10.1016/j.ijhm.2020.102682 PMID:33071426

Buffon, M., & Bassani, M. (2013). Benefícios fiscais: uma abordagem à luz da cidadania fiscal e da legitimação constitucional da extrafiscalidade [Tax benefits: an approach in the light of fiscal citizenship and the constitutional legitimation of extrafiscality]. *Revista da AJURIS, 40*(130), 245-280. http://ajuris.kinghost.net/OJS2/index.php/REVAJURIS/article/view/298

Buhalis, D., & Amaranggana, A. (2015). Smart tourism destinations enhancing tourism experience through personalisation of services. In I. Tussyadiah & A. Inversini (Eds.), *Information and Communication Technologies in Tourism* (pp. 377–389). Springer International Publishing. doi:10.1007/978-3-319-14343-9_28

Bundesinstitut für Risikobewertung. (2020). *Can The New Type Of Coronavirus Be Transmitted Via Food And Objects?* https://www.bfr.bund.de/en/can_the_new_type_of_coronavirus_be_ transmitted_via_food_and_objects_-244090.html

Buonincontri, P., Morvillo, A., Okumus, F., & van Niekerk, M. (2017). Managing the experience co-creation process in tourism destinations: Empirical findings from Naples. *Tourism Management, 62*, 264–277. doi:10.1016/j.tourman.2017.04.014

Bureau of Economic Analysis. (2020, April 20). *Gross Domestic Product, 1st Quarter 2020 (advance Estimate)*. https://www.bea.gov/news/2020/gross-domestic-product-1st-quarter-2020-advance-estimate

Burlingame, B., Gitz, V., & Meybeck, A. (2015). *Mediterranean food consumption patterns: diet, environment, society, economy and health* [White Paper]. FAO.

Burlingame, B., & Dernini, S. (2011). Sustainable diets: The Mediterranean diet as an example. *Public Health Nutrition, 14*(12A), 2285–2287. doi:10.1017/S1368980011002527 PMID:22166185

Burns, J., & Scapens, R. (2000). Conceptualizing management accounting change: An institutional framework. *Management Accounting Research, 11*, 3–25. doi:10.1006/mare.1999.0119

Buzzell, L., & Chalquist, C. (2009). Psyche and nature in a circle of healing. In Ecotherapy: Healing with Nature in Mind. Catapult.

Byrd, K., Her, E., Fan, A., Almanza, B., Liu, Y., & Leitch, S. (2021). Restaurants and COVID-19: What are consumers' risk perceptions about restaurant food and its packaging during the pandemic? *International Journal of Hospitality Management, 94*, 102821. doi:10.1016/j.ijhm.2020.102821 PMID:34866742

CAC. (2020). *General Principles of Food Hygiene CAC/RCP 1-1969*. https://www.fao.org/fao-who-codexalimentarius/sh-proxy/en/?lnk=1&url=https%253A%252F%252Fworkspace.fao.org%252Fsites%252Fcodex%252FStandards%252FCXC%2B1-1969%252FCXC_001e.pdf

Caldwell, M. L. (2007). Feeding the body nourishing the soul. *Food, Culture, & Society, 10*(1), 43–71.

Calıcıoglu, O., Flammini, A., Bracco, S., Bellù, L., & Sims, R. (2019). The future challenges of food and agriculture: An integrated analysis of trends and solutions. *Sustainability, 11*(1), 222. doi:10.3390u11010222

Callén, Y. F., & Gadea, A. L. (2004). Deficiencias y alternativas de la información contable actual [Deficiencies and alternatives of current accounting information]. *Partida Doble, 152*, 62–73.

Campbell, D. (2000). Legitimacy theory or managerial reality construction? corporate social disclosure in marks and spencer plc corporate reports, 1969-1997. *Accounting Forum, 24*(1), 80–100. doi:10.1111/1467-6303.00030

Campos, A. C., Mendes, J., Valle, P. O., & Scott, N. (2016). Co-creation of tourist experiences: A literature review. *Current Issues in Tourism, 21*(4). Advance online publication. doi:10.1080/13683500.2015.1081158

Cao, M. T., & Zhu, X. H. (2018). Research on the current situation of tourism IP. *China Market*, (35), 70–78. doi:10.13939/j.cnki.zgsc.2018.35.070

Cao, X. (2020). COVID-19: Immunopathology and its implications for therapy. *Nature Reviews. Immunology, 20*(5), 269–270. doi:10.103841577-020-0308-3 PMID:32273594

Cao, Z. H. (2021). Research on the marketing strategy of food e-commerce under the guidance *of E-Commerce and Network Marketing. Food Industries, 512*(4).

Carlsen, J., & Boksberger, P. (2015). Enhancing Consumer Value in Wine Tourism. *Journal of Hospitality & Tourism Research (Washington, D.C.), 39*(1), 132–144. https://journals.sagepub.com/doi/abs/10.1177/1096348012471379

Carvalho, A. F. (2015). Domesticação e difusão: as origens da Dieta Mediterrânica [Domestication and dissemination: the origins of the Mediterranean Diet]. In Dimensões da Dieta Mediterrânica: Património Cultural Imaterial da Humanidade. Universidade do Algarve.

Carvalho, M., Kastenholz, E., & Carneiro, M. J. (2021). Interaction as a Central Element of Co-Creative Wine Tourism Experiences—Evidence from Bairrada, a Portuguese Wine-Producing Region. *Sustainability, 13*, 9374. doi:10.3390u13169374

Çaycı, A. E. (2019). *The role of social media in the transformation of globalized food culture: Instagram gourmets* [Doctoral Thesis]. Istanbul University Institute of Social Sciences.

Çelik, K., & Ergönül, B. (2020). Determination of Microbiological Attributes of Unpackaged and Packaged Raisin Samples Sold in Manisa Markets. *Acta Scientific Nutritional Health, 4*(5), 10–13.

Centers for Disease Control and Prevention. (2020). *How Coronavirus Spreads*. CDC.

Centers for Disease Control and Prevention. (2020a). *Updates COVID19 Transmission Webpage to Clarify Information About Types of Spread*. https://www.cdc.gov/media/releases/2020/s0522-cdc-updates-covid-transmission.html

Central People's Government of the People's Republic of China. (2019, February 19). *Suggestions on giving priority to the issues of agriculture, rural areas and farmers*. http://www.moa.gov.cn/ztzl/jj2019zyyhwj/2019zyyhwj/201902/t20190220_6172154.htm

Cetin, G., Alrawadieh, Z., Dincer, M. Z., Dincer, F. I., & Ioannides, D. (2017). Willingness to pay for tourist tax in destinations: Empirical evidence from Istanbul. *Economies*, *5*(2), 15. doi:10.3390/economies5020021

Çevik, N. K. (1997). *Türk Mutfağının Akdeniz Mutfak Kültürünün Genel Özellikleri Yönünden Değerlendirilmesinin Önemi* [Importance of Evaluating Turkish Cuisine in Terms of General Characteristics of Mediterranean Culinary Culture]. V. Milletlerarası Türk Halk Kültürü Kongresi.

Chan, J. F. W., Yuan, S., Kok, K. H., To, K. K. W., Chu, H., & Yang, J. (2020). A familial cluster of pneumonia associated with the 2019 novel coronavirus indicating person-to-person transmission: A study of a family cluster. The Lancet, 395, 514–523. , 10223. doi:10.1016/S0140-6736(20)30154-9

Chaney, S., & Ryan, C. (2012). Analyzing the evolution of singapore's world gourmet summit: An example of gastronomic tourism. *International Journal of Hospitality Management*, *31*(2), 309–318. doi:10.1016/j.ijhm.2011.04.002

Chang, H. S., Capuozzo, B., Okumus, B., & Cho, M. (2021). Why cleaning the invisible in restaurants is important during COVID-19: A case study of indoor air quality of an open-kitchen restaurant. *International Journal of Hospitality Management*, *94*, 102854. doi:10.1016/j.ijhm.2020.102854 PMID:33776189

Chathoth, U. G. R., Harrington, R. J., Altinay, L., Okumus, F., & Chan, E. S. W. (2013). Conceptualization of value co-creation in the tourism context. In N. K. Prebensen, J. S. Chen, & M. S. Uysal (Eds.), Creating experience value in tourism. CABI.

Chaudhry, A., Fuller, C., Coetsee, D., Rands, E., Bakker, E., Merwe, M., ... Balasubramanian, T. (2016). Interpretation and application of international financial reporting standards. Wiley.

Cheng, J. (2020). *China Records First Ever Contraction in Quarterly GDP on Coronavirus*. https://www.wsj.com/articles/china-set-to-report-plunge-in-first-quarter-gdp-11587086697

Chen, L., Lu, Y. H., Li, D. L., & Jiang, X. T. (2018). An empirical study on culinary tourism destination brand personality and its impact in the context of Confucian culture. *Luyou Xuekan*, *1*, 25–41. https://www.cabdirect.org/cabdirect/abstract/20183093032

Chen, S., Brahma, S., Mackay, J., Cao, C., & Aliakbarian, B. (2020). The role of smart packaging system in food supply chain. *Journal of Food Science*, *85*(3), 517–525. doi:10.1111/1750-3841.15046 PMID:32056210

Cherblanc, J., Bergeron, C., Maltais, D., Cadell, S., Gauthier, G., & Labra, O. (2021). Predictive factors of spiritual quality of life during the COVID - 19 Pandemic : A multivariate analysis. *Journal of Religion and Health*, *60*(3), 1475–1493. doi:10.100710943-021-01233-6 PMID:33759072

Chernev, A. H. R., & Gal, D. (2011). Competing for consumer identity: Limits to self-expression and the perils of lifestyle branding. *Journal of Marketing*, *75*, 66–82.

Chhabra, D. (2020). Transformational wellness tourism system Model in the pandemic era. *International Journal of Health Management and Tourism*, *5*(2), 76–101. doi:10.31201/ijhmt.788628

China Banking and Insurance Regulatory Commission. (2020, April 9). *Notice on doing well the key work of banking and insurance services in the field of agriculture, rural areas and farmers in 2020.* http://www.cbirc.gov.cn/cn/view/pages/ItemDetail.html?docId=899357&itemId=878&generaltype=1

China National Tourism Administration. (2015, December 17). *China tourism statistics bulletin 2014.* http://www.xinhuanet.com/travel/2015-12/17/c_128541973.htm

China National Tourism Administration. (2016, October 18). *China tourism statistics bulletin 2015.* http://zwgk.mct.gov.cn/zfxxgkml/tjxx/202012/t20201204_906456.html

China National Tourism Administration. (2017, June 12). *Guidelines for the establishment of all-for-one tourism demonstration areas.* http://zwgk.mct.gov.cn/zfxxgkml/zykf/202012/t20201213_919347.html

China National Tourism Administration. (2017, November 8). *China tourism statistics bulletin 2016.* http://zwgk.mct.gov.cn/zfxxgkml/tjxx/202012/t20201215_919598.html

China National Tourism Administration. (2018, February 6). *2017 Annual tourism market and comprehensive contribution data report.* http://zwgk.mct.gov.cn/zfxxgkml/tjxx/202012/t20201204_906468.html

China Tourism Academy. (2020, October 25). *Report on China's night economy 2020.* http://www.ctaweb.org.cn/cta/xsjl/202103/c730a94cae6a4847b63679756926f851.shtml

China Tourism Academy. (2021, February 22). *From overall recovery to high-quality development: Analysis of tourism economy in 2020 and its development forecast in 2021.* http://www.ctaweb.org.cn/cta/ztyj/202103/80532de9677f4c8685428b5a362e5ed9.shtml

China Tourism Academy. (2021, July 13). *Survey report on the recovery of urban leisure market in the first half of 2021.* http://www.ctaweb.org.cn/cta/ztyj/202107/08e09c680cfe4172ae635cdeb05dcbdd.shtml

Chirico, F. (2016). Spiritual well-being in the 21st century : It is time to review the current WHO ' s health definition. *Journal of Health and Social Sciences*, *1*(1), 11–16. doi:10.19204/2016prt2

Cho, R. (2019, June 20). *How Climate Change Impacts the Economy, How Climate Change Impacts the Economy.* State of the Planet. https://news.climate.columbia.edu/2019/06/20/climate-change-economy-impacts/

CIHEAM. (2012). *Mediterra 2012. The Mediterranean Diet for Sustainable Regional Development.* Presses de Sciences Po.

CIHEAM. (2020). *The Covıd-19 Pandemıc: Threats on Food Securıty In The Medıterranean Regıon.* https://www.ciheam.org/wp-content/uploads/2020/07/COVID-rapport-FINAL.pdf

Činjarević, M., & Pestek, A. (2014). Tourist perceived image of local cuisine: The case of Bosnian food culture. *British Food Journal*, *116*(11), 1821–1838. doi:10.1108/BFJ-01-2014-0046

Cityloops. (n.d.). https://cityloops.eu/cities/porto

Clancy, K. (2017). Digging deeper: Transdisciplinary and systems approaches to food security. *Journal of Agriculture, Food Systems, and Community Development*, *7*(4), 13–16. doi:10.5304/jafscd.2017.074.012

Clark, P. (2018). *The world's 10 most popular foodie destinations.* https://www.travelpulse.com/gallery/destinations/the-world-s-10-most-popular-foodie-destinations.html?image=2

Classen, A. (2007). The symbolic function of food as iconic representation of culture and spirituality in Wolfram von Eschenbach's Parzival. *Orbis Litterarum*, *62*(4), 315–335.

Co2 earth. (n.d.). *Página inicial de CO2 da Terra* [Earth CO2 homepage.] https://www.co2.earth

Coelho, M. F., Goslingb, M. S., & Almeida, A. S. (2018). Tourism experiences: Core processes of memorable trips. *Journal of Hospitality and Tourism Management*, •••, 11–22. doi:10.1016/j.jhtm.2018.08.004

Cohen, E. (1988). Authenticity and commoditization in tourism. *Annals of Tourism Research*, *15*(3), 371–386. doi:10.1016/0160-7383(88)90028-X

Cohen, E., & Avieli, N. (2004). Food in tourism: Attraction and impediment. *Annals of Tourism Research*, *31*, 755–778. doi:10.1016/j.annals.2004.02.003

Cohen, M. J. (2020). Does the Covıd-19 outbreak mark the onset of a sustainable consumption transition? *Sustainability: Science Practice and Policy*, *16*(1), 1–3. doi:10.1080/15487733.2020.1740472

Colares, L. G. T., Figueiredo, V. O., Ferreira, A. A., & Oliveira, A. G. M. (2018). Good environmental practices check list for food services: Elaboration, content validation and inter-rater reliability. *Brazilian Journal of Food Technology*, *21*(e2017066).

Cömert, M. (2014). Turizm Pazarlamasında Yöresel Mutfakların Önemi ve Hatay Mutfağı Örneği [Importance of Regional Cuisines in Tourism Marketing and Hatay Cuisine Example]. *Journal of Tourism and Gastronomy Studies*, 64-70.

Comigo, F. (2021). *Portugal de norte a sul pela mítica estrada nacional 2* [Portugal from north to south by the mythical national road 2]. Foge Comigo.

Contento, I. R. (2011). *Nutrition education linking research.* Jones and Bartlett Publishers.

Cook, P. S. (2010). Constructions and experiences of authenticity in medical tourism: The performances of places, spaces, practices, objects and bodies. *Tourist Studies*, *10*(2), 135–153. doi:10.1177/1468797611403048

Correia, R. (2006). *The relevance of collective action in tourism* [Conference session]. Cutting Edge Research in Tourism Conference, University of Surrey. http://hdl.handle.net/10198/5837

Correia, A., Kozak, M., & Del Chiappa, G. (2020). Examining the meaning of luxury in tourism: A mixed-method approach. *Current Issues in Tourism*, *23*(8), 952–970. doi:10.1080/13683500.2019.1574290

Cosmescus, N. (2005). Formula 1: scena şi culisele marilor premii [Formula 1: the stage and the backstage of the big prizes]. Piteşti.

Costa, A. P. (2011). *Depreciações e amortizações no SNC: alterações contabilísticas e impacto fiscal* [Depreciation and amortization in the CNS: accounting changes and fiscal impact]. Almedina.

Coulthard, H., Sharps, M., Cunliffe, L., & van den Tol, A. (2021). Eating in the lockdown during the Covid 19 pandemic; self-reported changes in eating behaviour, and associations with BMI, eating style, coping and health anxiety. *Appetite*, *161*, 105082. doi:10.1016/j.appet.2020.105082 PMID:33476651

Couret, C. (2016). Managing experiences within the field of creative tourism: Best practices and guidelines. In M. Sotiriadis & D. Gursoy (Eds.), *The Handbook of Managing and Marketing Tourism Experiences* (pp. 235–259). Emerald Group Publishing Limited. doi:10.1108/978-1-78635-290-320161034

Courinha, G. L. (2019). *Manual do Imposto sobre o rendimento das pessoas coletivas* [Corporate Income Tax Manual]. Almedina.

Coveney, J. (1999). The science and spirituality of nutrition. *Critical Public Health*, *9*(1), 23–37. doi:10.1080/09581599908409217

Crespi-Vallbona, M., & Mascarilla-Miró, O. (2020). Wine lovers: Their interests in tourist experiences. *International Journal of Culture, Tourism and Hospitality Research*, *14*(2), 239–258. doi:10.1108/IJCTHR-05-2019-0095

Cronin, C. J., & Evans, W. N. (2021). Total shutdowns, targeted restrictions, or individual responsibility: How to promote social distancing in the COVID-19 Era? *Journal of Health Economics*, *79*, 102497. doi:10.1016/j.jhealeco.2021.102497 PMID:34325285

Cross, T. D. (2021). *Frequent extreme weather events are increasing crop losses in Europe.* Sustainability Times. https://www.sustainability-times.com/environmental-protection/frequent-extreme-weather-events-are-increasing-crop-losses-in-europe/

Crossley, T. F., Griffith, R., Jin, W. M., & Lechene, V. (2018). *A structural analysis of the decline of home-cooked food* [Conference session]. LISER Conference, Luxembourg. https://www.bc.edu/content/dam/bc1/schools/mcas/economics/pdf/seminars/CGJL.pdf

Cui, M., & Zheng, H. D. (2019). The construction and countermeasure of China's tourism food safety guarantee system. *Food Industries*, *40*(10), 275–279. https://www.cabdirect.org/cabdirect/abstract/20183205065

Dai, M., Li, H., Yan, N., Huang, J., & Zhao, L. (2020). Long-term survival of salmon-attached SARS-CoV-2 at 4oC as a potential source of transmission in seafood markets. *bioRxiv*. Advance online publication. doi:10.1101/2020.09.06.284695

Danylyshyn, B. (2020). The peculiarities of economic crisis due to COVID-19 pandemic in a developing country: Case of Ukraine. *Problems and Perspectives in Management*, *18*(2), 13–22. doi:10.21511/ppm.18(2).2020.02

De Faria Coelho-Ravagnani, C., Corgosinho, F. C., Sanches, F. L. F. Z., Prado, C. M. M., Laviano, A., & Mota, J. F. (2021). Dietary recommendations during the Covıd-19 pandemic. *Nutrition Reviews*, *79*(4), 382–393. doi:10.1093/nutrit/nuaa067 PMID:32653930

De Silva, S. S., & Soto, D. (2009). Climate change and aquaculture: Potential impacts, adaptation and mitigation. *FAO Fisheries and Aquaculture Technical Paper*, *530*, 151–212.

De Souza, T. S. P., Miyahira, R. F., Matheus, J. R. V., de Brito Nogueira, T. B., Maragoni-Santos, C., Barros, F. F. C., Costa Antunes, A. E., & Fai, A. E. C. (2022). Food services in times of uncertainty: Remodeling operations, changing trends, and looking into perspectives after the COVID-19 pandemic. Trends in Food Science & Technology. *Trends in Food Science & Technology*, *120*, 301–307. doi:10.1016/j.tifs.2022.01.005 PMID:35035090

Dedeoğlu, B. B., & Boğan, E. (2021). The motivations of visiting upscale restaurants during the COVID-19 pandemic: The role of risk perception and trust in government. *International Journal of Hospitality Management*, *95*, 102905. doi:10.1016/j.ijhm.2021.102905

Del Castillo, F. (2020). Health, spirituality and Covid-19: Themes and insights. *Journal of Public Health Health*, *43*(2), 254–255. doi:10.1093/pubmed/fdaa185 PMID:33044544

Deloitte & China Cuisine Association. (2020, March). *Impact of COVID-19 on finance and operations in the Chinese catering industry.* https://www2.deloitte.com/content/dam/Deloitte/global/Documents/About-Deloitte/COVID-19/gx-covid-19-impact-on-finance-operations-in-the-chinese-catering-industry.pdf

Deng, L. Z., Mujumdar, A. S., Pan, Z., Vidyarthi, S. K., Xu, J., & Zielinska, M. (2019). Emerging chemical and physical disinfection technologies of fruits and vegetables: A comprehensive review. *Critical Reviews in Food Science and Nutrition*, *60*, 2481–2508. doi:10.1080/10408398.2019.1649633 PMID:31389257

Denmark Government. (2003). *Culture and experience economy-5 new steps The Danish Growth Strategy.* Author.

Dernini, S., & Berry, E. M. (2015). Mediterranean diet: From a healthy diet to a sustainable dietary pattern. *Frontiers in Nutrition*, *2*, 15. doi:10.3389/fnut.2015.00015 PMID:26284249

Dernini, S., Berry, E. M., Serra-Majem, L., La Vecchia, C., Capone, R., & Medina, F. X. (2017). Med Diet 4.0: The Mediterranean diet with four sustainable benefits. *Public Health Nutrition*, *20*(7), 1322–1330. doi:10.1017136898001600317 7 PMID:28003037

Dernini, S., Meybeck, A., Burlingame, B., Gitz, V., Lacirignola, C., & Debs, P. (2013). Developing a methodological approach for assessing the sustainability of diets: The Mediterranean diet as a case study. *New Medit*, *12*(3), 28–36.

Deschasaux-Tanguy, M., Druesne-Pecollo, N., Esseddik, Y., de Edelenyi, F. S., Allès, B., & Andreeva, V. A. (2021). Diet and physical activity during the coronavirus disease 2019 (COVID-19) lockdown (March-May 2020): Results from the French NutriNet-Santé cohort study. *The American Journal of Clinical Nutrition*, *113*(4), 924–938. doi:10.1093/ajcn/nqaa336 PMID:33675635

DGS,, PNPAS, DGC, & FCNAUP. (2016). *Roda da alimentação Mediterrânica: cartaz da roda da alimentação Mediterrânica* [Mediterranean food wheel: Mediterranean food wheel poster] [Infographic]. https://nutrimento.pt/activeapp/wp-content/uploads/2016/06/Cartaz-A3-13_med.pdf

DGS. (2020a). *Estudo de adesão ao padrão alimentar mediterrânico* [Study of mediterranean food pattern adhering]. https://nutrimento.pt/activeapp/wp-content/uploads/2020/10/Estudo-de-adesa%CC%83o-ao-padra%CC%83o-alimentar-mediterra%CC%82nico.pdf

DGS. (2020b). *Inquérito sobre alimentação e atividade física em contexto de contenção social* [Survey on food and physical activity in a context of social restraint]. https://alimentacaosaudavel.dgs.pt/activeapp2020/wp content/uploads/2021/01/REACTCOVID_Survey_Retificado.pdf

Dhar, R., & Wertenbroch, K. (2000). Consumer Choice between Hedonic and Utilitarian Goods. *JMR, Journal of Marketing Research*, *37*(1), 60–71. https://journals.sagepub.com/doi/abs/10.1509/jmkr.37.1.60.18718

Di Renzo, L., Gualtieri, P., Pivari, F., Soldati, L., Attinà, A., & Cinelli, G. (2020). Eating habits and lifestyle changes during COVID-19 lockdown: An Italian survey. *Journal of Translational Medicine*, *18*(1), 229. doi:10.118612967-020-02399-5 PMID:32513197

Diario El Expreso. (2021, March 20). Samborondón crea una playa artificial de agua dulce, al pie del río Babahoyo [Samborondón creates an artificial freshwater beach, at the foot of the Babahoyo River]. *Diario El Expreso*. https://www.expreso.ec/guayaquil/samborondon-crea-playa-artificial-agua-dulce-pie-rio-babahoyo-100958.html

Diario El Universo. (2019, November 7). El malecón de Samborondón tiene más atractivos [The boardwalk of Samborondón has more attractions]. *Diario El Universo*. https://www.eluniverso.com/guayaquil/2019/11/07/nota/7591352/malecon-tiene-mas-atractivos/

Dias, R. R., & Nogueira, M. (2018). *Agri-Food trends in Portugal for 2027 according to the principles of slow food.* Academic Press.

Dias, R. R., & Nogueira, M. (2019). Agri-food trends for Portugal based on the slow food principles: a new approach. *13th International European Forum on System Dynamics and Innovation in Food Networks.*

Dias, M. G., & Viegas, E. J. (2005). *Projecto do Museu da Oliveira e do Azeite, Mirandela* [Olive and Olive Oil Museum Project, Mirandela]. Ouro Virgem - Revista do Museu da Oliveira e do Azeite - Mirandela.

Dias, R. R., Nogueira, M., & Azinheira, F. (2021). Spirituality in agri-food sector : A complementary vision of slow food principles. *International Journal of Food System Dynamics. Proceedings in System Dynamics and Innovation in Food Networks*, *2021*, 4–13.

Dinu, M., Pagliai, G., Casini, A., & Sof, F. (2018). Mediterranean diet and multiple health outcomes: An umbrella review of meta-analyses of observational studies and randomised trials. *European Journal of Clinical Nutrition*, *72*, 30–43. doi:10.1038/ejcn.2017.58 PMID:28488692

Dirsehan, T., & Cankat, E. (2021). Role of mobile food-ordering applications in developing restaurants' brand satisfaction and loyalty in the pandemic period. *Journal of Retailing and Consumer Services*, *62*, 102608. doi:10.1016/j.jretconser.2021.102608

Donini, L. M., Serra-Majem, L., Bulló, M., Gil, Á., & Salas-Salvadó, J. (2015). The Mediterranean diet: Culture, health, and science. *British Journal of Nutrition*, *113*(2), 1–3. doi:10.1017/S0007114515001087 PMID:26148911

du Rand, G. E., & Heath, E. (2006). Towards a framework for food tourism as an element of destination marketing. *Current Issues in Tourism*, *9*(3), 206–234.

Dumas, Y., Dadomo, M., Di Lucca, G., & Grolier, P. (2003). Effects of environmental factors and agricultural techniques on antioxidant content of tomatoes. *Journal of the Science of Food and Agriculture*, *83*(5), 369–382. doi:10.1002/jsfa.1370

Dunford, D., Dale, B., Stylianou, N., Lowther, E., Ahmed, M., & De la Torres Arenas, I. (2020). *Coronavirus: The World in Lockdown in Maps and Charts*. https://www.bbc.com/news/world-52103747

Dunkley, D. (2021, February 14). *A nation of meat-eaters: How Kiwis swapped lamb for chicken*. Stuff. https://www.stuff.co.nz/business/300224037/a-nation-of-meateaters-how-kiwis-swapped-lamb-for-chicken

Dunn, K. I., Mohr, P. B., Wilson, C. J., & Wittert, G. A. (2008). Beliefs about fast food in Australia: A qualitative analysis. *Appetite*, *51*(2), 331–334. doi:10.1016/j.appet.2008.03.003 PMID:18430490

Dürdaroğlu, O., Barboroz, M., Mercan, Ş. O., & Avcıkurt, C. (2020). *Turistlerin Sağlıklı Beslenme ve Diyet Danışmanlığı Hizmeti Alma Motivasyonuyla Gerçekleştirdikleri Turizm Hareketlerinde Çekici ve İtici Faktörler: Sosyal Medya Destekli Bir Araştırma* [Attractive and Repulsive Factors in Tourism Movements carried out by tourists with the motivation to receive healthy nutrition and dietary consultancy services: a social media-supported research]. Seyahat ve Otel İşletmeciliği Dergisi.

Dvorak, G. (2008). *Disinfection 101*. Center for Food Security and Public Health. http://www.cfsph.iastate.edu/Disinfection/Assets/ Disinfection101.pdf

DW. (2020). *How climate change paved the way to war in Syria*. DW Made for minds. https://www.dw.com/en/how-climate-change-paved-the-way-to-war-in-syria/a-56711650

Dwyer, L. (2015). Globalization of tourism: Drivers and outcomes. *Tourism Recreation Research*, *40*(3), 326–339. doi:10.1080/02508281.2015.1075723

Edsor, B. (2017, November 30). 26 European cities where you can live on less than £600 a month. *The Independent*. https://www.independent.co.uk/travel/news-and-advice/european-cities-live-less-ps600-month-dnipro-lviv-kharkiv-pristina-craiova-a8084766.html

El Telegrafo. (2020, September 27). La vuelta ancestral del río Babahoyo, una nueva ruta fluvial [The ancestral return of the Babahoyo River, a new river route]. *Diario El Telegrafo*. https://www.eltelegrafo.com.ec/noticias/sociedad/6/samborondon-ruta-fluvial-rio-babahoyo

Elias, S. R., & Barbero, A. C. (2017). Situación del oleoturismo y lineamientos para su desarrollo en la región del sudoeste bonaerense, Argentina [Situation of oleotourism and guidelines for its development in the southwestern region of Buenos Aires, Argentina]. *Revista Interamericana de Ambiente y Turismo*, 91-104. https://riat.utalca.cl/index.php/test/article/view/338/pdf

Elkington, J. (1998). *Cannibals with Forks: the triple bottom line of 21st century business.* New Society Publishers.

Ellis, A., Park, E., Kim, S., & Yeoman, I. (2018). What is food tourism. *Tourism Management, 68*, 250–263. doi:10.1016/j.tourman.2018.03.025

Ellis, D., & Caruana, A. (2018). Consumer wine knowledge: Components and segments. *International Journal of Wine Business Research, 30*(3), 277–291. doi:10.1108/IJWBR-03-2017-0016

Engel, Ü. (2021). COVID-19 and "New Normal" tourism: Reconstructing tourism. *Journal of Tourism and Development, 35*, 217–226. https://proa.ua.pt/index.php/rtd/article/view/24652

Entorno turístico. (2020, February 18). Turismo gastronómico: Una forma diferente de hacer turismo [Gastronomic tourism: A different way of doing tourism]. *Entorno turístico.* https://www.entornoturistico.com/turismo-gastronomico-una-forma-diferente-turismo/

Environmental Protection Agency. (2020). *List N: Disinfectants for Use Against SARS-CoV-2.* https://www.epa.gov/coronavirus/about-list-n-disinfectants-coronavirus-covid-19-0

Epstein, L., & King, G. (2002). The rules of inference. *The University of Chicago Law Review. University of Chicago. Law School, 69*(1), 1–133. doi:10.2307/1600349

Eraqi, M. I. (2011). Co-creation and the new marketing mix as an innovative approach for enhancing tourism industry competitiveness in Egypt. *International Journal of Services and Operations Management, 8*(1), 76–91. https://www.inderscienceonline.com/doi/abs/10.1504/IJSOM.2011.037441

Erkuş-Öztürk, H., & Terhorst, P. (2016). Innovative restaurants in a mass-tourism city: Evidence from Antalya. *Tourism Management, 54*, 477–489. doi:10.1016/j.tourman.2016.01.003

Erman, Y. (2007). *Erkek ve Kadınların Diyet-Kanser İlişkisi Hakkında Bilgi ve İnanışları* [Knowledge and Beliefs of Men and Women about the Diet-Cancer Relationship]. Yüksek Lisans Tezi. Ankara Üniversitesi, Fen Bilimleri Enstitüsü, Ev Ekonomisi (Beslenme Bilimleri) Anabilim Dalı.

Ertaş, Y., & Karadağ, M. G. (2013). *Sağlıklı Beslenmede Türk Mutfak Kültürünün Yeri* [Place of Turkish Culinary Culture in Healthy Eating]. Gümüşhane Üniversitesi Sağlık Bilimleri Dergisi.

Eser, B. (2018). Where are the migrant restaurants on the gastronomic field? A theoretical introduction to immigrant neighborhood, refugee cuisine and spatial segregation. *ViraVerita E-Journal: Interdisciplinary Encounters, 8*, 172–195.

Esposito, K., Ciotola, M., & Giugliano, D. (2007). Mediterranean diet and the metabolic syndrome. *Molecular Nutrition & Food Research, 51*(10), 1268–1274. doi:10.1002/mnfr.200600297 PMID:17879992

Esteki, M., Regueiro, J., & Simal-Gándara, J. (2019). Tackling fraudsters with global strategies to expose fraud in the. *Food Chain, 18*(2), 425–440. doi:10. 1111/1541-4337.12419 PMID:33336950

Estell, M., Hughes, J., & Grafenauer, S. (2021). Plant Protein and Plant-Based Meat Alternatives: Consumer and Nutrition Professional Attitudes and Perceptions. *Sustainability, 13*(3), 1478.

Estruch, R., & Bach-Faig, A. (2018). Mediterranean diet as a lifestyle and dynamic food pattern. *European Journal of Clinical Nutrition.* Advance online publication. doi:10.103841430-018-0302-z PMID:30487557

Etcheverria, O. (2020). *The Restaurant, a Geographical Approach: from invention to gourmet tourist destinations.* John Wiley & Sons.

Euractiv. (2020). *No evidence of COVID-19 transmission through food, says EFSA.* https ://www.euractiv.com/section/coronavirus/news/no-evidence-of-COVID-19-transmission-through-food-says-efsa/

European Bank Report. (2020). *The COVID-19 pandemic and implications in the tourism sector in Ukraine*. https://www.ntoukraine.org/assets/files/EBRD-COVID19-Report-UKR.pdf

European Commission. (2020). *A new Circular Economy Action PlanFor a cleaner and more competitive Europe*. https://eur-lex.europa.eu/resource.html?uri=cellar:9903b325-6388-11ea-b735-01aa75ed71a1.0017.02/DOC_1&format=PDF

European Parliament and Council Regulation (EC) No 178/2002 (2002). https:// publications.europa.eu/en/publication-detail/-/publication/b18e2753-3f89-4e5e8b03-6c6a9d340ac1/language-en

Eurostat. (2019, March 1). Olive trees cover 4.6 million hectares in the EU. *Eurostat*. https://ec.europa.eu/eurostat/web/products-eurostat-news/-/DDN-20190301-1?inheritRedirect=true

Fachin, R. C., & Mendonça, J. R. C. (2003). O conceito de profissionalização e da teoria institucional [The concept of professionalization and institutional theory]. In M. F. Vieira & C. A. Carvalho (Eds.), *Organizações, instituições e poder no Brasil* (pp. 19–41). FGV.

Fang, Y., Li, L. L., & Xu, P. (2020). Chengdu tourism and characteristic snack industry fusion development status analysis based on SWOT. *Shanxi Agricultural Economy*, *13*, 28–29. doi:10.16675/j.cnki.cn14-1065/f.2020.13.017

FAO. (2011). *Global Food Losses and Waste. Extent, Causes and Prevention*. https://www.fao.org/docrep/014/mb060e/mb060e00.pdf

FAO. (2017). The future of food and agriculture – Trends and challenges. FAO.

FAO. (n.d.). http://www.fao.org/faostat/en/?#data/QC

FAO. (n.d.a). *What is meant by the term "sustainability"?* https://www.fao.org/3/ai388e/AI388E05.htm

FAO/Bioversity. (2012). *Sustainable diets and biodiversity*. FAO: Directions and Solutions for Policy, Research and Action.

Farmery, A. K., Alexander, K., Anderson, K., Blanchard, J. L., Carter, C. G., & Evans, K. (2021). Food for all: Designing sustainable and secure future seafood systems. *Reviews in Fish Biology and Fisheries*, 1–21. doi:10.100711160-021-09663-x(0123456789) PMID:34092936

Faustino, M. (2015). Sobre a reforma do IRS (I) [About the reform of the IRS (I)]. *Revista de finanças públicas e direito fiscal*, *4*(7).

Faustino, M. (2017). Imposto sobre o rendimento das pessoas singulares [Personal income tax]. In *Lições de Fiscalidade. Princípios gerais e fiscalidade interna*. Almedina.

Federal Institute for Risk Assessment. (2012). *EHEC outbreak 2011 Investigation of the outbreak along the food chain*. https://mobil.bfr.bund.de/cm/350/ehec-outbreak-2011-investigation-of-the-outbreak-along-the-food-chain.pdf

Feng, J. Y., Mu, W. S., & Fu, Z. T. (2006). A review of consumer purchase intention. *Modern Management Science*, *11*, 7-9.

Ferguson, T. W., & Tamburello, J. A. (2015). The natural environment as a spiritual resource: A theory of regional variation in religious adherence. *Sociology of Religion*, *76*(3), 295–314.

Fernández, J. I. P., & Rivero, M. S. (2010). Competitiveness versus growth in tourist destinations. An analysis using multivariate techniques. *Cuadernos de Economía (Santiago, Chile)*, *33*(91), 159–182. doi:10.1016/S0210-0266(10)70061-3

Fernández, J. I., Montilla, J. C., & Hidalgo, I. C. (2019). Introducing olive-oil tourism as a special interest tourism. *Heliyon*, 1–8. doi:10.1016/j.heliyon.2019.e02975 PMID:31872132

Fernández, J., Montilla, J., Hidalgo, I., & Fernández, M. (2022). Evaluating olive oil tourism experiences based on the segmentation of demand. *International Journal of Gastronomy and Food Science*, *27*. Advance online publication. doi:10.1016/j.ijgfs.2021.100461

Fields, K. (2002). Demand for the gastronomy tourism product: Motivational factors. In A. M. Hjalager & G. Richards (Eds.), *Tourism and gastronomy* (pp. 37–50). Routledge.

Filipic, A., Gutierrez-Aguirre, I., Primc, G., Mozetic, M., & Dobnik, D. (2020). Cold plasma, a new hope in the field of virus inactivation. *Trends in Biotechnology*, *38*, 1278–1291. doi:10.1016/j.tibtech.2020.04.003 PMID:32418663

Fischler, C. (2021). Commensality, society and culture. *Social Sciences Information. Information Sur les Sciences Sociales*, *50*, 528–548. doi:10.1177/0539018411413963

Fisher, D., Reilly, A., Kang Eng Zheng, A., Cook, A. R., & Anderson, D. E. (2020). Seeding of outbreaks of COVID-19 by contaminated fresh and frozen food. *bioRxiv*. Advance online publication. doi:10.1101/2020.08.17.255166

Floarea, A., & Sgârciu, V. (2016). Smart refrigerator: A next generation refrigerator connected to the IoT [Conference session]. 8th international conference on electronics, computers and artificial intelligence (ECAI), Ploiesti. 10.1109/ECAI.2016.7861170

Florek, M., & Gazda, J. (2021). Traditional food products— Between place marketing, economic importance and sustainable development. *Sustainability*, *13*(3), 1277. doi:10.3390u13031277

Food and Agriculture Organization of the United Nations & World Health Organization (2020). *COVID-19, and food safety. Guidance for food businesses*. Author.

Food and Agriculture Organization of the United Nations & World Health Organization. (2020). *COVID-19, And Food Safety. Guidance For Food Businesses*. https://www.who.int/publications/i/item/covid-19-and-food-safety-guidance-for-food-businesses

Food and Agriculture Organization. (2020). *Q&A: COVID-19 pandemic–impact on food and agriculture*. https://www.fao.org/2019-ncov/q-and-a/impact-on-food-and-agriculture/en/

Food and Drug Administration (2020). *Shopping for food during the COVID-19 pandemic - information for consumers*. FDA.

Food and Drug Administration. (2020). *Shopping for food during the COVID-19 pandemic -information for consumers*. Author.

Food Programe. (n.d.). *Recomendações* [Recommendations]. https://www.food-programme.eu/pt/ferramentas/article/recomendacoes

FoodSave. (n.d.). http://www.foodsave.org/wp-content/uploads/2014/07/FoodSaveFacts.pdf

Ford, R. C., & Sturman, M. C. (2020). *Managing Hospitality Organizations: Achieving Excellence in the Guest Experience*. Sage Publications Inc.

Fossgard, K., & Fredman, P. (2019). Dimensions in the nature-based tourism experiencescape: An explorative analysis. *Journal of Outdoor Recreation and Tourism*, *28*, 100219. doi:10.1016/j.jort.2019.04.001

Fourat, E., Fournier, T., & Lepiller, O. (2021). Snatched commensality: To eat or not to eat together in times of Covid-19 in France. *Food & Foodways*, *29*(2), 204–212. doi:10.1080/07409710.2021.1901383

Fragkiadakis, G. A. (2007). *Combining tourism business and Mediterranean diet: An innovative approach*. Animal Hemagglutinins Technology View Project Archimedes Protect- Primary School Children Nutrition View Project.

Freeman, R. E. (1994). The politics of stakeholder theory: Some future directions. *Business Ethics Quarterly*, *4*(4), 409–421. doi:10.2307/3857340

Freightos. (2022, March 3). *Shipping & Freight Cost Increases, Freight Capacity, and Shipping Container Shortage.* https://www.freightos.com/freight-resources/coronavirus-updates/

Fuentes, R. C., Moreno-Gil, S., González, C. L., & Ritchie, J. B. (2015). La creación y promoción de experiencias en un destino turístico. Un análisis de la investigación y necesidades de actuación [The creation and promotion of experiences in a tourist destination. An analysis of research and needs for action]. *Cuadernos de Turismo*, *35*, 71–94. doi:10.6018/turismo.35.221511

Fuertes, G., Soto, I., Carrasco, R., Vargas, M., Sabattin, J., & Lagos, C. (2016). Intelligent packaging systems: Sensors and nanosensors to monitor food quality and safety. *Journal of Sensors*, *8*. Advance online publication. doi:10.1155/2016/4046061

Fung, T. T., Rexrode, K. M., Mantzoros, C. S., Manson, J. E., Willett, W. C., & Hu, F. B. (2009). Mediterranean diet and incidence of and mortality from coronary heart disease and stroke in women. *Circulation*, *119*, 1093–1100. doi:10.1161/CIRCULATIONAHA.108.816736 PMID:19221219

Fuste-Forne, F. (2021). Robot chefs in gastronomy tourism: What's on the menu. *Tourism Management Perspectives*, *37*, 100774. doi:10.1016/j.tmp.2020.100774

Fu, W. Y. (2014). Review of the "Six Elements of Tourism". *Journal of Zhongzhou University*, *31*(5), 14–17. doi:10.13783/j.cnki.cn41-1275/g4.2014.05.005

Galanakis, C. M. (2020). The food systems in the era of the coronavirus (COVID-19) pandemic crisis. *Foods*, *9*(4), 523. doi:10.3390/foods9040523 PMID:32331259

Gall, T. L., Malette, J., & Guirguis-Younger, M. (2011). Spirituality and religiousness : A diversity of definitions. *Journal of Spirituality in Mental Health*, *13*(3), 158–181. doi:10.1080/19349637.2011.593404

Galvez, J. F., Mejuto, J. C., & Simal-Gandara, J. (2018). Future challenges on the use of blockchain for food traceability analysis. *Trends in Analytical Chemistry*, *107*, 222–232. doi:10.1016/j.trac.2018.08.011

Ganzaroli, A., De Noni, I., & van Baalen, P. (2017). Vicious advice: Analyzing the impact of TripAdvisor on the quality of restaurants as part of the cultural heritage of Venice. *Tourism Management*, *61*, 501–510. doi:10.1016/j.tourman.2017.03.019

Garcês, S., Pocinho, M., & de Jesus, S. N. (2018). Review of optimism, creativity and spirituality in tourism research. *Tourism and Hospitality Management*, *24*(1), 107–117. doi:10.20867/thm.24.1.6

Garcia-Fernandez, E., Rico-Cabanas, L., Rosgaard, N., Estruch, R., & Bach-Faig, A. (2014). Mediterranean diet and cardiodiabesity: A review. *Nutrients*, *6*(9), 3474–3500. doi:10.3390/nu6093474 PMID:25192027

Garibaldi, R., & Pozzi, A. (2020). Gastronomy tourism and Covid-19: technologies for overcoming current and future restrictions. In F. Burini (Ed.), *Tourism facing a pandemic: From Crisis to recovery.* https://aisberg.unibg.it/handle/10446/160698#.XxlqKp4zbIX

Garraturo, F., del Guidice, C., Morelli, M., Cerullo, V., Libralato, G., & Galdiero, E. (2020). Persistence of SARs-CoV-2 in the environment and COVID-19 transmission risk from environmental matrices and surfaces. *Environmental Pollution*, *265*, 115010. doi:10.1016/j.envpol.2020.115010 PMID:32570023

Garrod, B. (2003). Local participation in the planning and management of ecotourism: A revised model approach. *Journal of Ecotourism*, 33–53. doi:10.1080/14724040308668132

Garssen, B., Visser, A., Pool, G., & Garssen, B. (2021). Does spirituality or religion positively affect mental health? *The International Journal for the Psychology of Religion*, *31*(1), 4–20. doi:10.1080/10508619.2020.1729570

Garvey, N. L. (2010). *À table: The impacts of globalization on French cuisine* [Doctoral dissertation]. Texas A & M University.

Gavilan, D., Balderas-Cejudo, A., Fernández-Lores, S., & Martinez-Navarro, G. (2021). Innovation in online food delivery: Learnings from COVID-19. *International Journal of Gastronomy and Food Science*, *24*, 100330. doi:10.1016/j.ijgfs.2021.100330 PMID:34745390

Gephart, J. A., Golden, C. D., Asche, F., Belton, B., Brugere, C., & Froehlich, H. E. (2020). Scenarios for global aquaculture and its role in human nutrition. *Reviews in Fisheries Science & Aquaculture*, 1–17. doi:10.1080/23308249.2020.1782342

Gerritsen, S., Sing, F., Lin, K., Martino, F., Backholer, K., Culpin, A., & Mackay, S. (2021). The timing, nature and extent of social media marketing by unhealthy food and drinks brands during the COVID-19 pandemic in New Zealand. *Frontiers in Nutrition*, *8*, 1–13. doi:10.3389/fnut.2021.645349 PMID:33748177

Getz, D., & Robinson, R. N. S. (2014). Foodies and food events. *Scandinavian Journal of Hospitality and Tourism*, *14*(3), 315–330. doi:10.1080/15022250.2014.946227

Gezon, L. L. (2018). Global scouts: Youth engagement with spirituality and wellness through travel, Lake Atitlán, Guatemala. *Journal of Tourism and Cultural Change*, *16*(4), 365–378. doi:10.1080/14766825.2017.1310217

Ghahremanzadeh, M., & Ziaei, M. B. (2014). Food Price Change and its welfare impact on Iranian Households. *International Journal of Agricultural Management and Development*, *4*, 313–323. doi:10.22004/ag.econ.234396

Gillespie, C. (2001). *European gastronomy into the 21st century*. Butterworth-Heinemann.

Ginter, E., & Simko, V. (2015). Recent data on Mediterranean diet, cardiovascular disease, cancer, diabetes and life expectancy. *Bratislavské Lekárske Listy*, *116*(6), 346–348. doi:10.4149/BLL_2015_065 PMID:26084734

Gnonlonfoun, R. (2017). *Restaurants Owner Strategies for Financial Sustainability Beyond 5 Years* [Doctoral dissertation]. Walden University.

Gonçalves, C., Santos, D., Rodrigo, J., & Fernandes, S. A. (2017). *Financial accounting explained - Practical Manual*. Vida Económica.

González, J. A., Armenteros, E. M., & Ruiz, F. J. (2020). Characterization of olive oil tourism as a type of Special Interest Tourism: An analysis from the tourist experience perspective. *Sustainability*, 1–10. doi:10.3390u12156008

Goodman, S., Lockshin, L., & Cohen, E. (2005). Best-worst scaling: a simple method to determine drinks and wine style preferences. *International Wine Symposium*, 1-16.

Goodwin, R., Schley, D., Lai, K. M., Ceddia, G. M., Barnett, J., & Cook, N. (2012). Interdisciplinary approaches to zoonotic disease. *Infectious Disease Reports*, *4*(2), e37. doi:10.4081/idr.2012.e37 PMID:24470951

GoRaymi. (2020). Balao. *Go Raymi*. https://www.goraymi.com/es-ec/guayas/balao/ ciudades/balao-a87d9782e

Gorbalenya, A. E., Baker, S. C., Baric, R. S., Groot, R. J. D., Drosten, C., & Gulyaeva, A. A. (2020). Severe acute respiratory syndrome-related coronavirus: The species and its viruses-a statement of the coronavirus study group. *bioRxiv*. Advance online publication. doi:10.1101/2020.02.07.937862

Gorji, S., & Gorji, A. (2021). COVID-19 pandemic: The possible influence of the long-term ignorance about climate change. *Environmental Science and Pollution Research International*, *28*(13), 15575–15579. doi:10.100711356-020-12167-z PMID:33403640

Gössling, S., Scott, D., & Hall, C. M. (2020). Pandemics, tourism and global change: A rapid assessment of COVID-19. *Journal of Sustainable Tourism*, *29*(1), 1–20. doi:10.1080/09669582.2020.1758708

Gotsis, E., Anagnostis, P., Mariolis, A., Vlachou, A., Katsiki, N., & Karagiannis, A. (2015). Health benefits of the Mediterranean diet: An update of research over the last 5 years. *Angiology*, *66*(4), 304–318. doi:10.1177/0003319714532169 PMID:24778424

Goyal, A. (2021). Consumer Perceived Associations of Masstige Brands: An Indian Perspective. *Journal of Promotion Management*, *27*(3), 399–416. doi:10.1080/10496491.2020.1838027

Graça, P. (2015). Dieta Mediterrânica: uma realidade multifacetada [Mediterranean Diet: a multifaceted reality]. In *Dimensões da Dieta Mediterrânica: Património Cultural Imaterial da Humanidade*. Universidade do Algarve. http://hdl.handle.net/10400.1/6728

Graça, P. (2014). Dieta Mediterrânica: um desafio para a cooperação entre a cultura e a saúde nos próximos anos [Mediterranean Diet: a challenge for cooperation between culture and health in the coming years]. In A. Romano (Ed.), *A Dieta Mediterrânica em Portugal: Cultura, Alimentação e Saúde* (pp. 332–335). Universidade do Algarve.

Granitto, Y. (2020, January 31). *O negócio* [The business.] OliveOilTimes: https://www.oliveoiltimes.com/pt/briefs/oleotourism-becomes-law-in-italy/79917

Granot, E., Russell, L. T. M., & Brashear-Alejandro, T. G. (2013). Populence: Exploring Luxury for the Masses. *Journal of Marketing Theory and Practice*, *21*(1), 31–44. doi:10.2753/MTP1069-6679210102

Green Globe. (n.d.). https://greenglobe.com/standard/

Green Key. (n.d.). https://greenkey.abae.pt/

Green, G. P., & Dougherty, M. L. (2009). Localizing linkages for food and tourism: Culinary tourism as a community development strategy. *Community Development (Columbus, Ohio)*, 148–158. doi:10.1080/15575330809489674

Green, R., Cornelsen, L., Dangour, A. D., Turner, R., Shankar, B., Mazzocchi, M., & Smith, R. D. (2013). The effect of rising food prices on food consumption: Systematic review with meta-regression. *British Medical Journal*, *346*. Advance online publication. doi:10.1136/bmj.f3703 PMID:23775799

Grimaldi, P., Fassino, G., & Porporato, D. (2019). Introduction. In C. Petrini (Ed.), Culture, heritage, identity and food: a methodological approach. Stampa: Geca Industrie Grafiche.

Gruda, N. (2005). Impact of environmental factors on product quality of greenhouse vegetables for fresh consumption. *Critical Reviews in Plant Sciences*, *24*(3), 227–247. doi:10.1080/07352680591008628

Grunert, K. G., De Bauw, M., Dean, M., Lähteenmäki, L., Maison, D., Pennanen, K., Sandell, M. A., Stasiuk, K., Stickel, L., Tarrega, A., Vainio, A., & Vranken, L. (2021). No lockdown in the kitchen: How the COVID-19 pandemic has affected food-related behaviours. *Food Research International*, *150*, 110752. doi:10.1016/j.foodres.2021.110752 PMID:34865770

Guan, W., Ni, Z., Hu, Y., Liang, W., Ou, C., & He, J. (2020). Clinical characteristics of coronavirus disease 2019 in China. *The New England Journal of Medicine*, *382*, 1708–1720. doi:10.1056/NEJMoa2002032 PMID:32109013

Guerra, I. R., Molina, V., & Quesada, J. M. (2018). Multidimensional research about oleotourism attraction from the demand point of view. *Journal of Tourism Analysis: Revista de Análisis Turístico*, *25*(2), 114–128. doi:10.1108/JTA-06-2018-0017

Guerreiro, R., Pereira, C., & Rezende, A. (2006). In search of understanding the formation of habits and routines of management accounting. *Revista de Administração Mackenzie*, *7*(2), 78–101. doi:10.1590/1678-69712006/administracao.v7n2p78-101

Gürhan, N. (2017). A Sociological Analysis on Social Change and Food Culture: Mardin Sample. *Journal of International Social Research*, *10*(54), 561–570.

Gussow, J. (1995). Mediterranean diets: Are they environmentally responsible? *The American Journal of Clinical Nutrition*, 1383S–1389S. doi:10.1093/ajcn/61.6.1383S PMID:7754992

Gustafson, C. R., Lybbert, T. J., & Sumner, D. A. (2016). Consumer knowledge affects valuation of product attributes: Experimental results for wine. *Journal of Behavioral and Experimental Economics*, *65*, 85–94. doi:10.1016/j.socec.2016.08.004

Guzel, B., & Apaydin, M. (2016). Gastronomy tourism, motivations and destinations. In C. Avcıkurt, M. S. Dinu, N. Hacıoğlu, R. Efe, A. Soykan, & N. Tetik (Eds.), *Global Issues and Trends in Tourism* (pp. 394–404). St. Kliment Ohridski University Press.

Hakimi, M. A. (2013, August 31). *Kebab, the symbol of Iranian hospitality*. www.hotelmagazine.ir

Hakim, M. P., Zanetta, L. D., & da Cunha, D. T. (2021). Should I stay, or should I go? Consumers' perceived risk and intention to visit restaurants during the COVID-19 pandemic in Brazil. *Food Research International*, *141*, 110152. doi:10.1016/j.foodres.2021.110152 PMID:33642018

Hall, C. M., & Sharples, L. (2003). The consumption of experiences or the experience of consumption: an introduction to the tourism of taste. In C. M. Hall, L. Sharples, R. Mitchell, N. Macionis & B. Cambourne (Eds.), Food tourism around the world: Development, management and markets. Butterworth-Heinemann.

Hall, C. M., & Mitchell, R. (2001). Wine and food tourism. In N. Douglas, N. Douglas, & R. Derrett (Eds.), *Special interest tourism* (pp. 307–329). John Wiley and Sons.

Hall, C. M., Sharples, L., Mitchell, R., Macionis, N., & Cambourne, B. (2003). *Food tourism Around the World: Development, management and markets*. Butterworth-Heinemann.

Hall, M., & Mitchell, R. (2003). Tourism as a force for gastronomic globalization and localization. In A. M. Hjalager & G. Richards (Eds.), *Tourism and gastronomy*. Routledge.

Hall, M., & Mitchell, R. (2007). *Wine Marketing*. Routledge., doi:10.4324/9780080553283

Hamming, I., Timens, W., Bulthuis, M. L., Lely, A. T., Navis, G., & van Goor, H. (2004). Tissue distribution of ACE2 protein, the functional receptor for SARS coronavirus. A first step in understanding SARS pathogenesis. *The Journal of Pathology*, *203*, 631–637. doi:10.1002/path.1570 PMID:15141377

Hancock, E. (2016, December 23). The 12 cheapest holiday destinations in the world. *Business Insider*. https://www.businessinsider.com/cheapest-holiday-destinations-to-visit-in-2017-2016-12

Han, S., Kumar Roy, P., Iqbal Hossain, M., Bun, K. H., Choi, C., & Ha, S. D. (2021). COVID-19 pandemic crisis and food safety: Implications and inactivation strategies. *Trends in Food Science & Technology*, *109*, 25–36. doi:10.1016/j.tifs.2021.01.004 PMID:33456205

Harapan, H., Itoh, N., Yufika, A., Winardi, W., Keam, S., & Te, H. (2020). Coronavirus disease 2019 (COVID-19): A literature review. *Journal of Infection and Public Health*, *13*, 667–673. doi:10.1016/j.jiph.2020.03.019 PMID:32340833

Harms, R., Alfert, C., Cheng, C. F., & Kraus, S. (2021). Effectuation and causation configurations for business model innovation: Addressing COVID-19 in the gastronomy industry. *International Journal of Hospitality Management*, *95*, 102896. doi:10.1016/j.ijhm.2021.102896

Harrington, R. J. (2005). Defining gastronomic identity: The impact of environment and culture on prevailing components, texture and flavors in wine and food. *Journal of Culinary Science & Technology*, *4*(2-3), 129–152. doi:10.1300/J385v04n02_10

Harvard. (n.d.). *Coronavírus, Mudanças Climáticas e Meio Ambiente Uma Conversa sobre COVID-19 com Dr. Aaron Bernstein, Diretor de Harvard Chan C-CHANGE* [Coronavirus, Climate Change and Environment A Conversation on COVID-19 with Dr. Aaron Bernstein, Director of Harvard Chan C-CHANGE]. https://www.hsph.harvard.edu/c-change/subtopics/coronavirus-and-climate-change/

Haven-Tang, C., & Jones, E. (2005). Using local food and drink to differentiate tourism destinations through a sense of place: A story from Wales-dining at Monmouthshire's great table. *Journal of Culinary Science & Technology*, *4*(4), 69–86. doi:10.1300/J385v04n04_07

Haven-Tang, C., & Jones, E. (2006). Learning provision, the labour market and skills needs of the tourism and related sectors in Wales. *Journal of Human Resources in Hospitality & Tourism*, *5*(2), 13–35. doi:10.1300/J171v05n02_02

He, M. L. (2020). *Research on the construction of experience value evaluation system of food tourism* [Master's dissertation]. Soochow University. https://kns.cnki.net/KCMS/detail/detail.aspx?dbname=CMFD202101&filename=1020152831.nh

Hecht, A. A., Biehl, E., Barnett, D. J., & Neff, R. A. (2019). Urban food supply chain resilience for crises threatening food security: A qualitative study. *Journal of the Academy of Nutrition and Dietetics*, *119*(2), 211–224. doi:10.1016/j.jand.2018.09.001

He, J. T., Zhu, N., Feng, W. L., & Huang, Y. Y. (2020). Research on the economic development platform of Shunde food culture. *China Journal of Commerce*, *9*, 7–8. doi:10.19699/j.cnki.issn2096-0298.2020.09.007

Henderson, E. (2019, February). Why Lviv makes the best weekend break this winter. *The Independent*. https://www.independent.co.uk/travel/48-hours-in/iviv-what-do-travel-guide-winter-weekend-break-away-ukraine-cheap-holiday-a7955531.html

Henderson, J. C. (2009). Food tourism reviewed. *British Food Journal*, *111*(4), 317–326.

Henwood, A. F. (2020). Coronavirus disinfection in histopathology. *Journal of Histotechnology*, *43*(2), 102–104. doi:10.1080/01478885.2020.1734718 PMID:32116147

Hernandez-Rojas, R. D., Folgado-Fernandez, J. A., & Palos-Sanchez, P. R. (2021). Influence of the restaurant brand and gastronomy on tourist loyalty: A study in Cordoba (Spain). *International Journal of Gastronomy and Food Science*, *23*, 100305. doi:10.1016/j.ijgfs.2021.100305

Herrera, D. B. (2021). Estimation of the tourism crisis in Ecuador by covıd-19 with a view to a Post-Pandemic Reactivation. *Journal of Contemporary Issues in Business and Government*, *27*(4), 91–110.

Hidalgo, I., Montilla, J., & Fernández, J. (2021). Desarrollo sostenible del oleoturismo: Un análisis de la colaboración entre actores [Sustainable development of oleotourism: An analysis of collaboration between actors]. *Revista de Ciencias Sociales*, *27*(2), 19–38. doi:10.31876/rcs.v27i2.35896

Hill, C. W. L., & Jones, T. M. (1992). Stakeholder-agency theory. *Journal of Management Studies*, *29*(2), 131–154. doi:10.1111/j.1467-6486.1992.tb00657.x

Hodos, T. (2010). Globalization and colonization: A view from Iron Age Sicily. *Journal of Mediterranean Archaeology*, *23*(1), 81–106. doi:10.1558/jmea.v23i1.81

Holding of Emotions! (2021). *FEST*. https://www.fest.lviv.ua

Hong, Q. H. (2018). *Taking the IP, command the world*. China Travel & Tourism Press.

Honoré, C. (2010). *In praise of slow: How a worldwide movement is challenging the cult of speed*. Hachette.

Honoré, C. (2013). *The slow fix: solve problems, work smarter and live better in a world addicted to speed*. Knopf Canada.

Hoogendoorn, G., & Fitchett, J. M. (2018). Tourism and climate change: A review of threats and adaptation strategies for Africa. *Current Issues in Tourism*, *21*(7), 742–759. doi:10.1080/13683500.2016.1188893

Horng, J. S., Liu, C. H., Chou, H. Y., & Tsai, C. Y. (2012). Understanding the impact of culinary brand equity and destination familiarity on travel intentions. *Tourism Management*, *33*(4), 815–824. doi:10.1016/j.tourman.2011.09.004

Hou, B., & Chen, Z. Y. (2008). Study on the development pattern of catering street on the basis of urban tourism. *Culinary Science Journal of Yangzhou University*, *4*, 26-31.

Huang, X. Y., & Liang, R. Q. (2020, March 3). How can the tourism industry turn "crisis" into "opportunity" in the post-pandemic? *China Travel News*. https://www.traveldaily.cn/article/136134

Hughes, L. (2020). *The restaurant chains which collapsed, restructured and cut jobs in 2020*. InYourArea. https://www.inyourarea.co.uk/news/the-restaurant-chains-which-collapsed-restructured-and-slashed-jobs-in-2020/

Hughes, G. (1995). Authenticity in tourism. *Annals of Tourism Research*, *22*(4), 781–803. doi:10.1016/0160-7383(95)00020-X

Huhn, S., Kharabian Masouleh, S., Stumvoll, M., Villringer, A., & Witte, A. V. (2015). Components of a Mediterranean diet and their impact on cognitive functions in aging. *Frontiers in Aging Neuroscience*, *7*, 132. doi:10.3389/fnagi.2015.00132 PMID:26217224

Hui, D. S., Azhar, E. I., Madani, T. A., Nitoumi, F., Kock, R., & Dar, O. (2020). The continuing 2019-nCoV epidemic threat of novel coronaviruses to global health – the latest 2019 novel coronavirus outbreak in Wuhan, China. *International Journal of Infectious Diseases*, *91*, 264–266. doi:10.1016/j.ijid.2020.01.009 PMID:31953166

Hummel, M., Hallahan, B. F., Brychkova, G., Ramirez-Villegas, J., Guwela, V., Chataika, B., & … . (2018). Reduction in nutritional quality and growing area suitability of common bean under climate change induced drought stress in Africa. *Scientific Reports*, *8*(1), 1–11. doi:10.103841598-018-33952-4 PMID:30385766

Hu, X. F. (2010). A deep reflection over Guangzhou's loss of "Gourmet Capital". *Journal of Guangdong University of Finance & Economics*, *25*(6), 77–80.

Hu, X., Yan, H., Casey, T., & Wu, C.-H. (2021). Creating a safe haven during the crisis: How organizations can achieve deep compliance with COVID-19 safety measures in the hospitality industry. *International Journal of Hospitality Management*, *92*, 102662. doi:10.1016/j.ijhm.2020.102662 PMID:32904503

Hwang, L. J. J., van Westering, J., & Chen, H. H. (2005). Exploration of the linkages between the gastronomy and heritage of Tainan city, Taiwan. In *Advances in hospitality and leisure*. Emerald Group Publishing Limited.

Hwang, Y., & Felitti, D. (2021). A quest for the healthy lifestyle and learning experiences in olive oil tourism. *Journal of Tourism and Cultural Change*. Advance online publication. doi:10.1080/14766825.2021.1917586

ICAEW. (2020). *Financial Accounting and Reporting (IFRS) Q&A 2020. ACA Simplified*. Great Britain Amazon.

Ichijo, A., & Ranta, R. (2016). *Food, national identity and nationalism: From everyday to global politics.* Palgrave Macmillan.

İçöz, O. (2014). *Seyahat Acentaları ve Tur Operatörlüğü Yönetimi* [Agencies and Tour Operator Management]. Turhan Kitabevi.

Igamberdievna, A. D. (2021). Promising measures for the development of sustainable tourism in the context of the COVID-19 pandemic. *Academic Journal of Digital Economics and Stability, 4,* 1–6.

iiMedia Research. (2020, April 12). *China's catering and food delivery merchants during COVID-19 report.* https://www.iimedia.cn/c400/70742.html

iiMedia Research. (2021, January 4). *2020-2021 China mobile social industry report.* https://www.iimedia.cn/c400/76205.html

Ilincic, M. (2014). Benefits of creative tourism - The tourist perspective I Tourism I Creativity. *Academia.Eu,* 99–113. https://pt.scribd.com/document/309796523/Benefits-of-Creative-Tourism-The-Tourist-Perspective

International Monetary Fund. (2020). *World Economic Outlook, April 2020: The Great Lockdown.* https://www.imf.org/en/Publications/WEO/Issues/2020/04/14/weo-april-2020

International Olive Council. (2020). www.internationaloliveoil.org/

International Organization for Migration. (2020). *World Migration Report 2020.* https://publications.iom.int/system/files/pdf/wmr_2020.pdf

IPCC. (2019). *IPCC Special Report on the Ocean and Cryosphere in a Changing Climate. IPCC.*

İplikçi, G., & Ayhar, N. Y. (2017). Beslenme Alzheimerdan Koruyucu Bir Faktör Müdür? [Is Nutrition a Protective Factor For Alzheimer's?] *Ankara Sağlık Bilimleri Dergisi, 1-2-3,* 1-11.

Iranwonders. (2022). https://iranwonders.com/fa/articles-fa/165-iran-cuisine

İrek, A. (1991). *Turizm İşletmelerinde Örgütlenme* [Organizing in Tourism Enterprises]. Yüksek Lisans Tezi. İstanbul Üniversitesi Sosyal Bilimler Enstitüsü.

iResearch. (2020, April 20). *2020 outlook for the recovery of hospitality and tourism industries in China after COVID-19 pandemic.* http://www.iresearchchina.com/content/details8_61202.html

IRNA. (2022). https://www.irna.ir/news/81130204

Ivanov, D., & Dolgui, A. (2020). Viability of intertwined supply networks: Extending the supply chain resilience angles towards survivability. A position paper motivated by COVID-19 outbreak. *International Journal of Production Research, 58*(10), 2904–2915. doi:10.1080/00207543.2020.1750727

Ivanov, S., Gavrilina, M., Webster, C., & Ralko, V. (2016). Impacts of political instability on the tourism industry in Ukraine. *Journal of Policy Research in Tourism, Leisure & Events.* Advance online publication. doi:10.1080/19407963.2016.1209677

IVV/Viniportugal. (2021). *Statistics. Portugal in the world.* https://www.viniportugal.pt/Statistics

Jabs, J., & Devine, C. M. (2006). Time scarcity and food choices: An overview. *Appetite, 47*(2), 196–204. doi:10.1016/j.appet.2006.02.014 PMID:16698116

Jackie. (2018, December 16). *Top 12 UNESCO creative cities of gastronomy foodie destinations to visit.* https://lifeofdoing.com/unesco-creative-cities-of-gastronomy-foodie-destinations/

Jacobsen, J. K. S., & Haukeland, J. V. (2002). A lunch with a view: Motor tourists' choices and assessments of eating-places. *Scandinavian Journal of Hospitality and Tourism*, *2*(1), 4–16.

Jangid, K. K., & Dwivedi, P. (2016). Physiological responses of drought stress in tomato: A review. *International Journal of Agriculture Environment and Biotechnology*, *9*(1), 53–61. doi:10.5958/2230-732X.2016.00009.7

Jang, S., & Cai, L. A. (2009). Travel motivations and destination choice: A study of British outbound market. *Journal of Travel & Tourism Marketing*, *13*(3), 111–133. doi:10.1080/10548400209511570

Jarrett, S., & Jarvis, W. (2016). *Defining the luxury wine category via an assessment of price-tier perceptions* [Conference session]. 9th Academy of Wine Business Research Conference, Adelaide.

Jentsch, B., & Simard, M. (2016). *International Migration and Rural Areas, Cross-National Comparative Perspectives - Studies in Migration and Diaspora*. Taylor and Francis.

Jepson, D. (2015). The lure of the countryside: the spiritual dimension of rural spaces of leisure. In S. Gammon & S. Elkington (Eds.), *Landscapes of leisure* (pp. 202–219). Palgrave Macmillan. doi:10.1057/9781137428530_15

Johnson, R. (2014). *Food fraud and "economically motivated adulteration" of food and food ingredients*. Congressional Research Service. https://fas.org/sgp/crs/misc/R43358.pdf

Jones, A., & Jenkins, I. (2002). A taste of Wales-Blas Ar Gymru: institutional mailaise in promoting Welsh food tourism product. In A.-M. Hjalager & G. Richards (Eds.), Tourism and gastronomy. Routledge.

Kala, D. (2021). Thank you, God. You saved us' - examining tourists' intention to visit religious destinations in the post COVID. *Current Issues in Tourism*, *0*(0), 1–7. doi:10.1080/13683500.2021.1876643

Kampf, G., Todt, D., Pfaender, S., & Steinmann, E. (2020). Persistence of coronaviruses on inanimate surfaces and their inactivation with biocidal agents. *The Journal of Hospital Infection*, *104*(3), 246–251. doi:10.1016/j.jhin.2020.01.022 PMID:32035997

Kampf, G., Voss, A., & Scheithauer, S. (2020a). Inactivation of coronaviruses by heat. *The Journal of Hospital Infection*, *105*, 348–349. doi:10.1016/j.jhin.2020.03.025 PMID:32243951

Kapferer, J.-N. (2014). The future of luxury: Challenges and opportunities. *Journal of Brand Management*, *21*(9), 716–726. doi:10.1057/bm.2014.32

Karatepe, O. M., Rezapouraghdam, H., & Hassannia, R. (2020). Job insecurity, work engagement and their effects on hotel employees' non-green and nonattendance behaviors. *International Journal of Hospitality Management*, *87*, 102472. doi:10.1016/j.ijhm.2020.102472

Karatepe, O. M., Rezapouraghdam, H., & Hassannia, R. (2021a). Sense of calling, emotional exhaustion and their effects on hotel employees' green and non-green work outcomes. *International Journal of Contemporary Hospitality Management*, *33*(10), 3705–3728.

Karatepe, O. M., Rezapouraghdam, H., & Hassannia, R. (2021b). Does employee engagement mediate the influence of psychological contract breach on pro-environmental behaviors and intent to remain with the organization in the hotel industry? *Journal of Hospitality Marketing & Management*, *30*(3), 326–353. doi:10.1080/19368623.2020.1812142

Kastorini, C. M., Milionis, H. J., Esposito, K., Giugliano, D., Goudevenos, J. A., & Panagiotakos, D. B. (2011). The effect of Mediterranean diet on metabolic syndrome and its components: A meta-analysis of 50 studies and 534,906 individuals. *Journal of the American College of Cardiology*, *57*(11), 1299–1313. doi:10.1016/j.jacc.2010.09.073 PMID:21392646

Kastorini, C. M., Milionis, H. J., Goudevenos, J. A., & Panagiotakos, D. B. (2010). Mediterranean diet and coronary heart disease: Is obesity a link? - A systematic review. *Nutrition, Metabolism, and Cardiovascular Diseases*, *20*, 536–551. doi:10.1016/j.numecd.2010.04.006 PMID:20708148

Keller, K. L. (2010). Brand Equity Management in a Multichannel, Multimedia Retail Environment. *Journal of Interactive Marketing*, *24*(2), 58–70. doi:10.1016/j.intmar.2010.03.001

Kemer, A. K. (2011). *Otellerde Çalışan Mutfak Personelinin ve Aşçılık Alanında Yüksek Öğrenim* [Kitchen Staff working in Hotels and Higher Education in Cooking]. Basılmamış Yüksek Lisans Tezi. Gazi Üniversitesi Eğitim Bilimleri Enstitüsü.

Kernel, P. (2005). Creating and implementing a model for sustainable development in tourism and enterprises. *Journal of Cleaner Production*, *13*(2), 151–164. doi:10.1016/j.jclepro.2003.12.023

Keshavarzi, P., & Akbari, M. (2012). *The role of food tourism in urban and regional development: the case of Bonab city* [Conference session]. 1st National Conference on Tourism, Geography and Environmental Sustainability, Hamadan, Iran.

Kesse-Guyot, E., Ahluwalia, N., Lassale, C., Hercberg, S., Fezeu, L., & Lairon, D. (2013). Adherence to Mediterranean diet reduces the risk of metabolic syndrome: A 6-year prospective study. *Nutrition, Metabolism, and Cardiovascular Diseases*, *23*(7), 677–683. doi:10.1016/j.numecd.2012.02.005 PMID:22633793

Keys, A. (1970). Coronary heart disease in seven countries. Summary. *Circulation*, *41*(4, Suppl), I186–I195. PMID:5442782

Keys, A. (1980). *Seven countries: a multivariate analysis of death and coronary heart disease*. Harvard University Press.

Keys, A., & Keys, M. (1975). *How to eat well and stay well. The Mediterranean way*. Doubleday Publishing.

Keys, A., Menotti, A., Aravanis, C., Blackburn, H., Djordevic, B., & Buzina, R. (1984). The seven countries study: 2,289 deaths in 15 years. *Preventive Medicine*, *13*(2), 141–154. doi:10.1016/0091-7435(84)90047-1 PMID:6739443

Kiatkawsin, K., & Han, H. (2019). What drives customers' willingness to pay price premiums for luxury gastronomic experiences at michelin-starred restaurants? *International Journal of Hospitality Management*, *82*, 209–219. doi:10.1016/j.ijhm.2019.04.024

Kielkiewicz, K., & Dalzell, T. (2015). Towards comprehension of spirituality through its semantics. *A Journal of Religion, Education and the Arts, 9*(January).

Kimball, A., Hatfield, K. M., Arons, M., James, A., Taylor, J., Spicer, K., & … . (2020). Asymptomatic and presymptomatic SARS-COV-2 infections in residents of a long-term care skilled nursing facility. *Morbidity and Mortality Weekly Report*, *69*, 377–381. doi:10.15585/mmwr.mm6913e1 PMID:32240128

Kim, H. M., & Laskowski, M. (2018). Agriculture on the Blockchain: Sustainable Solutions for Food, Farmers, and Financing. In D. Tapscott (Ed.), *Supply Chain Revolution, Barrow Books 2018*. doi:10.2139srn.3028164

Kim, J.-H. (2014). The antecedents of memorable tourism experiences: The development of a scale to measure the destination attributes associated with memorable experiences. *Tourism Management*, *44*, 34–45. doi:10.1016/j.tourman.2014.02.007

Kim, J., Kim, J., Lee, S. K., & Tang, L. (2020). Effects of epidemic disease outbreaks on financial performance of restaurants: Event study method approach. *Journal of Hospitality and Tourism Management*, *43*, 32–41. doi:10.1016/j.jhtm.2020.01.015

Kim, K., Bonn, M. A., & Cho, M. (2021). Clean safety message framing as survival strategies for small independent restaurants during the Covid-19 pandemic. *Journal of Hospitality and Tourism Management*, *46*, 423–431. doi:10.1016/j.jhtm.2021.01.016

Kim, S., Park, E., & Xu, M. (2020). Beyond the authentic taste: The tourist experience at a food museum restaurant. *Tourism Management Perspectives*, *36*, 100749. doi:10.1016/j.tmp.2020.100749

Kim, Y. (2018). Power moderates the impact of desire for exclusivity on luxury experiential consumption. *Psychology and Marketing*, *35*(4), 283–293. doi:10.1002/mar.21086

Kim, Y. G., & Eves, A. (2012). Construction and validation of a scale to measure tourist motivation to consume local food. *Tourism Management*, *33*(6), 1458–1467. doi:10.1016/j.tourman.2012.01.015

Kim, Y. G., Eves, A., & Scarles, C. (2009). Building a model of local food consumption on trips and holidays: A grounded theory approach. *International Journal of Hospitality Management*, 423–431. doi:10.1016/j.ijhm.2008.11.005

Kim, Y.-I., & Mason, J. B. (1996). Nutrition chemoprevention of gastrointestinal cancers: A critical review. *Nutrition Reviews*, *54*(9), 259–279. doi:10.1111/j.1753-4887.1996.tb03947.x PMID:9009668

Kiptenko, V., & Doan, P. (2015). Destination Ukraine: Tourism litmus of the transition to a green economy. In M. Reddy & K. Wilkes (Eds.), *Tourism in the Green Economy*. Routledge. doi:10.4324/9781315885681

Kivela, J., & Crotts, J. C. (2005). Gastronomy tourism: A meaningful travel market segment. *Journal of Culinary Science & Technology*, *4*(2/3), 39–55. doi:10.1300/J385v04n02_03

Kivela, J., & Crotts, J. C. (2006). Tourism and gastronomy: Gastronomy's influence on how tourists experience a destination. *Journal of Hospitality & Tourism Research (Washington, D.C.)*, *30*(3), 354–377. doi:10.1177/1096348006286797

Klenosky, D. B. (2003). The pull of tourism destinations: A means-end investigation. *Journal of Travel Research*, 385–395. doi:10.1177/004728750204000405

Kocaman, S. (2011). *Edirne İlinin Turizm Coğrafyası* [Tourism Geography of Edirne Province]. Doktora Tezi. Atatürk Üniversitesi Sosyal Bilimler Enstitüsü.

Kolawole Olayiwola, A. S. (2007). Impact of globalization on food consumption, health and nutrition in Nigeria. In FAO (Ed.), Globalization of Food Systems in Developing Countries: impact on food security and nutrition. FAO.

Koloverou, E., Esposito, K., Giugliano, D., & Panagiotakos, D. (2014). The effect of Mediterranean diet on the development of type 2 diabetes mellitus: A meta-analysis of 10 prospective studies and 136,846 participants. *Metabolism: Clinical and Experimental*, *63*(7), 903–911. doi:10.1016/j.metabol.2014.04.010 PMID:24931280

Korinth, A., Schiess, S., & Westenhoefer, J. (2010). Eating behaviour and eating disorders in students of nutrition sciences. *Public Health Nutrition*, 221–232. doi:10.1017/S1368980009005709 PMID:19433007

Korkmaz, M., & Başaran, G. (2021). The domestic transformation of coffee consumption habits in the Covid-19 pandemic process. *Journal of Original Studies*, *2*(1), 43–56.

Kostynets, V., Kostynets, I., & Olshanska, O. (2021). Pent-up Demand's Realization in the Hospitality Sector in the Context of COVID-19. *Journal of International Students*, *14*(1), 89–102. doi:10.14254/2071-8330.2021/14-1/6

Kotler, P., & Keller, K. L. (2016). *A framework for marketing management*. Pearson.

Koyuncu, M. (2017). Global Climate Change and Animal Husbandry. *Selcuk Journal of Agriculture and Food Sciences*, *31*(2), 98–106. doi:10.15316/SJAFS.2017.26

Kubiszewski, I., Costanza, R., Anderson, S., & Sutton, P. (2020). The future value of ecosystem services: Global scenarios and national implications. In Environmental Assessments. In K. N. Ninan (Ed.), *Environmental Assessments: Scenarios, Modelling and Policy* (pp. 81–108). Edward Elgar Publishing. doi:10.4337/9781788976879.00016

Kumar, A., Mangla, S. K., Kumar, P., & Song, M. (2021). Mitigate risks in perishable food supply chains: Learning from COVID-19. *Technological Forecasting and Social Change*, *166*, 120643. doi:10.1016/j.techfore.2021.120643

Kumar, A., & Nayar, K. R. (2021). COVID 19 and its mental health consequences. *Journal of Mental Health (Abingdon, England)*, *30*(1), 1–2. doi:10.1080/09638237.2020.1757052 PMID:32339041

Kumar, A., Paul, J., & Starčević, S. (2021a). Do brands make consumers happy?- A masstige theory perspective. *Journal of Retailing and Consumer Services*, *58*, 102318. doi:10.1016/j.jretconser.2020.102318

Kumar, A., Paul, J., & Unnithan, A. B. (2020). 'Masstige' marketing: A review, synthesis and research agenda. *Journal of Business Research*, *113*, 384–398. doi:10.1016/j.jbusres.2019.09.030

Kündeş, H. (2019). *Tip 2 Diyabetli Hastalarda Akdeniz Diyeti ve Dash Diyetine Uyumun Glisemik Kontrol Üzerine Etkisi* [Effect of Mediterranean Diet and Dash Diet Compliance on Glycemic Control in Patients with Type 2 Diabetes]. Yüksek Lisans Tezi. Hacettepe Üniversitesi Sağlık Bilimleri Enstitüsü, Toplu Beslenme Sistemleri Programı.

Kürklü, B. B. (2021). Akdeniz Diyeti ve Bileşenlerinin Osteoartrit ile İlişkisi [Relationship of Mediterranean Diet and Ingredients with Osteoarthritis]. *Beslenme ve Diyetetik Güncel Konular, 10*, 211–221.

Kurtdaş, M. Ç. (2020). Some evaluations on the social effects of Covid-19. *Journal of City and Civilization,* 530-545.

Kuswandi, B., Wicaksono, Y., Jayus, A. A., Heng, L. Y., & Ahmad, M. (2011). Smart packaging: Sensors for monitoring of food quality and safety. *Sensing and Instrumentation for Food Quality and Safety*, *5*(3), 137–146. doi:10.1007/s11694-011-9120-x

Kuyumcu, A., & Kuyumcu, M. S. (2019). Akdeniz Diyeti ve Karotis Arter Hastalığı Arasındaki İlişki [The Relationship Between Mediterranean Diet and Carotis Artery Disease]. *Süleyman Demirel Üniversitesi Sağlık Bilimleri Dergisi*, *10*(2), 99–104.

Kyriakaki, A., Zagkotsi, S., & Trihas, N. (2016). Gastronomy, tourist experience and location. The case of the greek breakfast. Tourismos. *An International Multidisciplinary Journal of Tourism*, *11*(3), 227–261. doi:10.26215/tourismos.v11i3.489

L'viv – the Ensemble of the Historic Centre. (2021). *UNESCO: Culture: World Heritage Centrest: World Heritage List.* https://whc.unesco.org/en/list/865

Lagiso, D. (2020). Challenges and opportunities of Covid-19 in agricultural economy: The case of Ethiopia. *International Journal of African and Asian Studies*, *65*, 24–29. https://iiste.org/Journals/index.php/JAAS/article/view/52989

Lai, M. Y., Wang, Y., & Khoo-Lattimore, C. (2020). Do food image and food neophobia affect tourist intention to visit a destination? The case of Australia. *Journal of Travel Research*, *59*(5), 928–949. doi:10.1177/0047287519867144

Lapatza, J. J. F. (2006). Simplificación; base imponible y renta empresarial. *Quincena fiscal: Revista de actualidad fiscal, 21,* 13-159.

Lapatza, J. J. F. (2007). Resultado contable y base imponible: Deslegalización, autodeterminación y delito fiscal [Accounting result and tax base: Delegalization, self-determination and tax crime]. *Quincena fiscal: Revista de actualidad fiscal, 8,* 9-12.

Laranjeiro, A. (2020). Hotéis continuam a investir apesar da pandemia. *Dinheiro Vivo.* https://www.dinheirovivo.pt/economia/hoteis-continuam-a-investir-apesar-da-pandemia-12895854.html

Laudan, R. (2004). Slow Food: The french terroir strategy, and culinary modernism. *Food, Culture, & Society*, *7*(2), 133–144. doi:10.2752/155280104786577833

Lawler, O. K., Allan, H. L., Baxter, P. W., Castagnino, R., Tor, M. C., & Dann, L. E. (2021). The COVID-19 pandemic is intricately linked to biodiversity loss and ecosystem health. *The Lancet. Planetary Health*, *5*(11), e840–e850. doi:10.1016/S2542-5196(21)00258-8 PMID:34774124

Lee, C., Hallak, R., & Sardeshmukh, S. (2019). Creativity and innovation in the restaurant sector: Supply-side processes and barriers to implementation. *Tourism Management Perspectives*, *31*, 54–62. doi:10.1016/j.tmp.2019.03.011

Lee, K. H., & Scott, N. (2015). Food tourism reviewed using the paradigm funnel approach. *Journal of Culinary Science & Technology*, *13*(2), 95–115. doi:10.1080/15428052.2014.952480

Lee, S., Bee-Lia Chua, & Han, H. (2020). Variety-seeking motivations and customer behaviors for new restaurants: An empirical comparison among full-service, quick-casual, and quick-service restaurants. *Journal of Hospitality and Tourism Management*, *43*, 220–231. doi:10.1016/j.jhtm.2020.04.004

Lehtinen, U. (2012). Sustainability and local food procurement: A case study of finnish public catering. *British Food Journal*, *114*(8), 1053–1071. doi:10.1108/00070701211252048

Lehtolainen, M. (2003). Public infrastructure investments and their role in tourism development in the Finnish Lake Region. In *Proceedings of Lake Tourism Conference 2003*. Savonlinna Institute for Regional Development and Research International.

Levitt, T. (1972). Production-line approach to service. *Harvard Business Review*, *50*(5), 41–52.

Li, H. N. (2021, August 18). Not afraid of the epidemic, the catering industry actively seeks change and looks forward to a bright future. *China Economic Times*. http://jjsb.cet.com.cn/

Li, S. B. (2009). *Research on the Development of the Tourism of Macao* [Master's dissertation]. Huazhong University of Science & Technology. https://kns.cnki.net/KCMS/detail/detail.aspx?dbname=CMFD2012&filename=1011276171.nh

Li, B., Zhong, Y., Zhang, T., & Hua, N. (2021). Transcending the COVID-19 crisis: Business resilience and innovation of the restaurant industry in China. *Journal of Hospitality and Tourism Management*, *49*, 44–53. doi:10.1016/j.jhtm.2021.08.024

Li, J. (2021). Current situation and countermeasures of tourism food industry in Yunnan, *Food. Research for Development*, *42*(6), 229–230. https://doi.org/CNKI:SUN:SPYK.0.2021-06-041

Lima, E. V. (2003). *VAT: Value added tax, commented and annotated* (9th ed.). Porto Editora.

Linares, E. C., Cortés, V. S., Rodríguez, M. M., & Padrón, B. R. (2018). Patrimonio gastronómico y desarrollo local en dos Pueblos Mágicos de México [Gastronomic heritage and local development in two Magical Towns of Mexico]. *Revista Iberoamericana de Viticultura, Agroindustria y Ruralidad*, *5*(15), 22–38. https://www.redalyc.org/jatsRepo/4695/469565683002/469565683002.pdf

Lin, Y. C., Pearson, T. E., & Cai, L. A. (2011). Food as a form of destination identity: A tourism destination brand perspective. *Tourism and Hospitality Research*, *11*(1), 30–48. doi:10.1057/thr.2010.22

Lipor. (n.d.). *Dose Certa* [Certain Dose]. https://www.lipor.pt/pt/sensibilizar/100-desperdicio/desperdicio-alimentar/

Lipor. (n.d.a). *Embrulha. e os seus restaurantes à distância de um clique!* [Wraps. and your restaurants just a click away!]. https://www.lipor.pt/pt/noticias/embrulha-e-os-seus-restaurantes-a-distancia-de-um-clique/

Little, J. (2012). Transformational tourism, nature and wellbeing: New perspectives on fitness and the body. *Sociologia Ruralis*, *52*(3), 257–271.

Liu, G. (2012). Landscape design of sanhe quezhu culture city based on regional culture. *Anhui Architecture*, 2. doi:10.16330/j.cnki.1007-7359.2012.02.060

Liu, Y., Ning, Z., Chen, Y., Guo, M., Liu, Y., Gali, N. K., Sun, L., Duan, Y., Cai, J., Westerdahl, D., Liu, X., Xu, K., Ho, K. F., Kan, H., Fu, Q., & Lan, K. (2020). Aerodynamic analysis of SARS-CoV-2 in two Wuhan hospitals. *Nature, 1–4*. Advance online publication. doi:10.103841586-020-2271-3 PMID:32340022

Li, X., He, X. D., & Liu, S. Y. (2019). The development trend of culinary tourism at home and abroad. *Tourism Research, 11*(4), 5–9.

Lockshin, L., Cohen, E., & Zhou, X. (2011). What Influences Five-star Beijing Restaurants in Making Wine Lists? *Journal of Wine Research, 22*(3), 227–243. doi:10.1080/09571264.2011.596200

Lockshin, L., Jarvis, W., Hauteville, F., & Perrouty, J.-P. (2006). Using simulations from discrete choice experiments to measure consumer sensitivity to brand, region, price and awards in wine choice. *Food Quality and Preference, 17*(3/4), 166–178.

Loco. (n.d.). *Projeto*. http://www.in-loco.pt/pt/projectos/20190301/100-local/

Long, L. (2004). *Culinary tourism (material worlds)*. The University Press of Kentucky.

Long, L. M. (1998). Introduction to the special issue on culinary tourism. *Southern Folklore, 55*(3), 179.

Long, L. M. (2004). *Culinary tourism*. The University Press of Kentucky.

Long, M. (2004). *Culinary Tourism*. University Press of Kentucky.

Longue, A. W. (1991). *The psychology of eating and drinking: An introduction*. WH Freeman.

Lopes, C. T. D., Oliveira, A., Severo, M., Alarcão, V., Guiomar, S., & Mota, J. (2017). *Inquérito Alimentar Nacional e de Atividade Física* [National Food and Physical Activity Survey]. IAN-AF 2015-2016: Relatório de Resultados. www.ian-af.up.pt

Lopes, A. S., Sargento, A., & Carreira, P. (2021). Vulnerability to COVID-19 unemployment in the Portuguese tourism and hospitality industry. *International Journal of Contemporary Hospitality Management, 33*(5), 1850–1869. doi:10.1108/IJCHM-11-2020-1345

López-Guzmán, T., & Sánchez-Cañizares, S. (2012). Gastronomy, tourism and destination differentiation: A case study in Spain. *Revue d'Economie Financiere, 1*, 63–72.

Lu, C. B., & Xu, T. R. (2021). Dimensions and focus of perceived value of Internet celebrity restaurant visitors —— Empirical research based on network text analysis. *Journal of Researches on Dietetic Science and Culture, 1*, 33–42. https://doi.org/10.19913/j.cnki.2095-8730msyj.2021.0003

Lück, M., & Aquino, R. S. (2021). Domestic nature-based tourism and wellbeing - A roadmap for the new domestic nature. In J. Wilks, D. Pendergast, P. Leggat, & D. Morgan (Eds.), *Tourist health, safety and wellbeing in the new normal*. Springer., doi:10.1007/978-981-16-5415-2

Luczaj, L., Pieroni, A., Tardío, J., Pardo-de-Santayana, M., Sõukand, R., Svanberg, I., & Kalle, R. (2012). Wild food plant use in 21 st century Europe, the disapperance of old traditions and the search for new ciusines involving wild edibles. *Acta Societatis Botanicorum Poloniae, 81*(4), 359–370. doi:10.5586/asbp.2012.031

Lugares turísticos. (2020). 3 Lugares Turísticos en Naranjal [Tourist Places in Naranjal]. *Lugares turísticos*. https://www.lugaresturisticos.org/naranjal/

Lu, L., & Li, P. L. (2021). Research on the development of food culture tourism under the guidance of Food Culture and Tourism. *Food Industries*, *3*, 374–375. https://doi.org/CNKI:SUN:SPGY.0.2021-03-110

Luo, Y., & Xu, X. (2021). Comparative study of deep learning models for analyzing online restaurant reviews in the era of the COVID-19 pandemic. *International Journal of Hospitality Management*, *94*, 102849. doi:10.1016/j.ijhm.2020.102849 PMID:34785843

Lu, R., Zhao, X., Li, J., Niu, P., Yang, B., Wu, H., & Wang, W. (2020). Genomic characterisation and epidemiology of 2019 novel coronavirus: Implications for virus origins and receptor binding. *Lancet*, *395*(10224), 565–574. doi:10.1016/S0140-6736(20)30251-8 PMID:32007145

Lviv Tourist Barometer. (2021). http://www.barometer.lviv.ua

Lynch, L. (2021, April 26). Why I travel for food and so should you. *The Planet D: Adventure Travel Blog*. https://theplanetd.com/culinary-travel-why-i-travel-for-food/

Ma, L. J., & Liu, C. (2021). Study on the preference and influencing factors of potential customers for city food in China — based on the perspective of network attentions. *Resource Development & Market, 37*(3), 365-371. doi:10.3969/j.issn.1005-8141.2021.03.016

MacCannell, D. (1973). Staged authenticity: Arrangements of social space in tourist settings. *American Journal of Sociology*, *79*(3), 589–603. https://www.jstor.org/stable/2776259

Macedo, V. (2020). The opportunity for a new tax paradigm. In Os Impostos e o Combate às Desigualdades (Na era do Covid-19). Rei dos livros.

Machado, A. F., Sousa, B. B., Rocha, A. M. A., & de Abreu, Z. H. L. (2020, Apr 01-02). *Communication Strategies to Combat the Negative Effects of Covid-19 On Tourism* [Conference session]. International Business Information Management Association Conference (IBIMA).

Mackenbach, J. P. (2007). The Mediterranean diet story illustrates that "why" questions are as important as "how" questions in disease explanation. *Journal of Clinical Epidemiology*, *60*(2), 105–109. doi:10.1016/j.jclinepi.2006.05.001 PMID:17208115

Major, M., Conceição, A., & Clegg, S. (2018). When institutional entrepreneurship failed: The case of a responsibility centre in a Portuguese hospital. *Accounting, Auditing & Accountability Journal*, *31*(4), 1199–1229.

Mak, A. H., Lumbers, M., & Eves, A. (2012). Globalization and food consumption in tourism. *Annals of Tourism Research*, *39*, 171–196. doi:10.1016/j.annals.2011.05.010

Makkar, S., & Singh, A. K. (2021). Development of a spirituality measurement scale. *Current Psychology (New Brunswick, N.J.)*, *40*(3), 1490–1497.

Makroo, H., Majid, D., Siddiqi, M. A., Greiner, R., & Dar, B. N. (2020). *COVID-19 pandemic and its implication on food system*. Preprints., doi:10.20944/preprints202008.0321.v1

Maldonado, J. K., Shearer, C., Bronen, R., Peterson, K., & Lazrus, H. (2013). The impact of climate change on tribal communities in the US: displacement, relocation, and human rights. In J. K. Maldonado, C. Shearer, R. Bronen, K. Peterson, & H. Lazrus (Eds.), *Climate change and indigenous peoples in the United States* (pp. 93–106). Springer.

Marino, E. (2012). The long history of environmental migration: Assessing vulnerability construction and obstacles to successful relocation in Shishmaref, Alaska. *Global Environmental Change*, *22*(2), 374–381. doi:10.1016/j.gloenvcha.2011.09.016

Marques, C. P., Guedes, A., & Bento, R. (2021). Rural tourism recovery between two COVID-19 waves: The case of Portugal. *Current Issues in Tourism*. Advance online publication. doi:10.1080/13683500.2021.1910216

Marshall, B., Cardon, P., Poddar, A., & Fontenot, R. (2013). Does sample size matter in qualitative research? A review of qualitative interviews in IS research. *Journal of Computer Information Systems*, *54*(1), 11–22. doi:10.1080/08874417.2013.11645667

Martinez-Gonzalez, M. A., de la Fuente-Arrillaga, C., Nunez-Cordoba, J. M., Basterra-Gortari, F. J., Beunza, J. J., & Vazquez, Z. (2008). Adherence to Mediterranean diet and risk of developing diabetes: Prospective cohort study. *BMJ (Clinical Research Ed.)*, *336*(7657), 1348–1351. doi:10.1136/bmj.39561.501007.BE PMID:18511765

Martínez-González, M. Á., Hershey, M. S., Zazpe, I., & Trichopoulou, A. (2017). Transferability of the Mediterranean diet to non-Mediterranean countries. What is and what is not the Mediterranean diet. *Nutrients*, *9*(11), 1226. doi:10.3390/nu9111226 PMID:29117146

Martinez-Lacoba, R., Pardo-Garcia, I., Amo-Saus, E., & Escribano-Sotos, F. (2018). Mediterranean diet and health outcomes: A systematic meta-review. *European Journal of Public Health*, *28*(5), 955–961. doi:10.1093/eurpub/cky113 PMID:29992229

Martín, J. C., Roman, C., Lopez-Guzman, T., & Moral-Cuadra, S. (2020). A fuzzy segmentation study of gastronomical experience. *International Journal of Gastronomy and Food Science*, *22*, 100248. https://doi.org/10.1016/j.ijgfs.2020.100248

Martins, A., Dinis, A., & Lopes, C. (2018). Interpretative complexity and tax law changes: The taxation of corporate expenses in Portugal. *Journal of Applied Accounting Research*, *19*(4), 537–551. doi:10.1108/JAAR-04-2017-0050

Martins, A., Sá, C., & Taborda, D. (2020). *A dedutibilidade de gasto no IRC. Uma análise económica-fiscal* [The deductibility of expenditure in irc. An economic-fiscal analysis.]. Almedina.

Martins, H. P. (2017). Corporation tax. In *Lições de Fiscalidade. Princípios gerais e fiscalidade interna* (5th ed., Vol. 1). Almedina.

Marty, L., de Lauzon-Guillain, B., Labesse, M., & Nicklaus, S. (2021). Food choice motives and the nutritional quality of diet during the COVID-19 lockdown in France. *Appetite*, *157*, 105005. PMID:33068666

Marujo, N. (2016). Turismo, turistas e experiências: abordagens teóricas [Tourism, tourists and experiences: theoretical approaches]. *Revista: Turydes Revista Turismo y Desarrollo*. http://hdl.handle.net/10174/20116

Mason, K., & Bequette, J. (1998). Product experience and consumer product attribute inference accuracy. *Journal of Consumer Marketing*, *15*(4), 343–357. doi:10.1108/07363769810225984

Matta, R. (2014). Gastronomic republic and country of cooks: Food, politics, media and a new idea of nation for Peru. *Colombian Journal of Anthropology*, *50*(2), 15–40. http://www.scielo.org.co/ pdf/rcan/v50n2/ v50n2a02.pdf

Maynard, D. C., Zandonadi, R. P., Nakano, E. Y., & Botelho, R. B. A. (2020). Sustainability indicators in restaurants: The development of a checklist. *Sustainability*, *12*, 4076. doi:10.3390u12104076

McDonald's Reports Fourth Quarter & Full Year. (2019). *Results and Quarterly Cash Dividend*. McDonald's Newsroom. McDonald's Corporation.

Mcguire, M. (2003). Why bodies matter: A sociological reflection on spirituality and materiality. *Journal of Christian Spirituality*, *3*(1), 1–18.

Mediterrânica. (n.d.). *D. Feira da Dieta Mediterrânica*. https://www.dietamediterranica.pt/?q=pt/node/759

Mehmetoglu, M., & Engen, M. (2011). Pine and Gilmore's concept of experience economy and its dimensions: An empirical examination in tourism. *Journal of Quality Assurance in Hospitality & Tourism, 12*(4), 237–255. doi:10.1080/1528008X.2011.541847

Melbourne University Report. (2015). *Appetite for Change Global Warming Impacts on Food and Farming Regions in Australia*, Melbourne Sustainable Society Institute.

Melville, A. (2019). *International Financial Reporting: A Practical Guide* (7th ed.). Person.

Mendez, M., & Popkin, B. (2004). Globalization, urbanization, and nutritional change in the developing world. *The Electronic Journal of Agricultural and Development Economics, 1*(2), 220–241.

Meneguel, C. R. D. A., Mundet, L., & Aulet, S. (2019). The role of a high-quality restaurant in stimulating the creation and development of gastronomy tourism. *International Journal of Hospitality Management, 83*, 220–228. https://doi.org/10.1016/j.ijhm.2018.10.018

Meyer, J. W., & Rowan, B. (1991). Institutionalized organizations: formal structure as myth and ceremony. In W. W. P. P. J. DiMaggio (Ed.), *The new institutionalism in organizational analysis.* University of Chicago Press.

Miao, Y. D., & Wu, C. Z. (2020). Practice and suggestions on the construction of "City of Gastronomy" Yangzhou from the perspective of Creative Cities Network. *Humanities World*, (13), 2-6. doi:10.16737/j.cnki.rwtx81281190.2020.13.001

Michopoulou, E., & Jauniškis, P. (2020). Exploring the relationship between food and spirituality: A literature review. *International Journal of Hospitality Management, 87*(February), 102494. doi:10.1016/j.ijhm.2020.102494

Miguel-Molina, M., Miguel-Molina, B., Santamarina, V., & Segarra-Oña, M. V. (2016). Intangible Heritage and Gastronomy: The Impact of UNESCO Gastronomy Elements. *Journal of Culinary Science & Technology, 14*(4), 293–310. doi:10.1080/15428052.2015.1129008

Millán, G., Arjona, J., & Amador, L. (2014). A new market segment for olive oil: Olive oil tourism in the South of Spain. *Agricultural Sciences, 5*(3), 179–185. doi:10.4236/as.2014.53022

Millán, G., Arjona, J., & Amador, L. (2017). Olive oil tourism: Promoting rural development in Andalusia. *Tourism Management Perspectives, 21*, 100–108. doi:10.1016/j.tmp.2016.12.003

Ministerio de Turismo. (2018). *Mapa gastronómico del Ecuador* [Gastronomic map of Ecuador]. https://files.goraymi.com/2020/04/01/60d71579ff1651d857a1a6c8f25af41c.pdf

Ministry of Culture and Tourism of the People's Republic of China. (2015, December 31). *Looking back to 2015: "Culture+" has become a new economic engine.* https://www.mct.gov.cn/whzx/bnsj/whcys/201512/t20151231_760240.htm

Ministry of Culture and Tourism of the People's Republic of China. (2018, April 16). *Implementation plan for the construction of Manzhouli in Inner Mongolia and Fangchenggang border tourism pilot zone in Guangxi.* http://zwgk.mct.gov.cn/zfxxgkml/scgl/202012/t20201213_919404.html

Ministry of Culture and Tourism of the People's Republic of China. (2018, September 10). *Reply to recommendation 3240 of the first session of the 13th National People's Congress.* http://zwgk.mct.gov.cn/zfxxgkml/fwzwhyc/202012/t20201206_916874.html

Ministry of Culture and Tourism of the People's Republic of China. (2019, February 12). *Basic situation of tourism market in 2018.* http://zwgk.mct.gov.cn/zfxxgkml/tjxx/202012/t20201204_906481.html

Ministry of Culture and Tourism of the People's Republic of China. (2019, July 26). *Notice on Agricultural Bank of China's financial support for the construction of national rural tourism key villages*. http://zwgk.mct.gov.cn/zfxxgkml/zykf/202012/t20201213_919373.html

Ministry of Culture and Tourism of The People's Republic of China. (2020, February 5). *Notice on temporarily refunding part of the tourism service quality deposit to support travel agencies in coping with operational difficulties*. http://www.gov.cn/zhengce/zhengceku/2020-02/07/content_5475676.htm

Ministry of Culture and Tourism of the People's Republic of China. (2020, July 17). *Notice on normalizing epidemic prevention in rural tourism and accelerating market recovery*. http://zwgk.mct.gov.cn/zfxxgkml/zykf/202012/t20201213_919386.html

Ministry of Culture and Tourism of the People's Republic of China. (2020, March 10). *Basic situation of tourism market in 2019*. https://www.mct.gov.cn/whzx/whyw/202003/t20200310_851786.htm

Ministry of Culture and Tourism of the People's Republic of China. (2021, April 29). *14th Five-Year Plan for culture and tourism development*. http://zwgk.mct.gov.cn/zfxxgkml/ghjh/202106/t20210602_924956.html

Ministry of Culture and Tourism of the People's Republic of China. (2021, February 18). *Domestic tourism data in 2020*. http://zwgk.mct.gov.cn/zfxxgkml/tjxx/202102/t20210218_921658.html

Ministry of Culture and Tourism of the People's Republic of China. (2021, February 22). *To build an internationally influential cultural and leisure Guangdong-Hong Kong-Macao Greater Bay Area*. http://zwgk.mct.gov.cn/zfxxgkml/zcfg/zcjd/202101/t20210106_920503.html

Ministry of Culture and Tourism of the People's Republic of China. (2021, July 19). *Domestic tourism data in the first half of 2021*. http://zwgk.mct.gov.cn/zfxxgkml/tjxx/202107/t20210729_926770.html

Ministry of Culture and Tourism of the People's Republic of China. (2021, July 19). *Notice on organizing the 2021 national culture and tourism consumption season*. http://zwgk.mct.gov.cn/zfxxgkml/cyfz/202107/t20210726_926673.html

Ministry of Culture and Tourism of the People's Republic of China. (2021, July 5). *Notice on building the first batch of state-level night culture and tourism consumption agglomeration zones*. http://zwgk.mct.gov.cn/zfxxgkml/cyfz/202107/t20210706_926242.html

Ministry of Culture and Tourism of the People's Republic of China. (2021, March 8). *Suggestions on promoting high-quality development of public cultural services*. http://zwgk.mct.gov.cn/zfxxgkml/ggfw/202103/t20210323_923230.html

Min, J., Kim, J., & Yang, K. (2021). How generations differ in coping with a pandemic: The case of restaurant industry. *Journal of Hospitality and Tourism Management*, *48*, 280–288. doi:10.1016/j.jhtm.2021.06.017

Mir, U. R., Hassan, S. M., & Qadri, M. M. (2014). Understanding globalization and its future: An analysis. *Pakistan Journal of Social Sciences*, *34*(2), 607–624. http://pjss.bzu.edu.pk/website/journal/article/5ea0309b5c9a2/page

Mishra, J., Mishra, P., & Arora, N. K. (2021). Linkages between environmental issues and zoonotic diseases: With reference to COVID-19 pandemic. *Environmental Sustainability*, *4*(3), 455–467. doi:10.100742398-021-00165-x

Mogollón, J. M., Clemente, E. D., Fernández, J. A., & Cerro, A. M. (2019). Olive oil tourism: State of the art. *Tourism and Hospitality Management*, 179–208. doi:10.20867/thm.25.1.5

Möhring, M. (2010). Döner Kebab and West German Consumer (Multi-) Cultures. *Cross/Cultures, 129*, 151-166.

Momtaz, N. J., Alizadeh, S., & Vaghefi, M. S. (2013). A new model for assessment fast food customer behavior case study: An Iranian fast-food restaurant. *British Food Journal*, *115*(4), 601–613. doi:10.1108/00070701311317874

Montarini, A., & Staniscia, B. (2009). Culinary tourism as a tool for regional re-equilibrium. *European Planning Studies*, *17*(10), 1463–1483. https://doi.org/10.1080/09654310903141656

Morais, R. D. (2016). *About the IRS*. Almedina.

Morales, P. M., Guzmán, T. J.-G., Cuadra, S. M., & Agüera, F. O. (2015). *Análisis de la demanda del oleoturismo en Andalucía* [Analysis of the demand for oleotourism in Andalusia.]. Revista de Estudios Regionales.

Moreira, P., Santos, S., Padrao, P., Cordeiro, T., Bessa, M., & Valente, H. (2010). Food patterns according to sociodemographics, physical activity, sleeping and obesity in Portuguese children. *International Journal of Environmental Research and Public Health*, *7*(3), 1121–1138. doi:10.3390/ijerph7031121 PMID:20617022

Moreno, I. (2015). Culturas Mediterrânicas e sistemas alimentares: continuidades, imaginários e novos desafios [Mediterranean cultures and food systems: continuities, imaginaries and new challenges]. In *Dimensões da Dieta Mediterrânica: Património cultural imaterial da humanidade*. Universidade do Algarve. http://hdl.handle.net/10400.1/6728

Morgan, M., Lugosi, P., & Ritchie, J. R. B. (2010). *The tourism and leisure experience: Consumer and managerial perspectives*. Channel View Publications.

Morgan, N., Pritchard, A., & Pride, R. (2004). *Destination branding: Creating the unique destination proposition*. Butterworth Heinemann.

Moro, E. (2016). The Mediterranean Diet from Ancel keys to the UNESCO cultural heritage. A pattern of sustainable development between myth and reality. *Procedia: Social and Behavioral Sciences*, *223*(Supplement C), 655–661. doi:10.1016/j.sbspro.2016.05.380

Morris, H. (2017, December 14). Where you should fly in 2018: The 12 most exciting new routes. *The Telegraph*. https://www.telegraph.co.uk/travel/news/new-flight-routes-2018-holiday-destinations/

Morrisey, L. (2010). Slow Food: An Interview with Carlo Petrini. *Development*, *53*(3), 447–448. doi:10.1057/dev.2010.51 PMID:20056678

Mozas-Moral, A. (2019). *Contribución de las Cooperativas Agrarias al Cumplimiento de los Objetivos de Desarrollo Sostenible. Especial Referencia al Sector Oleícola* [Contribution of Agricultural Cooperatives to the Fulfillment of the Sustainable Development Goals. Special Reference to the Olive Sector]. Centro internacional de investigación e información sobre la economía pública, social y cooperativa.

Muller, C. (2020, May 19). Will Dine-In Restaurants Survive the Pandemic? *Futurity*. https://www.futurity.org/restaurants-after-COVID-19-pandemic-2369632/

Mullis, L., Saif, L. J., Zhang, Y., Zhang, X., & Azevedo, M. S. (2012). Stability of bovine coronavirus on lettuce surfaces under household refrigeration conditions. *Food Microbiology*, *30*(1), 180–186. doi:10.1016/j.fm.2011.12.009 PMID:22265299

Murgado, E. M. (2013). Turning food into a gastronomic experience: Olive oil tourism. In N. Arcas, F. N. Arroyo López, J. Caballero, R. D'Andria, M. Fernández, R. Fernandez Escobar, A. Garrido, J. López-Miranda, M. Msallem, M. Parras, L. Rallo, & R. Zanoli (Eds.), *Present and future of the Mediterranean olive sector* (pp. 97-109). CIHEAM. https://om.ciheam.org/om/pdf/a106/00006809.pdf

Nabais, J. C. (2018). *Introduction to corporate tax law* (3rd ed.). Almedina.

Nabais, J. C. (2019). *Tax Law* (11th ed.). Almedina.

Nápoles, N., Nobre, H., & Meneses, R. (2020). The Portuguese Luxury Wine Market: A Study on the Perceptions of Wine Brand Managers. In V. Chkoniya, A. Madsen, & P. Bukhrashvili (Eds.), *Anthropological Approaches to Understanding Consumption Patterns and Consumer Behavior* (pp. 303–324). IGI Global. doi:10.4018/978-1-7998-3115-0.ch017

Naranjal. (2020). Historia del Centro Shuar [History of the Shuar Center]. *Naranjal Turistico*. https://naranjal-turistico.es.tl/Centro-Shuar.htm

Narayanan, L., Pandit, M., Basu, S., Karmakar, A., Bidhan, V., & Kumar, H. (2020). *Impact of lockdown due to COVID-19 outbreak: Lifestyle changes and public health concerns in India*. Preprints., doi:10.20944/preprints202006.0129.v1

Nardell, E. A., & Nathavitharana, R. R. (2020). Airborne spread of SARS-CoV2 and a potential role for air disinfection. *Journal of the American Medical Association, 324*(2), 141–142. doi:10.1001/jama.2020.7603 PMID:32478797

Naserghandi, A., Allameh, S. F., & Saffarpour, R. (2020). All about COVID-19 in brief. *New Microbes and New Infections, 35*, 100678. doi:10.1016/j.nmni.2020.100678 PMID:32292590

Naska, A., Oikonomou, E., Trichopoulou, A., Psaltopoulou, T., & Trichopoulos, D. (2007). Siesta in healthy adults and coronary mortality in the general population. *Archives of Internal Medicine, 167*(3), 296–301. doi:10.1001/archinte.167.3.296 PMID:17296887

National Bureau of Statistics of China. (n.d.). *Number of domestic tourists in China from 2010 to 2019 (in billions)*. Author.

National Geographic. (2018). 5 países que destacan por su gastronomía a nivel mundial [5 countries that stand out for their gastronomy worldwide]. *National Geographic*. https://www.ngenespanol.com/gastronomia/5-paises-gastronomia-mundial/

National Restaurant Association. (2020). *New Research Details Early Impact of Coronavirus Pandemic on Restaurant Industry*. https://restaurant.org/articles/news/study-details-impact-of-coronavirus-on-restaurants

Neise, T., Verfürth, P., & Franz, M. (2021). Rapid responding to the COVID-19 crisis: Assessing the resilience in the German restaurant and bar industry. *International Journal of Hospitality Management, 96*, 102960. doi:10.1016/j.ijhm.2021.102960

Nemati, M., Ghorbanpour, H., Razavieh, S. V., & Hoseini, M. (2008). Chemical composition and microbiological quality of the Bonab kebabs sold in Tabriz market. *Journal of Food Safety, 28*(3), 315–323. doi:10.1111/j.1745-4565.2008.00112.x

Nestle, M. (1995). Mediterranean diets: Historical and research overview. *The American Journal of Clinical Nutrition, 61*(6, Suppl), 1313s–1320s. doi:10.1093/ajcn/61.6.1313S PMID:7754981

Newbert, S. L. (2008). Value, rareness, competitive advantage, and performance: A conceptual-level empirical investigation of the resource-based view of the firm. *Strategic Management Journal, 29*(7), 745–768. doi:10.1002mj.686

Nhamo, G., Dube, K., & Chikodzi, D. (2020). Restaurants and COVID-19: A focus on sustainability and recovery pathways. In G. Nhamo, K. Dube, & D. Chikodzi (Eds.), *Counting the cost of COVID-19 on the global Tourism Industry* (pp. 205–224). Springer Nature Switzerland AG. doi:10.1007/978-3-030-56231-1_4

Nicodemo, M., Spreghini, M. R., Manco, M., Wietrzykowska Sforza, R., & Morino, G. (2021). Childhood obesity and COVID-19 lockdown: Remarks on eating habits of patients enrolled in a food-education program. *Nutrients, 13*(2). Advance online publication. doi:10.3390/nu13020383 PMID:33530570

Nicola, M., Alsafi, Z., Sohrabi, C., Kerwan, A., Al-Jabir, A., Iosifidis, C., & Agha, R. (2020). The socio-economic implications of the coronavirus pandemic (COVID-19): A review. *International Journal of Surgery, 78*, 185–193. doi:10.1016/j.ijsu.2020.04.018 PMID:32305533

Nistoreanu, P. (2020). *New Trends and Opportunities for Central and Eastern European Tourism.* IGI Global. doi:10.4018/978-1-7998-1423-8

Nolla Restaurant. (n.d.). https://www.restaurantnolla.com/

Nor, N. M., Sharif, M. S. M., Zahari, M. S. M., Salleh, H. M., Isha, N., & Muhammad, R. (2012). The transmission modes of Malay traditional food knowledge within generations. *Procedia: Social and Behavioral Sciences*, *50*, 79–88. doi:10.1016/j.sbspro.2012.08.017

North, D. C. (1990). *Institutions, institutional change and economic performance.* Cambridge University Press.

North, D. C. (1994). *Structure and change in economic history.* Shanghai People's Publishing House.

North, D. C., & Thomas, R. P. (2009). *The rise of the western world: A new economic history.* Huaxia Publishing House.

Nummedal, M., & Hall, M. (2006). Local food and tourism: An investigation of the New Zealand South Island's bed and breakfast section's use and perception of local food. *Tourism Review International*, *9*, 365–378. doi:10.3727/154427206776330571

Nutrimento. (2019, July 15). *Alimentação saudável no ensino superior -selo excelência* [Healthy eating in higher education-seal excellence]. Nutrimento. https://nutrimento.pt/noticias/alimentacao-saudavel-no-ensino-superior-selo-excelencia/

Oberholzer, M., & Febbo, P. (2020). *What we know today about coronavirus SARS-CoV-2 and where do we go from here.* Genetic Engineering and Biotechnology News. https://www.genengnews.com/insights/what-we-know-today-about-coronavirus-sars-cov-2-and-where-do-we-go-from-here/

OECD. (2020). *COVID-19 and international trade: Issues and actions.* https://www.oecd.org/coronavirus/policy-responses/covid-19-and-international-trade-issues-and-actions-494da2fa/

Ogundijo, D. A., Tas, A. A., & Onarinde, B. A. (2021). Exploring the impact of COVID-19 pandemic on eating and purchasing behaviours of people living in England. *Nutrients*, *13*(5). Advance online publication. doi:10.3390/nu13051499 PMID:33946799

Oh, H.-C., Uysal, M., & Heung, C. (1995). Product bundles and market segments based on travel motivations: A canonical correlation approach. *International Journal of Hospitality Management*, *14*(2), 123–137. doi:10.1016/0278-4319(95)00010-A

Oh, H., Fiore, A. M., & Jeoung, M. (2007). Measuring experience economy concepts: Tourism applications. *Journal of Travel Research*, *46*(2), 119–132. doi:10.1177/0047287507304039

OIV. (2021). *State of the Vitivinicultural World In 2020.* https://www.oiv.int/public/medias/7880/oiv-state-of-the-vitivinicultural-world-2020-ppt.pdf

Ojelel, S., Mucunguzi, P., Katuura, E., Kakudidi, E. K., Namaganda, M., & Kalema, J. (2019). Wild edible plants used by communities in and around selected forest reserves of Teso-Karamoja region, Uganda. *Journal of Ethnobiology and Ethnomedicine*, *15*(1), 1–14. doi:10.118613002-018-0278-8 PMID:30626418

Oliveira, B. S., Tricárico, L. T., Sohn, A. P. L., & Pontes, N. (2018). The culinary intangible cultural heritage of UNESCO: A review of journal articles in EBSCO platform. *Journal of Culinary Science & Technology.* Advance online publication. doi:10.1080/15428052.2018.1513882

Oliveira-Brochado, A., & Silva, R. (2014). The Wine List Design by Upscale Restaurants. World Academy of Science, Engineering and Technology, Open Science Index 88. *International Journal of Humanities and Social Science*, *8*(4), 1052–1056.

Oliveira, N. M. P. (2020). COViD-19, credit, lease and tourism-connected contracts in Portugal. *Revista De Derecho Civil*, *7*(2), 89–102.

Oliveira, T. C., Abranches, M. V., & Lana, R. M. (2020). Food security in Brazil in the context of the SARS-CoV-2 pandemic. *Cadernos de Saude Publica*, *36*(4), e00055220. doi:10.1590/0102-311X00055220 PMID:32267383

Omar, S. R., Ab Karim, S., Isa, S. S., & Omar, S. N. (2020). Enlivening the effect of tourists' knowledge and involvement on Malaysian heritage food (MHF) towards food tourism image in Malaysia. *International Journal of Tourism and Hospitality Management in the Digital Age*, 268–290.

Öney, H. (2013). *Gastronomi turizmi. H. Öney içinde, Alternatif Turizm* [Gatronomy tourism. H. In Öney, Alternative Tourism]. Anadolu Üniversitesi Açıköğretim Fakültesi Yayınları.

Orkun, N. D. (2009). *Food culture changed by globalization: Istanbul-Beyoğlu 2002-2009* [Doctoral Thesis]. Marmara University Institute of Social Sciences.

Oussedik, S. (2012). *Food and cuisine: Part of the migration process.* Institut Europeu de la Mediterrània.

Özgüç, N. (1996). *Turizm Coğrafyası* [Tourism Geography]. İstanbul Üniversitesi Edebiyat Fakültesi Coğrafya Dergisi.

OziliP. K.ArunT. (2020). SpilloverofCOVID-19: impact on the Global Economy. SSRN 3562570. doi:10.2139/ssrn.3562570

Paeratakul, S., Ferdinand, D. P., Champagne, C. M., Ryan, D. H., & Bray, G. A. (2003). Fast-food consumption among US adults and children: Dietary and nutrient intake profile. *Journal of the American Dietetic Association*, *103*(10), 1332–1338. doi:10.1016/S0002-8223(03)01086-1 PMID:14520253

Palma, C. (2015). *Public authorities and value added tax: A break in the principle of neutrality.* Almedina.

Pamukcu, H., Sarac, O., Aytugar, S., & Sandikci, M. (2021). The effects of local food and local products with geographical indication on the development of tourism gastronomy. *Sustainability*, *13*(12), 6692. doi:10.3390u13126692

Panagiotakos, D. B., Chrysohoou, C., Pitsavos, C., & Stefanadis, C. (2006). Association between the prevalence of obesity and adherence to the Mediterranean diet: The ATTICA study. *Nutrition (Burbank, Los Angeles County, Calif.)*, *22*, 449–456. doi:10.1016/j.nut.2005.11.004 PMID:16457990

Pan, X., Chen, D., Xia, Y., Wu, X., Li, T., & Ou, X. (2020). Asymptomatic cases in a family cluster with SARS-CoV-2 infection. *The Lancet. Infectious Diseases*, *20*, 410–411. doi:10.10161473-3099(20)30114-6 PMID:32087116

Panzone, L. A., Larcom, S., & She, P.-W. (2021). Estimating the impact of the first COVID-19 lockdown on UK food retailers and the restaurant sector. *Global Food Security*, *28*, 100495. doi:10.1016/j.gfs.2021.100495

Park, C. W., Sutherland, I., & Seul, K. L. (2021). Effects of online reviews, trust, and picture-superiority on intention to purchase restaurant services. *Journal of Hospitality and Tourism Management*, *47*, 228–236. doi:10.1016/j.jhtm.2021.03.007

Parkins, W., & Craig, G. (2015). Culture and the politics of alternative food networks. *Food, Culture, & Society*, *12*(1), 77–103. doi:10.2752/155280109X368679

Parsian, N., & Dunning, T. (2009). Developing and validating a questionnaire to measure spirituality : A psychometric process. *Global Journal of Health Science*, *1*(1), 2–11.

Pascual, L. A. M., & Zamora, M. P. M. (1998). *El Corporate income tax, general regime*. Centro de Estudios Financieros.

Paul, J. (2015). Masstige marketing redefined and mapped. *Marketing Intelligence & Planning*, *33*(5), 691–706. doi:10.1108/MIP-02-2014-0028

Paul, J. (2018a). Toward a'masstige'theory and strategy for marketing. *European Journal of International Management, 12*, 722–745.

Paul, J. (2018b). Toward a 'masstige' theory and strategy for marketing. *European Journal of International Management, 12*(5-6), 722–745. https://www.inderscienceonline.com/doi/abs/10.1504/EJIM.2018.094466

Paul, J. (2019). Masstige model and measure for brand management. *European Management Journal, 37*(3), 299–312. doi:10.1016/j.emj.2018.07.003

Peci, A. (2006). *The new institutional theory in organizational studies: a critical approach* (Vol. 4). Cadernos EBAPE.

Peeri, N. C., Shrestha, N., Rahman, M. D. S., Zaki, R., Tan, Z., & Bibi, S. (2020). The SARS, MERS and novel coronavirus (COVID-19) epidemics, the newest and biggest global health threats. What lessons have we learned? *International Journal of Epidemiology, 49*, 717–726. doi:10.1093/ije/dyaa033 PMID:32086938

Pereira, F. A. M. (2012). The evolution of institutional theory in organizational studies: a field of research to be explored. *Revista Organizações em Contexto (ROC)*, 275-295. https://www.metodista.br/revistas/revistas-metodista/index.php/OC/article/view/3396

Pereira, P. R. (2019). *Manual de IRS* (2nd ed.). Edições Almedina.

Pérez, A. F. (2017). Análisis de la demanda turística del cantón Naranjal- Provincia del Guayas, para el diseño de un plan promocional turístico [Analysis of the tourist demand of the canton Naranjal- Province of Guayas, for the design of a tourist promotional plan]. *Repositorios UG*, 1-113. https://rraae.cedia.edu.ec/Record/UG_cb87e2a3ebef54452b5940726579f9f9

Petrini, C. (2017). *Food Trends – Agri-Food Trends in Portugal to 2027 according to the Principles of Slow Food: Scientific Handbook*. Slow food.

Petrini, C. (2001). *Slow food: the case for taste*. Columbia University Press.

Petrini, C. (2006). *Slow Food Revolution*. Rizzoli.

Petrini, C. (2007). *Slow food nation*. Rizzoli.

Petrini, C. (2013). *Slow food nation: Why our food should be good, clean, and fair*. Rizzoli Publications.

Pezzi, E., & Vianna, S. L. (2015). Experiência turística e o turismo de experiência: Um estudo sobre as dimensões da experiência memorável [Tourist experience and experience tourism: a study on the dimensions of the memorable experience]. *Revista Turismo em Análise, 26*(1), 165–187. doi:10.11606/issn.1984-4867.v26i1p165-187

Pezzotti, G., Ohgitani, E., Shin-Ya, M., Adachi, T. D., Marin, E., & Boschetto, F. (2020). Rapid inactivation of SARS-CoV-2 by silicon nitride, copper, and aluminum nitride. [Preprint]. *bioRxiv*. Advance online publication. doi:10.1101/2020.06.19.159970

Phull, S. (2015). The Mediterranean diet: Socio-cultural relevance for contemporary health promotion. *The Open Public Health Journal, 8*, 35–40. doi:10.2174/1874944501508010035

Piers, L. S., Walker, K. Z., Stoney, R. M., Soares, M. J., & O'Dea, K. (2003). Substitution of saturated with monounsaturated fat in a 4-week diet affects body weight and composition of overweight and obese men. *British Journal of Nutrition, 90*(3), 717–727. doi:10.1079/BJN2003948 PMID:13129479

Pigini, D., & Conti, M. (2017). NFC-based traceability in the food chain. *Sustainability, 9*(10), 1910–1930. doi:10.3390u9101910

Pine, B. J., & Gilmore, J. H. (1999). *The experience economy*. Harvard Business Press.

Pine, B. J., & Gilmore, J. H. (2011). *The Experience Economy, Uptade Edition*. Harvard Business Review Press.

Pine, J., & Gilmore, J. H. (1998). Welcome to the experience economy. *Harvard Business Review*, *76*(4), 97–105. PMID:10181589

Pomfret, G. (2012). Personal emotional journeys associated with adventure activities on packaged mountaineering holidays. *Tourism Management Perspectives*, *4*, 145–154.

Pordata. (n.d.). https://www.pordata.pt/

Portugal Wine Landscapes. (2021). *Encouraging Signs for Portugal's wine market in 2021*. https://www.wineintelligence.com/encouraging-signs-for-portugals-wine-market-in-2021/

PortugalFoods. (2021). *Trends 2021 – Setor Agroalimentar* [Trends 2021 - Agri-Food Sector]. https://qualifica.portugalfoods.org/wp-content/uploads/2021/02/trends2021-portugalfoods.pdf

Portuguese Association of Distribution Companies. (n.d.). http://aped.pt/

Post Office Celebration Break Report. (2019). https://www.postoffice.co.uk/travel-money/city-cost-celebration-report

Poudel, P. B., Poudel, M. R., Gautam, A., Phuyal, S., Tiwari, C. K., Bashyal, N., & Bashyal, S. (2020). COVID-19 and its global impact on food and agriculture. *Journal of Biology and Today's World*, *9*(5), 221–225.

Prahalad, C. K., & Ramaswamy, V. (2004). Co-creation experiences: The next practice in value creation. *Journal of Interactive Marketing*, *18*(3), 5–14. doi:10.1002/dir.20015

Pramezwary, A., Pramono, R., Situmorang, J. M. H., Salim, S. L., & Christina, F. (2021). Impulsive buying of five star hotel Jakarta: effect new normal era mediated of price and promotion. *Academy of Strategic Management Journal, 20*(2), 1–16.

Prescott, J., Young, O., O'Neill, L., Yau, N. J. N., & Stevens, R. (2002). Motives for food choice: A comparison of consumers from Japan, Taiwan, Malaysia and New Zealand. *Food Quality and Preference*, *13*, 489–495. doi:10.1016/S0950-3293(02)00010-1

Presenza, A., & Cipollina, M. (2010). Analysing tourism stakeholders networks. *Tourism Review*, *65*, 17–30. doi:10.1108/16605371011093845

Pressman, P., Naidu, A. S., & Clemens, R. (2020). COVID-19 and Food Safety: Risk Management and Future Considerations. *Wolters Kluwer Health*, *55*(3), 125–128.

Prime, H., Wade, M., & Browne, D. T. (2020). Risk and resilience in family well-being during the COVID-19 pandemic. *The American Psychologist*, *75*(5), 631–643. doi:10.1037/amp0000660 PMID:32437181

Pung, R., Chiew, C. J., Young, B. E., Chin, S., Chen, M. I., & Clapham, H. E. (2020). Investigation of three clusters of COVID-19 in Singapore: Implications for surveillance and response measures. *Lancet*, *395*(1029), 1039–1046. doi:10.1016/S0140-6736(20)30528-6 PMID:32192580

Qi, E. Z., & Chen, J. (2018). Research on brand promotion strategy of "City of Gastronomy" Shunde. *Market Forum*, *2*, 37-41. doi:https://doi.org/CNKI:SUN:JHTS.0.2018-02-011

Qi, L. L. (2019). The construction of characteristic town of gastronomy of Leliu, Shunde: Based on the development of industry. *Journal of Guangdong Industry Polytechnic*, *18*(1), 27–30. https://doi.org/10.13285/j.cnki.gdqgxb.2019.0006

Qiu, M., Sha, J., & Scott, N. (2021). Restoration of visitors through nature-based tourism: A systematic review, conceptual framework, and future research directions. *International Journal of Environmental Research and Public Health*, *18*(5), 1–19. doi:10.3390/ijerph18052299 PMID:33652652

Qu, Q. K., & Chen, H. B. (2009, December 3). Food has become a cultural feature of Macao tourism. *China Trade News*, 6.

Quendler, E., Magnini, V., & Driouech, N. (2021). Strategic positioning of the farm holiday's post-pandemic competitive advantages: Fresh air and sunlight. *Zeitschrift für Tourismuswissenschaft*, *13*(1), 22–51. https://econpapers.repec.org/RePEc:bpj:touwis:v:13:y:2021:i:1:p:22-51:n:1

Quevedo-Leon, R., Bastías-Montes, J. M., Espinoza-Tellez, T., Ronceros, B., Balic, I., & Munoz, O. (2020). Inactivation of coronaviruses in food industry: The use of inorganic and organic disinfectants, ozone, and UV radiation. *Scientia Agropecuaria*, *11*(2), 257–266. doi:10.17268ci.agropecu.2020.02.14

Radd-Vagenas, S., Kouris-Blazos, A., Singh, M. F., & Flood, V. M. (2017). Evolution of Mediterranean diets and cuisine: Concepts and definitions. *Asia Pacific Journal of Clinical Nutrition*, *26*(5), 749–763. doi:10.6133/apjcn.082016.06 PMID:28802282

Raitsos, D. E., Beaugrand, G., Georgopoulos, D., Zenetos, A., Pancucci-Papadopoulou, A. M., Theocharis, A., & Papathanassiou, E. (2010). Global climate change amplifies the entry of tropical species into the Eastern Mediterranean Sea. *Limnology and Oceanography*, *55*(4), 1478–1484. doi:10.4319/lo.2010.55.4.1478

Ramos, P. M. G., Martins, F. V., & Barandas, H. G. (2011). Differences in the perception of wine attributes: A comparative view between consumers, producers and intermediaries. *Bulletin de l'OIV*, *84*(1), 271–306.

Ram, U. (2004). Glocommodification: How the global consumes the local-McDonald's in Israel. *Current Sociology*, *52*(1), 11–31. doi:10.1177/0011392104039311

Randall, E., & Sanjur, D. (1981). Food preferences: Their conceptualization and relationship to consumption. *Ecology of Food and Nutrition*, *11*(3), 151–161. doi:10.1080/03670244.1981.9990671

Randazzo, W., Fabra, M. J., Falcó, I., López-Rubio, A., & Sánchez, G. (2018). Polymers and biopolymers with antiviral activity: Potential applications for improving food safety. *Comprehensive Reviews in Food Science and Food Safety*, *17*, 754–768. doi:10.1111/1541- 4337.12349 PMID:33350126

Rasul, G. (2021). Twin challenges of COVID-19 pandemic and climate change for agriculture and food security in South Asia. *Environmental Challenges*, *2*, 100027. doi:10.1016/j.envc.2021.100027

Ratnesar-Shumate, S., Williams, G., Green, B., Krause, M., Holland, B., & Wood, S. (2020). Simulated sunlight rapidly inactivates SARS-CoV-2 on surfaces. *The Journal of Infectious Diseases*, *222*, 214–222. doi:10.1093/infdis/jiaa274/5841129 PMID:32432672

Real, H., Dias, R. R., & Graça, P. (2021a). Mediterranean Diet conceptual model and future trends of its use in Portugal. *Health Promotion International*, *36*(2), 548–560. doi:10.1093/heapro/daaa056 PMID:32696052

Real, H., & Graça, P. (2019). Marcos da História da Dieta Mediterrânica, desde Ancel Keys [Landmarks of the History of the Mediterranean Diet, from Ancel Keys]. *Acta Portuguesa de Nutrição*, *17*, 6–14.

Real, H., & Graça, P. (2020). Perceções sobre o conceito de Dieta Mediterrânica e a sua evolução - perspetivas de profissionais do Algarve-Portugal [Perceptions about the concept of Mediterranean Diet and its evolution - perspectives of professionals from the Algarve-Portugal]. *Revista Española de Comunicación en Salud*, *11*(2), 203–216. doi:10.20318/recs.2020/5306

Real, H., Soares, S., Ferreira, C., Marques, L., & Gonçalves, T. F. (2021). Referencial de critérios para checklist de avaliação da sustentabilidade em restaurantes [Benchmark criteria for checklist of sustainability assessment in restaurants]. *Acta Portuguesa de Nutrição*, *26*, 18–30. doi:10.21011/apn.2021.2604

Reguant-Aleix, J., Arbore, M. R., Bach-Faig, A., & Serra-Majem, L. (2009). Mediterranean heritage: An intangible cultural heritage. *Public Health Nutrition*, *12*(9a), 1591–1594. doi:10.1017/S1368980009990413 PMID:19911466

Reis, P. (2014). *O olival em Portugal dinâmicas, tecnologias e relação com o desenvolvimento rural* [The olive grove in Portugal dynamics, technologies and relationship with rural development]. Animar - Associação Portuguesa para o Desenvolvimento Local. https://www.animar-dl.pt/site/assets/files/1273/o_olival_em_portugal.pdf

Reisinger, Y., & Steiner, C. J. (2006). Reconceptualizing object authenticity. *Annals of Tourism Research*, *33*(1), 65–86. doi:10.1016/j.annals.2005.04.003

Renna, M. M. F., Serio, F., & Gonnella, M. (2021). The Mediterranean diet between traditional foods and human health through culinary examples. In G. Cm (Ed.), *Gastronomy and food science* (pp. 75–100). Elsevier.

Restaurante A Cozinha. (n.d.). https://restauranteacozinha.pt/

Restaurante Kitchen Dates. (n.d.). https://kitchendates.pt/

Restaurants And Mobile Food Services Global Market Report. (2021). https://www.thebusinessresearchcompany.com/global-market-reports

Rezapouraghdam, H., Behravesh, E., Ari, E., & Doh, W. L. (2018a). Cost-benefit analysis of educational tourism in North Cyprus: A qualitative study of the socio-economic impacts. *e-Review of Tourism Research, 15*(6), 457-479. https://ertr-ojs-tamu.tdl.org/ertr/index.php/ertr/article/view/259

Rezapouraghdam, H., & Akhshik, A. (2021). Tracing the complexity-sustainability nexus in a small Mediterranean island: Implications for hospitality and tourism education. *Worldwide Hospitality and Tourism Themes*, *13*(4), 476–487.

Rezapouraghdam, H., Akhshik, A., & Ramkissoon, H. (2021). Application of machine learning to predict visitors' green behavior in marine protected areas: Evidence from Cyprus. *Journal of Sustainable Tourism*. Advance online publication. doi:10.1080/09669582.2021.1887878

Rezapouraghdam, H., Alipour, H., & Arasli, H. (2019). Workplace spirituality and organization sustainability: A theoretical perspective on hospitality employees' sustainable behavior. *Environment, Development and Sustainability*, *21*(4), 1583–1601.

Rezapouraghdam, H., Alipour, H., & Darvishmotevali, M. (2018). Employee workplace spirituality and pro-environmental behavior in the hotel industry. *Journal of Sustainable Tourism*, *26*(5), 740–758. doi:10.1080/09669582.2017.1409229

Rezapouraghdam, H., & Karatepe, O. M. (2020). Applying health belief model to unveil employees' workplace COVID-19 protective behaviors: Insights for the hospitality industry. *International Journal of Mental Health Promotion*, *22*(4), 233–247. doi:10.32604/IJMHP.2020.013214

Rhazi, K. E., Nejjari, C., Romaguera, D., Feart, C., Obtel, M., Zidouh, A., & Gateau, R. B. (2012). Adherence to a Mediterranean diet in Morocco and its correlates: Cross-sectional analysis of a sample of the adult Moroccan population. *BMC Public Health*, *12*, 345. doi:10.1186/1471-2458-12-345 PMID:22578133

Richards, G. (2003). Gastronomy: an essential ingredient in tourism production and consumption? In Tourism and gastronomy. Routledge.

Richards, G. (2003). Gastronomy: an essential ingredient in tourism production and consumption? In *M.-A. Hjalager, & G. Richards, Tourism and gastronomy* (pp. 17–34). Routledge. doi:10.4324/9780203218617

Richards, G. (2013). Tourism development trajectories - From culture to creativity? In G. Richards & J. Wilson (Eds.), *Encontros Científicos - Tourism & Management Studies, 297–303*. Routledge. doi:10.4324/9780203933695

Richards, G., & Wilson, J. (2006). Developing creativity in tourist experiences: A solution to the serial reproduction of culture? *Tourism Management*, *27*, 1209–1223. doi:10.1016/j.tourman.2005.06.002

Riker, D. G. J., & Pozzetti, V. C., & B., C. (2018). The extratax nature of export and import taxes in environmental preservation. *Percurso*, *2*(25), 36–42. doi:10.21902/RevPercurso.2316-7521.v2i25.3095

Ristaino, J. B., Anderson, P. K., Bebber, D. P., Brauman, K. A., Cunniffe, N. J., & Fedoroff, N. V. (2021). The persistent threat of emerging plant disease pandemics to global food security. *Proceedings of the National Academy of Sciences of the United States of America*, *118*(23), e2022239118. doi:10.1073/pnas.2022239118 PMID:34021073

Ritchie, H., & Roser, M. (2019). *Seafood Production.* https://ourworldindata.org/seafood-production

Ritchie, J., Tung, S., & Ritchie, R. J. B. (2010). Tourism experience management research: Emergence, evolution and future directions. *International Journal of Contemporary Hospitality Management*, *23*(4), 419–438. doi:10.1108/09596111111129968

Ritzer, G. (2001b). Slow food versus McDonald's'. In C. Petrini & B. Watson (Eds.), Slow food: Collected thoughts on taste, tradition and the honest pleasures of food. Chelsea Green Publishing.

Ritzer, G. (1995). *The McDonaldization of society*. Sage Publication.

Ritzer, G. (2001a). *Explorations in the sociology of consumption: fast food, credit cards and casinos*. SAGE Publications.

Rivera-Ferre, M. G., López-i-Gelats, F., Ravera, F., Oteros-Rozas, E., di Masso, M., Binimelis, R., & El Bilali, H. (2021). The two-way relationship between food systems and the COVID19 pandemic: Causes and consequences. *Agricultural Systems*, *191*, 103134. doi:10.1016/j.agsy.2021.103134

Robertson, R. (2012). Globalisation or glocalisation? *Journal of International Communication*, *18*(2), 191–208. doi:10.1080/13216597.2012.709925

Rocha, R. G., & Pinheiro, P. G. (2021). Organizational spirituality: Concept and perspectives. *Journal of Business Ethics*, *171*(2), 241–252. doi:10.100710551-020-04463-y

Rodrigues, J. (2016). Accounting Standardization System. In SNC explicado (6ª ed.). Porto editora.

Rodriguez, E. (2020). Turismo gastronómico ¿Qué es y cómo es el turista gastronómico? [Gastronomic tourism What is and how is the gastronomic tourist?]. *Gastromarketing*. https://eloyrodriguez.com/turismo-gastronomico/

Rodríguez-Gutiérrez, P., Santa Cruz, F. G., Gallo, L. S. P., & López-Guzmán, T. (2020). Gastronomic satisfaction of the tourist: Empirical study in the creative city of Popayán, Colombia. *Journal of Ethnic Foods*, *7*(1), 1–12. doi:10.118642779-019-0044-0

Rodríguez-López, M.-E., Alcántara-Pilar, J.-M., Del Barrio-García, S., & Muñoz-Leiva, F. (2020). A review of restaurant research in the last two decades: A bibliometric analysis. *International Journal of Hospitality Management*, *87*, 102387. doi:10.1016/j.ijhm.2019.102387

Rodríguez-Pérez, C., Molina-Montes, E., Verardo, V., Artacho, R., García-Villanova, B., Guerra-Hernández, E. J., & Ruíz-López, M. D. (2020). Changes in dietary behaviours during the COVID-19 outbreak confinement in the Spanish COVIDiet study. *Nutrients*, *12*(6). Advance online publication. doi:10.3390/nu12061730 PMID:32531892

Romaguera, D., Norat, T., Mouw, T., May, A. M., Bamia, C., Slimani, N., ... Cottet, V. (2009). Adherence to the Mediterranean diet is associated with lower abdominal adiposity in European men and women. *The Journal of Nutrition*, *139*, 1728–1737. doi:10.3945/jn.109.108902 PMID:19571036

Rosa, R. (2017). *Food Trends – Agri-Food Trends in Portugal to 2027 according to the Principles of Slow Food: Scientific Handbook*. Academic Press.

Ross, L. (1995). The spiritual dimension: Its importance to patients' health, well-being and quality of life and its implications for nursing practice. *International Journal of Nursing Studies*, *32*(5), 457–468. doi:10.1016/0020-7489(95)00007-K PMID:8550306

Rota da Dieta Mediterrânica. (n.d.). http://www.dietamediterranica.pt/?q=pt/node/604

RotaN2. (2021, May). *Passaporte Rota da Estrada Nacional 2* [National Road Route Passport 2]. https://www.rotan2.pt/merchandise/4

Rozin, P., & Vollmecke, T. A. (1986). Food likes and dislikes. *Annual Review of Nutrition*, *6*, 433–456. PMID:3524623

Ruiz-Molina, M. E., Gil-Saura, I., & Berenguer-Contrí, G. (2010). Instruments for Wine Promotion in Upscale Restaurants. *Journal of Foodservice Business Research*, *13*(2), 98–113. doi:10.1080/15378021003784483

Russo, S., Sillmann, J., & Fischer, E. M. (2015). Top ten European heatwaves since 1950 and their occurrence in the coming decades. *Environmental Research Letters*, *10*(12), 124003. doi:10.1088/1748-9326/10/12/124003

Rutynskyi, M., & Kushniruk, H. (2020a). The impact of quarantine due to COVID-19 pandemic on the tourism industry in Lviv (Ukraine). *Problems and Perspectives in Management*, *18*(2), 194–205. doi:10.21511/ppm.18(2).2020.17

Rutynskyi, M., & Kushniruk, H. (2020b). Coffee Tourism in Lviv in the Context of World Coffee Tourism. *Annales Universitatis Mariae Curie-Skłodowska. Sec. B.*, *75*, 87–113. doi:10.17951/b.2020.75.0.87-113

Sabbağ, Ç. (2012). The seasonal BMI changes of rural women living in Anatolia. *International Journal of Environmental Research and Public Health*, *9*(4), 1159–1170. doi:10.3390/ijerph9041159 PMID:22690188

Sabbağ, Ç., & Mankan, E. (2012). Crosschecking Eastern and Western universities students food consumption attitudes in Turkey. *International Journal of Academic Research*, *4*(2), 23–27.

Saghai, Y. (2021). Subversive future seeks like-minded model: On the mismatch between visions of food sovereignty futures and quantified scenarios of global food futures. *Ethics & International Affairs*, *35*(1), 51–67.

Saint Akadiri, S., Alola, A. A., Olasehinde-Williams, G., & Etokakpan, M. U. (2020). The role of electricity consumption, globalization and economic growth in carbon dioxide emissions and its implications for environmental sustainability targets. *The Science of the Total Environment*, *708*, 134653. doi:10.1016/j.scitotenv.2019.134653 PMID:31806296

Sala, E., & Baldassar, L. (2017). Leaving Family to Return to Family: Roots Migration Among Second-Generation Italian-Australians. *Ethos (Berkeley, Calif.)*, *45*(3), 386–408. doi:10.1111/etho.12173

Sanabria-Díaz, J. M., Aguiar-Quintana, T., & Araujo-Cabrera, Y. (2021). Public strategies to rescue the hospitality industry following the impact of COVID-19: A case study of the European Union. *International Journal of Hospitality Management*, *97*, 102988. doi:10.1016/j.ijhm.2021.102988 PMID:34092893

Sanches, J. L. S. (2007). *Tax law manual* (3rd ed.). Coimbra Editora.

Sánchez-Cañizares, S. M., & López-Guzmán, T. (2012). Gastronomy as a tourism resource: Profile of the culinary tourist. *Current Issues in Tourism*, *15*(3), 229–245. https://doi.org/10.1080/13683500.2011.589895

Sánchez-Torres, J., Arroyo-Cañada, F., Solé-Moro, M., & Argila-Irurita, A. (2017). The Mediterranean Diet: Culture and its impact on consumer habits in Barcelona restaurant. *Revista CEA, 3*(6), 47–54. doi:10.22430/24223182.657

Sandoval, C. A. E. (2003). Vulnerability due to extremes: Droughts, Enzo Levi Lecture, 2002. *Ingeneria Hidraulica en Mexico, 18*(2), 133–155.

Sangpikul, A. (2008). Travel motivations of Japanese senior travelers to Thailand. *International Journal of Tourism Research, 10*(1), 81–94. doi:10.1002/jtr.643

Santos, V. R., Ramos, P., Almeida, N., & Santos-Pavón, E. (2019). Wine and wine tourism experience: A theoretical and conceptual review. *Worldwide Hospitality and Tourism Themes, 11*(6), 718–730. doi:10.1108/WHATT-09-2019-0053

Sarmento, J. M., Nunes, R., & Pinto, M. M. (2019). *Theoretical-Practical Manual de IRC* (2nd ed.). Almedina.

Saunders, M., Lewis, P., & Thornhill, A. (2000). *Research methods for business students*. Pearson Education Limited.

Scarmozzino, F. V. F. (2020). Covid-19 and the subsequent lockdown modified dietary habits of almost half the population in an Italian sample. *Foods, 9*, 675. doi:10.3390/foods9050675 PMID:32466106

Scarpato, R. (2002). Gastronomy as a tourist product: the perspective of gastronomy studies. In A. Hjalager & G. Richards (Eds.), *Tourism and Gastronomy*. Routledge.

Schlosser, E. (2001). *Fast Food Nation the Dark Side of the All-American Meal*. Mariner Books Houghton Mifflin Harcourt.

Schröder, H. (2007). Protective mechanisms of the Mediterranean diet in obesity and type 2 diabetes. *The Journal of Nutritional Biochemistry, 18*, 149–160. doi:10.1016/j.jnutbio.2006.05.006 PMID:16963247

Schulz, J. B., & Deuschl, G. (2015). Influence of lifestyle on neurodegenerative diseases. *Der Nervenarzt, 86*(8), 954–959. doi:10.100700115-014-4252-y PMID:26105162

Senbeto, D. L., & Hon, A. H. (2020). The impacts of social and economic crises on tourist behaviour and expenditure: An evolutionary approach. *Current Issues in Tourism, 23*(6), 740–755. doi:10.1080/13683500.2018.1546674

Şengül, S. (2015). *Akdeniz Bölgesi Mutfak Kültürü ve Yöresel Yemekleri* [Mediterranean Region Culinary Culture and Regional Dishes]. Beta Basım Yayım.

Şengül, S., & Türkay, O. (2014). *"Yöresel mutfaklar" eğitimi üzerinden kimlik, mutfak kültürü ve turizm gelişimine dair çıkarımlar: Mudurnu örneği* [Inferences on identity, culinary culture and tourism development through "regional cuisines" education: Mudurnu example]. Citta Slow Eko-Gastronomi Kongresi. Gökçeada.

Şengül, S., & Türkay, O. (2016). The evaluation of Mediterranean cuisine as a gastronomic tourism product. *Journal of Tourism and Gastronomy Studies, 4*(Special issue), 86–99.

Serra, F. (2015). *Turismo de experiência - Tempos de gestão* [Experience tourism - Management times]. http://www.temposdegestao.com/turismo/turismo-de-experiencia

Shang, W. W. (2015). *Research on tourist satisfaction and loyalty of time-honored catering brand from the perspective of food tourism motivation* [Master's dissertation]. Henan University. https://kns.cnki.net/KCMS/detail/detail.aspx?dbname=CMFD201601&filename=1015658405.nh

Sharma, P., & Nayak, J. K. (2019). Do tourists' emotional experiences influence images and intentions in yoga tourism? *Tourism Review, 74*(3). Advance online publication. doi:10.1108/TR-05-2018-0060

Sharpley, R., & Jepson, D. (2011). Rural tourism: A spiritual experience? *Annals of Tourism Research, 38*(1), 52–71. doi:10.1016/j.annals.2010.05.002

Sheldon, P. J. (2020). Designing tourism experiences for inner transformation. *Annals of Tourism Research, 83*, 102935.

Shetty, P. (2003, January 22-24). *Impact of globalization on food and agriculture from the farm to the plate* [Conference session]. Impacts of Globalization on Agricultural Production and Marketing with Focus on Food Quality, Tokyo.

Shi, Y. Y., & Li, Y. L. (2021). The influence of COVID-19 on the development of China's catering industry and its countermeasures. *Modern Business*, (1), 30-32. doi:10.14097/j.cnki.5392/2021.01.009

Shivakoti, A. (2021). Impact of COVID-19 on tourism in Nepal. *The Gaze: Journal of Tourism and Hospitality, 12*(1), 1–22. doi:10.3126/gaze.v12i1.35674

Shuarnum. (2020). *Gastronomía típica* [Typical Gastronomy]. https://shuarnum.com/gastronomia/

Sigala, M. (2020). Tourism and COVID-19: Impacts and implications for advancing and resetting industry and research. *Journal of Business Research, 117*, 312–321. doi:10.1016/j.jbusres.2020.06.015 PMID:32546875

Silo Restaurant. (n.d.). https://silolondon.com

Silva, E. S., Pereira, A., & Andrade, S. (2019). *Tax Benefits in irc*. Vida Económica.

Sims, R. (2009). Food, place and authenticity: Local food and the sustainable tourism experience. *Journal of Sustainable Tourism, 17*(3), 321–336.

Singh, P., Brown, D. M., Chelekis, J., Apostolidis, C., & Dey, B. L. (2021). Sustainability in the beer and pub industry during the COVID-19 period: An emerging new normal. *Journal of Business Research*. Advance online publication. doi:10.1016/j.jbusres.2021.11.066

Skara, T., & Rosnes, J. T. (2016). Emerging methods and principles in food contact surface decontamination/prevention environmental factors in infectious disease. In C. E. Leadley (Ed.), *Innovation and Future Trends in Food Manufacturing and Supply Chain Technologies* (pp. 151–172). Woodhead Publishing Series in Food Science. doi:10.1016/B978-1-78242-447-5.00006-X

Skegg, D., Gluckman, P., Boulton, G., Hackmann, H., Karim, S. S. A., Piot, P., & Woopen, C. (2021). Future scenarios for the COVID-19 pandemic. *Lancet, 397*(10276), 777–778. doi:10.1016/S0140-6736(21)00424-4 PMID:33607000

Skenderi, R. (2021). *Dissecting the Normal: The Effects of Covid-19 Pandemic in the Gastronomy Sector of Kosovo* [Thesis]. Rochester Institute of Technology. https://scholarworks.rit.edu/theses/10946

Slow Food. (n.d.). *Sobre nós.* https://www.slowfood.com/about-us/

Smith, M. (2003). Holistic holidays: Tourism and the reconciliation of body, mind and spirit. *Tourism Recreation Research, 28*(1), 103–108. doi:10.1080/02508281.2003.11081392

Smith, M. K., Ferrari, S., & László, P. (2016). Service innovations and experience creation in spas, wellness and medical tourism. In D. G. Marios Sotiriadis (Ed.), *Tourism Experiences* (pp. 299–319). Emerald Group Publishing Limited.

Smith, M., & Kelly, C. (2006). Holistic tourism: Journeys of the self? *Tourism Recreation Research, 31*(1), 15–24. doi:10.1080/02508281.2006.11081243

Soares, J., & Kupski, R. (2020). Covid-19 calls into question defense of the minimum state In *Os Impostos e o Combate às Desigualdades (Na era do Covid-19)*. Rei dos livros.

Sofronov, B. (2017). The economic impact on global tourism. *Annals of Spiru Haret University. Economic Series, 17*(2), 127-139. http://anale.spiruharet.ro/index.php/economics/article/view/1728/pdf

Solfrizzi, V., Panza, F., Frisardi, V., Seripa, D., Logroscino, G., Imbimbo, B. P., & Pilotto, A. (2011). Diet and Alzheimer's disease risk factors or prevention: The current evidence. *Expert Review of Neurotherapeutics*, *11*(5), 677–708. doi:10.1586/ern.11.56 PMID:21539488

Song, H. J., Yeon, J., & Lee, S. (2021). Impact of the COVID-19 pandemic: Evidence from the U.S. restaurant industry. *International Journal of Hospitality Management*, *92*, 102702. doi:10.1016/j.ijhm.2020.102702 PMID:33024347

Sorcaru, I. A. (2019). Gastronomy tourism - A sustainable alternative for local economic development. *Annals of the University Dunarea de Jos of Galati: Fascicle: I. Economics & Applied Informatics*, *25*(1). Advance online publication. doi:10.35219/eai1584040912

Sotiriadis, M. (2017). Pairing intangible cultural heritage with tourism: The case of Mediterranean diet. *Marios Sotiriadis*, *12*(3), 269–284. doi:10.1108/EMJB-07-2016-0019

Sotiriadis, M. D. (2015). Culinary tourism assets and events: Suggesting a strategic planning tool. *International Journal of Contemporary Hospitality Management*, *27*(6), 1214–1232. doi:10.1108/IJCHM-11-2013-0519

Speck, B. W. (2005). *What is spirituality?* Academic Press.

SPINS. (2020, May 26). *New Data Shows Plant-Based Food Outpacing Total Food Sales During COVID-19.* Plant Based Foods Association. https://www.spins.com/new-data-shows-plant-based-food-outpacing-total-food-sales-during-covid-19

Sportelli, M. C., Izzi, M., Kukushkina, E. A., Hossain, S. I., Picca, R. A., & Ditaranto, N. (2020). Can nanotechnology and materials science help the fight against SARS-CoV-2? *Nanomaterials (Basel, Switzerland)*, *10*(4), 802. doi:10.3390/nano1004 0802 PMID:32326343

Stahel, W. R. (2016). The circular economy. *Nature*, *531*, 435–438. doi:10.1038/531435a PMID:27008952

Stalmirska, A. M. (2021). Local food in tourism destination development: The supply-side perspectives. *Tourism Planning & Development*. Advance online publication. doi:10.1080/21568316.2021.1928739

State Council. (2016, December 7). *13th Five-Year Plan for tourism development.* http://www.gov.cn/zhengce/content/2016-12/26/content_5152993.htm

State Council. (2017, March 5). *2017 Government Work Report.* http://www.gov.cn/guowuyuan/2017zfgzbg.htm

State Statistics Service of Ukraine. (2021). http://www.ukrstat.gov.ua

State Taxation Administration. (2020, March 15). *Guidelines on preferential tax policies for epidemic control and economy development.* http://www.gov.cn/xinwen/2020-03/15/content_5491493.htm

Statista. (n.d.). *COVID-19/Coronavirus: Facts and Figures.* https://www.statista.com/page/covid-19-coronavirus

Stausberg, M. (2014). Religion and spirituality in tourism. Academic Press.

Stavrianea, A., Dipidis, C., & Siomkos, G. (2017). Gastronomy tourism: An examination of the "Greek Breakfast Initiative" potential. In N. Tsounis & A. Vlachvei (Eds.), *Advances in Applied Economic Research* (pp. 841–848). Springer.

Steardo, L., Steardo, L. Jr, Zorec, R., & Verkhratsky, A. (2020). Neuroinfection may attribute to pathophysiology and clinical manifestations of COVID-19. *Acta Physiologica (Oxford, England)*, *229*(3), e13473. Advance online publication. doi:10.1111/alph.13473 PMID:32223077

Ştefanovici, M., & Mazilu, M. (2021). The importance of gastronomic tourism in economic development of Romania. *Quaestus*, *18*, 365–376.

Strisciuglio, C., Cenni, S., Serra, M. R., Dolce, P., Martinelli, M. A. S., & Miele, E. (2020). Effectiveness of Mediterranean diet's adherence in children with inflammatory bowel diseases. *Nutrients*, *12*(10), 3206. doi:10.3390/nu12103206 PMID:33092159

Sudiyarto, S., Yuliati, N., Uhertiana, I., & Pratiwi, L. F. L. (2018). Exploring youngest consumers perceptions of traditional foods and fast foods product. In J. Island (Ed.), *AIP Publishing LLC.* AIP Conference Proceedings. doi:10.1063/1.5061859

Sugiura, T., Ogawa, H., Fukuda, N., & Moriguchi, T. (2013). Changes in the taste and textural attributes of apples in response to climate change. *Scientific Reports*, *3*(1), 1–7. doi:10.1038rep02418 PMID:23948740

Sun, G. (1994). Tourism, a sunrise industry forever. *China Reform*, *7*, 17-20. doi:https://doi.org/CNKI:SUN:ZGGG.0.1994-07-005

Sung, Y.-K., & Hu, H.-H., & King, B. (2021). Restaurant preventive behaviors and the role of media during a pandemic. *International Journal of Hospitality Management*, *95*, 102906. doi:10.1016/j.ijhm.2021.102906

Suntikul, W., & Jachna, T. (2016). The co-creation/place attachment nexus. *Tourism Management*, *52*, 276–286. doi:10.1016/j.tourman.2015.06.026

Szmigiera, M. (2021, February 21). *Impact of the coronavirus pandemic on the global economy - Statistics & Facts.* Statista. https://www.statista.com/topics/6139/covid-19-impact-on-the-global-economy/

Tang, J. Z. (2020, December 3). The Macanese cuisine is not repeated. *Chinese People's Political Consultative Conference Daily*, 8.

Taşdelen, A. H., Kara, A., Bozkurt, O., & Şaşmaz, Ö. (2020). Konaklama ve Seyahat Hizmetleri [Accommodation and Travel Services]. In Genel Turizm. Milli Eğitim Bakanlığı Yayınları.

Taş, M., & Zengin, B. (2019). A research on university students' perception of destination image: The case of Kemaliye. *Research in International Business and Finance*, *1*(3), 205–216.

Teixeira, D. J. (2005). *O Ecomuseu de Barroso. A Nova Museologia ao serviço do desenvolvimento local* [The Ecomuseum of Barroso. The New Museology at the service of local development] [Master Dissertation]. Universidade do Minho.

Tekin, H., Arpacı, S., Atlı, S., Açar, İ., Karadağ, S., Yükçeken, Y., & Yaman, A. (2001). *Cultivation of Pistachios. T.C. Ministry of Agriculture and Rural Affairs, General Directorate of Agricultural Research.* Pistachio Research Institute.

Telfer, D. J., & Wall, G. (2000). Strengthening backward economic linkages: Local food purchasing by three Indonesian hotels. *Tourism Geographies*, *2*(4), 421–447. doi:10.1080/146166800750035521

Tellström, R. (2016, November 28-29). *Inheriting Your Ancestor's Diasporic Food Culture–A Planned Project How Swedish Fourth and Fifth Generation Emigrants in USA Relate to their Ancestor's Former Swedish Food Culture, People Moving with Food: Food, Migration and Multiculturalism* [Conference session]. NAFS conference. https://www.diva-portal.org/smash/get/diva2:1119911/FULLTEXT01.pdf

Telukdarie, A., Munsamy, M., & Mohlala, P. (2020). Analysis of the Impact of COVID-19 on the Food and Beverages Manufacturing Sector. *Sustainability*, *12*(22), 9331. doi:10.3390u12229331

Testa, R., Galati, A., Schifani, G., Di Trapani, A. M., & Migliore, G. (2019). Culinary tourism experiences in agri-tourism destinations and sustainable consumption—Understanding Italian Tourists' Motivations. *Sustainability*, *11*(17), 4588. doi:10.3390u11174588

Thach, L., & Olsen, J. (2019). Luxury wine: Analyzing motivations of luxury wine buyers in the US market. *Applied Studies in Agribusiness and Commerce*, *13*(3-4), 51–58. doi:10.19041/APSTRACT/2019/3-4/6

The Economist. (2002, February 21). Thailand's gastro-diplomacy. *The Economist.* https://www.economist.com/asia/2002/02/21/thailands-gastro-diplomacy

Thornton, P. & Cramer, L. (2012). *Impacts of climate change on the agricultural and aquatic systems and natural resources within the CGIAR's mandate* [CCAFS Working Paper 23]. CGIAR Research Program on Climate Change, Agriculture and Food Security.

Tilman, D., & Clark, M. (2014). Global diets link environmental sustainability and human health. *Nature, 515*, 518–522. doi:10.1038/nature13959 PMID:25383533

TKD. (2021). https://sozluk.gov.tr/

Toker, B., & Rezapouraghdam, H. (2021). Intangible cultural heritage and management of educational tourism: An experiential learning approach. In *Handbook of Research on Human Capital and People Management in the Tourism Industry*. IGI Global., doi:10.4018/978-1-7998-4318-4

Tomczewska-Popowycz, N., & Quirini-Popławski, Ł. (2021). Political Instability Equals the Collapse of Tourism in Ukraine? *Sustainability, 13*(8), 4126. doi:10.3390u13084126

Tomlinson, J. (1991). *Cultural imperialism: A critical introduction.* Johns Hopkins University Press.

Tong, Z. D., Tang, A., Li, K. F., Li, P., Wang, H. L., & Yi, J. P. (2020). Potential presymptomatic transmission of SARS-CoV-2, Zhejiang Province, China. *Emerging Infectious Diseases, 26,* 1052–1054. doi:10.3201/eid2605.200198 PMID:32091386

Torre, M. M., Hidalgo, L. A., & Fuentes, A. J. (2015). El oleoturismo: Una alternativa para preservar los paisajes del olivar y promover el desarrollo rural y regional de Andalucía [Oleotourism: An alternative to preserve the landscapes of the olive grove and promote rural and regional development of Andalusia]. *Revista de Geografía Norte Grande, 60,* 195–214. doi:10.4067/S0718-34022015000100011

Tourism Office of Lviv City Council. (2021). https://city-adm.lviv.ua

Towler, G., & Shepherd, R. (1992). Application of Fishbein and Ajzen's expectancy-value model to understanding fat intake. *Appetite, 18,* 15–27. doi:10.1016/0195-6663(92)90207-M PMID:1562199

Travanca, D. (2009). *VAT on construction and real estate.* Vida Económica.

Tregua, M., D'Auria, A., & Marano-Marcolini, C. (2018). Oleotourism: Local actors for local tourism development. *Sustainability,* 1–20. doi:10.3390u10051492

Trichopoulou, A., Bamia, C., & Trichopoulos, D. (2005). Mediterranean diet and survival among patients with coronary heart disease in Greece. *Archives of Internal Medicine, 165*(8), 929–935. doi:10.1001/archinte.165.8.929 PMID:15851646

Trichopoulou, A., Bamia, C., & Trichopoulos, D. (2009). Anatomy of health effects of Mediterranean diet: Greek EPIC prospective cohort study. *BMJ (Clinical Research Ed.), 338,* b2337. doi:10.1136/bmj.b2337 PMID:19549997

Trichopoulou, A., & Lagiou, P. (1997). Healthy traditional Mediterranean diet: An expression of culture, history, and lifestyle. *Nutrition Reviews, 55*(11), 383–389. doi:10.1111/j.1753-4887.1997.tb01578.x PMID:9420448

Trichopoulou, A., Martínez-González, M. A., Tong, T. Y., Forouhi, N. G., Khandelwal, S., Prabhakaran, D., & Lorgeril, D. M. (2014). Definitions and potential health benefits of the Mediterranean diet: Views from experts around the world. *BMC Medicine, 12*(1), 112. doi:10.1186/1741-7015-12-112 PMID:25055810

Trihas, N., Kyriakaki, A., & Zagkotsi, S. (2015). *Local Cuisine and agricultural products as a means of enhancing tourist's gastronomic experiences in Greece* [Conference session]. 1st International Conference on Experimential Tourism, Santorini. https://imic2015.conferences.gr/wp-content/uploads/TRIHAS.pdf

Trilling, L. (1972). *Sincerity and authenticity*. Oxford University Press.

TripAdvisor. (2021a). *BBQ Restaurants for Large Groups in Lviv*. https://www.tripadvisor.com/Restaurants-g295377-c6-zfp9-Lviv_Lviv_Oblast.html

TripAdvisor. (2021b). *Cafés in Lviv*. https://www.tripadvisor.com/Restaurants-g295377-c8-Lviv_Lviv_Oblast.html

TripAdvisor. (2021c). *Restaurants near Old Town, Lviv, Ukraine*. https://www.tripadvisor.com/RestaurantsNear-g295377-d598874-Old_Town-Lviv_Lviv_Oblast.html

TripAdvisor: Lviv Delivery Restaurants. (2021). https://www.tripadvisor.com/Restaurants-g295377-zfp19-Lviv_Lviv_Oblast.html

TripAdvisor: Lviv Late Night Restaurants. (2021). https://www.tripadvisor.com/Restaurants-g295377-zfp8-Lviv_Lviv_Oblast.html

TripAdvisor: Lviv Romantic Restaurants. (2021). https://www.tripadvisor.com/Restaurants-g295377-zfp3-Lviv_Lviv_Oblast.html

Truong, Y., McColl, R., & Kitchen, P. J. (2009). New luxury brand positioning and the emergence of Masstige brands. *Journal of Brand Management*, *16*(5), 375–382. doi:10.1057/bm.2009.1

Trust, C. (2008). *Food preparation and catering Increase carbon savings without compromising on quality*. https://www.dcproducts.co.uk/documents/carbon_trust_food_preparation_and_catering-2.pdf

Tsai, C.-T., & Lu, P.-H. (2012). Authentic dining experiences in ethnic theme restaurants. *International Journal of Hospitality Management*, *31*(1), 304–306. doi:10.1016/j.ijhm.2011.04.010

Tuo, Y., Bai, C., & Chen, Z. (2014). Travel for meaning: the present and future of tourist wellbeing. *Seventh International Joint Conference on Computational Sciences and Optimization (IEEE)*, 479–483. 10.1109/CSO.2014.95

Tureac, C. E., & Turtureanu, A. (2008). Types and forms of tourism. *Economica*, *4*, 1.

Turismo de Portugal. (2017). *Estratégia turismo 2027, liderar o turismo do futuro*. https://www.turismodeportugal.pt/SiteCollectionDocuments/estrategia/estrategia-turismo-2027.pdf

Tyrovolas, S., & Panagiotakos, D. B. (2010). The role of Mediterranean type of diet on the development of cancer and cardiovascular disease, in the elderly: A systematic review. *Maturitas*, *65*(2), 122–130. doi:10.1016/j.maturitas.2009.07.003 PMID:19656644

UNESCO. (2010a). Representative list of the intangible cultural heritage of humanity. UNESCO.

UNESCO. (2010b). *Traditional Mexican cuisine - ancestral, ongoing community culture, the Michoacán paradigm*. https://ich.unesco.org/en/RL/traditional-mexican-cuisine-ancestral-ongoing-community-culture-the-michoacn-paradigm-00400

UNESCO. (2013a). ITH/13/8.COM/Decisions.

UNESCO. (2013b). *Mediterranean diet*. http://www.unesco.org/culture/ich/en/RL/mediterranean-diet-00884

United Nation World Tourism Organization. (2012). *Global report on food tourism*. https://www.e-unwto.org/doi/epdf/10.18111/9789284414819

United Nation. (1987). *Report of the World Commission on Environment and Development: Our Common Future.* http://www.un-documents.net/ocf-02.htm#I

United Nations Educational, Scientific and Cultural Organization. (n.d.a). *Chengdu.* https://en.unesco.org/creative-cities/chengdu

United Nations Educational, Scientific and Cultural Organization. (n.d.b). *Shunde.* https://en.unesco.org/creative-cities/shunde

United Nations Educational, Scientific and Cultural Organization. (n.d.c). *Yangzhou.* https://en.unesco.org/creative-cities/yangzhou

United Nations World Travel Organization. (2012). *Global report on food tourism.* http://cf.cdn.unwto.org/sites/all/files/docpdf/amreports4-foodtourism.pdf

United Nations. (2015). https://sdgs.un.org/goals

United Nations. (2017). *World Population Prospects 2017.* https://www.un.org/development/desa/pd/sites/www.un.org.development.desa.pd/files/files/documents/2020/Jan/un_2017_world_population_prospects-2017_revision_databooklet.pdf

United Nations. (2019). *World Population Prospects 2019.* https://www.un.org/development/desa/en/news/population/world-population-prospects-2019.html

Universo, E. (2015). *Comunidad* [Community]. El Universo. https://www.eluniverso.com/noticias/2015/ 09/27/nota/5148923/luis-aveiga-mete-camaron-rio-tortillas-verde/

Universo, E. (2017). *En Balao se preparó la ensalada de cangrejo más grande del Guayas* [In Balao the largest crab salad in Guayas was prepared]. https://www.eluniverso.com/guayaquil/2017/11/02/nota/ 6462787/balao-se-preparo-ensalada-cangrejo-mas-grande-guayas/

Urry, J. (1995). *Consuming Places.* Routledge.

Uysal, M., & Li, X. E. S.-T. (2008). Push–pull dynamics in travel decisions. In Handbook of Hospitality Marketing Management. Elsevier.

Vallone, G. (2018). *Turismo do azeite ganha adeptos no sul de Portugal* [Olive oil tourism wins fans in southern Portugal]. Folha de S. Paulo: https://www1.folha.uol.com.br

Van Doremalen, N., Bushmaker, T., Morris, D. H., Holbrook, M. G., Gamble, A., & Williamson, B. N. (2020). Aerosol and surface stability of SARS-CoV-2 as compared with SARS-CoV-1. *The New England Journal of Medicine, 382*(16), 1564–1567. doi:10.1056/NEJMc2004973 PMID:32182409

Vasconcelos, A. A., & Pinto, C. (2011). *Tax regime for depreciation and amortization.* Areal Editores.

Vasques, S. (2015). *Value added tax.* Almedina.

Veerman, C. L. (2013). Removing the GST exemption for fresh fruits and vegetables could cost lives. *The Medical Journal of Australia, 199*(8), 534–535. PMID:24138373

Velikova, N., Canziani, B., & Williams, H. (2019). Small winery-restaurant relationship building: Challenges and opportunities. *International Journal of Wine Business Research, 31*(1), 5–11. doi:10.1108/IJWBR-07-2018-0038

Verberne, L., Bach-Faig, A., Buckland, G., & Serra-Majem, L. (2010). Association between the Mediterranean diet and cancer risk: A review of observational studies. *Nutrition and Cancer, 62*(7), 860–870. doi:10.1080/01635581.2010.509834 PMID:20924961

Vik, F. N., Bjornara, H. B., Overby, N. C., Lien, N., Androutsos, O., & Maes, L. (2013). Associations between eating meals, watching TV while eating meals and weight status among children, ages 10-12 years in eight European countries: The ENERGY cross-sectional study. *The International Journal of Behavioral Nutrition and Physical Activity*, *10*, 58. doi:10.1186/1479-5868-10-58 PMID:23675988

Vilarnau, C., Stracker, D. M., Funtikov, A., da Silva, R., Estruch, R., & Bach-Faig, A. (2018). Worldwide adherence to Mediterranean Diet between 1960 and 2011. *European Journal of Clinical Nutrition*. Advance online publication. doi:10.103841430-018-0313-9 PMID:30487566

Wahlqvist, M. L. (2007). Regional food culture and development. *Asia Pacific Journal of Clinical Nutrition*, *16*(1), 2–7. PMID:17392068

Wang, H., Livingston, K. A., Fox, C. S., Meigs, J. B., & Jacques, F. P. (2013). Yogurt consumption is associated with better diet quality and metabolic profile in American men and women. *Nutrition Research (New York, N.Y.)*, *33*(1), 18–26. doi:10.1016/j.nutres.2012.11.009 PMID:23351406

Wang, W., Tang, J., & Wei, F. (2020). Updated understanding of the outbreak of 2019 novel coronavirus (2019-nCoV) in Wuhan, China. *Journal of Medical Virology*, *92*(4), 441–447. doi:10.1002/jmv.25689 PMID:31994742

Wang, Y. (2020). Research on IP commercialization operation of food tourism in Leshan city. *Journal of Sichuan Tourism University*, *2*, 52–56. https://doi.org/CNKI:SUN:SCPR.0.2020-02-013

Ward, S. (2021, March 5). *What is Internet marketing*? https://www.thebalancesmb.com/internet-marketing-2948348

Warnes, S., Little, Z. R., & Keevil, C. W. (2015). Human Coronavirus 229E Remains Infectious on Common Touch Surface Materials. *mBio*, *6*(6). Advance online publication. doi:10.1128/mBio.01697-15 PMID:26556276

Wei, W. E., Li, Z., Chiew, C. J., Yong, S. E., Toh, M. P., & Lee, V. J. (2020). Presymptomatic transmission of SARS-CoV-2. *Morbidity and Mortality Weekly Report*, *69*, 411–415. doi:10.15585/mmwr.mm6914e1 PMID:32271722

Wen, H., Leung, X., & Pongtornphurt, Y. (2020). Exploring the impact of background music on customers' perceptions of ethnic restaurants: The moderating role of dining companions. *Journal of Hospitality and Tourism Management*, *43*, 71–79. doi:10.1016/j.jhtm.2020.02.007

Wilkins, J., & Hill, S. (2009). *Food in the ancient world*. John Wiley & Sons.

Willett, W., Rockstrom, J., Loken, B., Springmann, M., Lang, T., & Vermeulen, S. (2019). Food in the anthropocene: The EAT-lancet commission on healthy diets from sustainable food systems. *Lancet*, *393*(10170), 447–492. doi:10.10160140-6736(18)31788-4 PMID:30660336

Williams, G., & Price, M. (2020, March 18). Businesses Pleased with Support. *Otago Daily Times*. https://www.odt.co.nz/regions/queenstown/businesses-pleased-support

Willis, C. (2015). The contribution of cultural ecosystem services to understanding the tourism–nature–wellbeing nexus. *Journal of Outdoor Recreation and Tourism*, *10*, 38–43.

Wisła, R., & Nowosad, A. (2020). *Economic Transformation in Poland and Ukraine. National and Regional Perspectives*. Routledge. doi:10.4324/9781003041375

Wolf, E. (2002). *Culinary tourism: a tasty economic proposition*.

Wolf, H. L., Morrish, S. C., & Fountain, J. (2016). A conceptualization of the perceptions and motivators that drive luxury wine consumption. *International Journal of Wine Business Research*, *28*(2), 120–133. doi:10.1108/IJWBR-09-2015-0038

Wong, I. A. (2015). A multimethod multilevel study of heritage transmission: The role of culture on tourist interest and authenticity. *Journal of Travel Research*, *54*(5), 672–685. doi:10.1177/0047287514532368

World Health Organization. (1998). *WHOQOL and spirituality, religiousness and personal beliefs (SRPB)*. https://apps.who.int/iris/bitstream/handle/10665/70897/WHO_MSA_MHP_98.2_eng.pdf?sequence=1&isAllowed=y

World Health Organization. (2012). *Spirituality, religiousness and personal Beliefs (SRPB) field-test instrument*. https://apps.who.int/iris/bitstream/handle/10665/77777/WHO_MSD_MER_Rev.2012.04_eng.pdf?sequence=1&isAllowed=y

World Health Organization. (2014a). *Zoonotic Disease: Emerging Public Health Threats in the Region*. http://www.emro.who.int/about-who /rc61/zoonotic-diseases.html

World Health Organization. (2014b). *Infection prevention and control of epidemic and pandemic Prone Acute respiratory infections in health care*. WHO Press.

World Health Organization. (2020a). *Infection Prevention and Control during Health Care When Novel Coronavirus (nCoV) Infection is Suspected: Interim Guidance*. https://www.who.int/publicationsdetail/infection-prevention-and-control-during-health-care-when-novelcoronavirus-(ncov)-infection-is-suspected-20200125

World Health Organization. (2020b). *Water, Sanitation, Hygiene and Waste Management for COVID- 19: Technical Brief*. https://www.who.int/publications-detail/watersanitation-hygiene-and-waste-management-for-COVID-19

World Health Organization. (2020b). *Water, Sanitation, Hygiene and Waste Management for COVID-19: Technical Brief*. https://www.who.int/publications-detail/watersanitation-hygiene-and-waste-management-for-covid-19

World Health Organization. (2020c). *Coronavirus disease (COVID-19) pandemic*. http ://www.who.int/emergencies/diseases/novel-coronavirus-2019/

World Health Organization. (2020d). *Cleaning and disinfection of environmental surfaces in the context of COVID-19: Interim guidance*. https://www.who.int/ publications/i/item/cleaning-and-disinfection-of-environmental-surfaces-inthecontext-of-COVID-19

World Health Organization. (2020e). *Responding to COVID-19 Community Transmission*. https://www.who.int/publications/i/item/responding-to-community-spread-of-covid-19

World Health Organization. (2021). *WHO Coronavirus (COVID-19) Dashboard*. https://covid19.who.int/

World Tourism Organization & Basque Culinary Center. (2019). *UNWTO.ORG. Guidelines for the Development of Gastronomy Tourism*. https://www.e-unwto.org/doi/pdf/10.18111/9789284420957

World Tourism Organization & Food and Agriculture Organization of the United Nations. (2020). *Gastronomy key to supporting hard-hit tourism industry*. https://unric.org/en/gastronomy-key-to-supporting-hard-hit-tourism-industry/

World Tourism Organization. (2017). *Second global report on gastronomy tourism*. https://www.e-unwto.org/doi/pdf/10.18111/9789284418701

World Tourism Organization. (2018). *Glosario de términos de turismo* [Glossary of tourism terms]. https://www.unwto.org/es/glosario-terminos-turisticos

World Tourism Organization. (2020). *World Tourism Barometer*. WTO.

World Tourism Organization. (2020a). *Conceptual guidance on tourism statistics in the COVID-19 Context*. UNWTO.

World Tourism Organization. (2020b). How are countries supporting tourism recovery? *UNWTO Briefing Note – Tourism and COVID-19, 1*, 1–28.

World Tourism Organization. (2020c). *Understanding domestic tourism and seizing its opportunities* (Issue 3). WTO.

World Tourism Organization. (2021a). *Gastronomy*. https://www.unwto.org/gastronomy

World Tourism Organization. (2021a). *How COVID-19 is changing the world : a statistical perspective Volume III*. WTO.

World Tourism Organization. (2021b). *Gastronomy and wine tourism*. https://www.unwto.org/gastronomy-wine-tourism#:~:text=The%20Committee%20on%20Tourism%20and%20Competitiveness%20%28CTC%29%20of,food%20and%20related%20products%20and%20activities%20while%20travelling

World Tourism Organization. (2021b). *One year ago today: WHO defined the COVID-19 outbreak as a pandemic*. https://www.unwto.org/one-year-ago-today-who-defined-the-covid-19-outbreak-as-a-pandemic

World Tourism Organization. (2021c). *Vaccines and Reopen Borders Driving Tourism's Recovery*. https://webunwto.s3.eu-west-1.amazonaws.com/s3fs-public/2021-10/211004-unwto-barometer-en.pdf?wLLYFuBNpS3FXCLZzMvreo8oOXC2UIbW

World Travel & Tourism Council. (2022). https://wttc.org/

World Travel and Tourism Council. (2020). *Contribution of China's travel and tourism industry to GDP 2014-2028*. https://www.statista.com/statistics/249794/contribution-of-chinas-travel-and-tourism-industry-to-gdp/

Wrap. (2013). *An overview of waste in the UK hospitality and food service sector*. https://www.wrap.org.uk/sites/files/wrap/Overview%20of%20Waste%20in%20the%20UK%20Hospitality%20and%20Food%20Service%20Sector%20FINAL.pdf

Wright, L. T., Nancarrow, C., & Kwok, P. M. H. (2001). Food taste preferences and cultural influences on consumption. *British Food Journal*, *103*(5), 348–357. doi:10.1108/00070700110396321

Wu, Y., Guo, C., Tang, L., Hong, Z., Zhou, J., & Dong, X. (2020). Prolonged persistence of SARS-CoV-2 viral RNA in faecal samples. *The Lancet. Gastroenterology & Hepatology*, *5*(5), 434–435. doi:10.1016/S2468-1253(20)30083-2 PMID:32199469

Xie, J. S. (2015). Research on the development of Shunde cultural and creative industry. *Culture Journal*, *12*, 130-136. doi:https://doi.org/CNKI:SUN:WHXU.0.2015-12-041

Xin, W. (2020). The reasons why Macao has gained "City of Gastronomy". *China Food Industry*, *1*, 65-69. doi:https://doi.org/CNKI:SUN:SPZG.0.2020-01-025

XINHUANET. (2018, April 25). *The white paper on the consumption power of China's tourism cuisine 2018*. http://www.xinhuanet.com/food/2018-04/25/c_1122738918.htm

XINHUANET. (2020, March 2). *Report of the current situation and trend of China's catering industry under COVID-19*. http://www.xinhuanet.com/food/2020-03/02/c_1125652997.htm

Xu, S. Y. (2021). Study on the strategy of "three beauties" to boost the local economy of Yangzhou under the impact of new epidemic situation. *Journal of Hubei Open Vocational College, 34*(14), 110-111+116. doi:https://doi.org/CNKI:SUN:HBHS.0.2021-14-050

Xu, J., Guo, S., Xie, D., & Yan, Y. (2020). Blockchain: A new safeguard for agri-foods. *Artificial Intelligence in Agriculture*, *4*, 153–161. doi:10.1016/j.aiia.2020.08.002

Yang, L. (2021). *Research on the development of Yangzhou urban tourism brand based on symbiosis theory* [Master's dissertation]. Yangzhou University. https://kns.cnki.net/KCMS/detail/detail.aspx?dbname=CMFDTEMP&filename=1021050858.nh

Yang, C. (2020). Does hand hygiene reduce SARS-CoV-2 transmission? *Graefes Archive for Clinical and Experimental Ophthalmology, 258*, 1133–1134. doi:10.100700417-020-04652-5 PMID:32221693

Yang, C. H., Feng, M. H., & Chen, Y. (2019). A study on Chengdu culinary image based on tourist perception. *Journal of Researches on Dietetic Science and Culture, 36*(4), 15–22. https://doi.org/CNKI:SUN:YZPX.0.2019-04-004

Yang, S., Guo, B., Ao, L., Yang, C., Zhang, L., Zhou, J., & Jia, P. (2020). Obesity and activity patterns before and during COVID-19 lockdown among youths in China. *Clinical Obesity, 10*(6), e12416. doi:10.1111/cob.12416 PMID:33009706

Yazaki, Y., Mizuo, H., Takahashi, M., Nishizawa, T., Sasaki, N., Gotanda, Y., & Okamoto, H. (2003). Sporadic acute or fulminant hepatitis E in Hokkaido, Japan, may be food-borne, as suggested by the presence of hepatitis E virus in pig liver as food. *The Journal of General Virology, 84*(9), 2351–2357. doi:10.1099/vir.0.19242-0 PMID:12917455

Yeoman, I., Brass, D., & McMahon-Beattie, U. (2007). Current issue in tourism: The authentic tourist. *Tourism Management, 28*, 1128–1138. doi:10.1016/j.tourman.2006.09.012

Yépiz-Gómez, M. S., Gerba, C. P., & Bright, K. R. (2013). Survival of respiratory viruses on fresh produce. *Food and Environmental Virology, 5*, 150–156. doi:10.100712560-013-9114-4 PMID:23681671

Yeung, P., & Thach, L. (2019). *Luxury Wine Marketing: The art and science of luxury wine branding*. Infinite Ideas.

Yi, K. G., & Hu, H. X. (2019, December 10). Construction path and value creation of cultural tourism IP. *China Tourism News*. http://www.ctnews.com.cn/

Yiannas, F. (2018). A new era of food transparency powered by blockchain. *Innovations: Technology, Governance, Globalization, 12*(1–2), 46–56. doi:10.1162/ inov_a_00266

Yılmaz, A. (2014). International Migration: Its types, reasons and effects. *Electronic Turkish Studies, 9*(2), 1685–1704. doi:10.7827/TurkishStudies.6274

Yin, C., Ling, W., Hong, S., & Yan, X. (2020). Vitro inactivation of SARS-CoV-2 by commonly used disinfection products and methods. Preprints. 10.2 1203/rs.3.rs-43489/v1

Yiqi, W., & Jewoo, K. (2021). Interconnectedness between online review valence, brand, and restaurant performance. *Journal of Hospitality and Tourism Management, 48*, 138–145. doi:10.1016/j.jhtm.2021.05.016

Yngve, A., Neuman, N., Haapala, I., & Scander, H. (2021). The Project Collection Food, Nutrition and Health, with a Focus on Eating Together. *International Journal of Environmental Research and Public Health, 18*(4), 1572. doi:10.3390/ijerph18041572 PMID:33562352

Yoon, Y., & Uysal, M. (2005). An examination of the effects of motivation and satisfaction on destination loyalty: A structural model. *Tourism Management*, 45–56. doi:10.1016/j.tourman.2003.08.016

Yoshimoto, J., Ono, C., Tsuchiya, Y., Kabuto, S., Kishi, M., & Matsuura, Y. (2020). *Virucidal effect of acetic acid and vinegar on SARS-CoV-2*. Preprints., doi:10.21203/rs.3.rs-37750/v1

Yost, E., & Cheng, Y. (2021). Customers' risk perception and dine-out motivation during a pandemic: Insight for the restaurant industry. *International Journal of Hospitality Management, 95*, 102889. doi:10.1016/j.ijhm.2021.102889

Yost, E., Kizildag, M., & Ridderstaat, J. (2021). Financial recovery strategies for restaurants during COVID-19: Evidence from the US restaurant industry. *Journal of Hospitality and Tourism Management, 47*, 408–412. doi:10.1016/j.jhtm.2021.04.012

Yousefi, M., & Marzuki, A. (2012). Travel motivations and the influential factors: The case of Penang, Malaysia. *Anatolia: An International Journal of Tourism and Hospitality Research, 23*(2), 169–176. doi:10.1080/13032917.2012.662906

Yu, Z. T. (2005, January). *Taste formation from the perspective of regional food culture.* http://www.dili360.com/cng/article/p5350c3d7eaea252.htm

Yuan, D. H. (1997). Discussion on the cognitive value of a product. In Chinese Psychological Society (Ed.), *Abstracts from the 8th National Conference on Psychology* (p. 321). Chinese Psychological Society.

Yu, C.-E., & Sun, R. (2018). The role of Instagram in the UNESCO's creative city of gastronomy: A case study of Macau. *Tourism Management, 75,* 257–268. doi:10.1016/j.tourman.2019.05.011

Yu, P., Zhu, J., Zhang, Z., Han, Y., & Huang, L. (2020). A familial cluster of infection associated with the 2019 novel coronavirus indicating potential person-to-person transmission during the incubation period. *The Journal of Infectious Diseases, 221,* 1757–1761. doi:10.1093/infdis/jiaa077 PMID:32067043

Zavaleta-Cortijo, C., Ford, J. D., Arotoma-Rojas, I., Lwasa, S., Lancha-Rucoba, G., García, P. J., & … . (2020). Climate change and COVID-19: Reinforcing Indigenous food systems. *The Lancet. Planetary Health, 4*(9), e381–e382. doi:10.1016/S2542-5196(20)30173-X PMID:32777205

Zazpe, I., Bes-Rastrollo, M., Ruiz-Canela, M., Sanchez-Villegas, A., Serrano-Martinez, M., & Martinez-Gonzalez, M. A. (2011). A brief assessment of eating habits and weight gain in a Mediterranean cohort. *British Journal of Nutrition, 105*(5), 765–775. doi:10.10170007114510004149 PMID:21138604

Zhai, S. (2015). Mapping the popularity of urban restaurants using social media data. *Applied Geography (Sevenoaks, England), 63,* 113–120. doi:10.1016/j.apgeog.2015.06.006

Zhang, F. Q. (2008). The concept and the study methods of regional culture. *Zhejiang Social Sciences, 4,* 63-66+50+127. doi:10.14167/j.zjss.2008.04.008

Zhang, S., Wang, S., Yuan, L., Liu, X., & Gong, B. (2020). The impact of epidemics on agricultural production and forecast of COVID-19. *China Agricultural Economic Review, 12*(3), 409–425. doi:10.1108/CAER-04-2020-0055

Zhan, L., & He, Y. (2012). Understanding luxury consumption in China: Consumer perceptions of best-known brands. *Journal of Business Research, 65*(10), 1452–1460. doi:10.1016/j.jbusres.2011.10.011

Zhong, Z. J., Lin, J. P., Zhou, M. Q., Zhang, H., & Han, Y. J. (2021a). "Food" tourism at home and abroad: Research comparison on "Food Tourism" "Gourmet Tourism" "Diet Tourism". *Resource Development & Market, 37*(4), 463-471. doi:10.3969/j.issn.1005-8141.2021.04.013

Zhong, Z. J., Zhou, M. Q., & Lin, J. (2021b). Research process of "food" tourism at home and abroad. *Yunnan Geographic Environment Research, 33*(2), 69-78. doi:https://doi.org/CNKI:SUN:YNDL.0.2021-02-010

Zhu, N., Zhang, D., Wang, W., Li, X., Yang, B., & Song, J. (2019). A Novel Coronavirus from Patients with Pneumonia in China. China Novel Coronavirus Investigating and Research Team. *The New England and Journal of Medicine, 382*(8), 727-733.

Zuber, S., & Brüssow, H. (2020). COVID-19: Challenges for virologists in the food industry. *Microbial Biotechnology,* 1–13. doi:10.1111/1751-7915.13638 PMID:32700430

About the Contributors

Ana Pinto Borges is PhD in Economics, Faculty of Economics, University of Porto. She is Coordinating Professor (since 2010), President of the Pedagogical Council (since 2015), Coordinator of the Master in Business Management (since 2015) and the Executive MBA (since 2016) at ISAG - European Business School. Scientific Coordinator of Research Center in Business Sciences and Tourism (CICET – FCVC) since 2021 and coordinated the ISAG Research Center (NIDISAG) between 2015 and 2021. She has published several articles and book chapters indexed in highly recognized international scientific journals. She also publishes books of international distribution, namely "Building Consumer-Brand Relationship in Luxury Brand Management" and "New Techniques for Brand Management in the Healthcare Sector". Participation in presentations in various national and international congresses and member of the Scientific Committees in academic events. Editor and one of the founding members of the European Journal of Applied Business and Management (EJABM). Former Accenture consultant in the financial area. Economist at the Health Regulatory Entity since 2010.

Paula Rodrigues holds a Ph.D. in Management, in the Faculty of Economics at the University of Porto, Portugal. She is an Associate Professor in Brand Management at the School of Economics and Management, Lusíada University - North, Portugal. Since 1995, she teaches Brand Management, Consumer Behaviour, Econometrics, statistics, and Quantitative Methods at the Universidade Lusíada - Norte. She has published several scientific papers, chapters, and books. Her on-going research projects concern consumer-brand relationships in the field of luxury, tourism and hospitality, and consumer goods.

* * *

Susana Aldeia is Assistant Professor at the European Business School (ISAG/EBS) and Polytechnic Institute of Maia (IPMAIA), Research Center in Business Sciences and Tourism (CICET-FCVC), Research on Economics, Management, and Information Technologies (REMIT).

Aida Maria Oliveira Carvalho is PhD in Cultural Science, by the University of Trás-os-Montes e Alto Douro, Master in History of Populations, by the University of Minho, post-graduate in Tourism and Religious Heritage, by the Catholic University of Lisbon and post-graduate in Cultural Management, by the Polytechnic Institute of Porto. Graduated in European Studies. President of the Board of Directors of the Côa Parque Foundation, since March 2021. Teacher at the Polytechnic Institute of Bragança, since September 2001. Coordinator of the Arts and Humanities Department and member of the Scientific Committee of the Tourism Marketing Master's Degree. Researcher at the Research, Development and

Innovation in Tourism Centre (CITUR). Guide Interpreter of the Côa Valley, between 1996 and 2001. Member of the National Council for Science, Technology and Innovation (CNCTI). Member of the jury of the "7 Maravilhas Doces de Portugal" contest promoted by 7 Maravilhas de Portugal® Member of the jury of the "7 Maravilhas da Nova Gastronomia" contest promoted by 7 Maravilhas de Portugal® Member of the Strategic Council of Turismo Porto e Norte. Member of ART&TUR Festival, International Tourism Film Festival, 2020 and 2021 edition. Member of the Founders Council, of Douro Museum Foundation, representing the Polytechnic Institute of Bragança, between 2009 and 2020.Member of the award jury of the contest Douro Entrepreneur Award 2012, representing the Polytechnic Institute of Bragança. He has published several articles in scientific journals specialising in the area of heritage and cultural tourism.

Catarina Correia is a Business Master Student of Universidade Lusíada.

Ricardo Correia, with an academic career started in 2003, is Associated Professor at the Polytechnic Institute of Bragança (Portugal), where he is the head of the Master programme in Tourism Marketing, and visiting Professor at ISM University of Management and Economics (Lithuania). Ricardo has served as an external consultant in programs such as the European Union Agris Program and Interreg and as consultant for Deloitte & Touche. We should also emphasize his regular participation in research programs, conferences and international scientific meetings, as well as the publication of several chapters in books and articles in international journals.

Palmira Felgueiras has a degree in Socio-Cultural Management from the Polytechnic Institute of Bragança. Master in Tourism Marketing by the Polytechnic Institute of Bragança. Works as Superior Technician of Socio-Cultural Management, in the Olive and Olive Oil Museum in Mirandela, Portugal.

Rahmat Hashim was born and raised in Penang. He is an associate professor at the Faculty of Social Sciences & Leisure Management. He holds a Bachelor's Degree in Hotel and Restaurant Management from Alliant International University in San Diego as well as a Master's Degree of Professional Services (MPS) in Travel and Tourism Management from the New School for Social Research, New York. Rahmat Hashim earned a PhD in Food and Leisure from Sheffield Hallam University in UK. He has authored and co-authored several books and book chapters such as "Restaurant Service Delivery and Customer Patronization", "New Product and Efficiency Strategy for Foodservice Business", "The Impact of Food Trends on Menu Innovation" and "Human Resource Management in Hospitality". He has also edited a book title "Positioning and Branding Tourism Destinations for Global Competitiveness". He has also authored and co-authored several journal articles and conference proceedings, besides serving as reviewers and editorial board committee members in Scopus and ISI indexed journals in relevant areas. Working with his colleagues in the faculty, he is one of the project leaders for the TU Flagship Research Grants. He, together with his colleagues, has won several awards in product innovation competitions such as the BioInno Awards Malaysia and ITEX Malaysia. Prior joining Taylor's he was an Associate Professor at UiTM Shah Alam. His research interests lie in the area of learning behavior, competence management, and branding and innovation. Throughout his career he has undertaken a variety of administrative duties at UiTM and served as the Associate Dean for Research for the School of Hospitality, Tourism & Culinary Arts, Taylor's University. He is a member of the Institute of Hospitality, United Kingdom, American Society of Quality and Tourism Educators Association of Malaysia (TEAM). He was the past president (2012-2014) of Tourism Educators Association of Malaysia (TEAM), chairperson for MQA's program standard for hospitality, Tourism and Culinary Arts and member of the Malaysian Qualification Agency's panel of assessors. Presently, he is a committee member, Malaysia Industry Standard (ISC-Tourism, Exhibition & Hospitality Services), Ministry of International Trade and Industry (MITI).

Mohd Raziff Jamaluddin is currently seconded in the Institute of Business Excellent, Universiti Teknologi MARA (UiTM), as the Fellow Researcher that heads the Retail Excellence and Consultancy unit. His primary affiliation is the Faculty of Hotel and Tourism Management, Universiti Teknologi MARA, currently in the Top 100 QS World ranking under hospitality and leisure subjects. He has been in the academic line for almost 15 years. He considered himself a proud alumnus of UiTM, where he completed his tertiary education from a Diploma to a Ph.D. from this institution. Currently, Dr. Raziff is the President of the Tourism Educators Association of Malaysia (TEAM) and the Editor-in-Chief for TEAM Journal for Hospitality and Tourism. He is an active International Resource Speaker for tourism events and destination topics with regular sharing in the webinar organized by the School of Hospitality and Tourism and Non-Government Association in the Asian countries. Dr. Raziff works closely with the industry, where he became the Advisor and consultant for companies and ministries. He is the Committee Member for the public and cultural event organized by the Department of Event Management, Malaysia Ministry of Tourism, Arts and Culture, where he was involved in the impact study for the cultural events organized by the Ministry since 2011. Dr. Raziff is one of UiTM's active consultants with more than 50 projects commissioned by government agencies and private institutions, mostly on impact study of major tourism events in Malaysia. Dr. Raziff is currently served as the Associate Consultant for Universiti Teknologi MARA business, consultant and training units, UiTM Techno Venture, for his active role in consultancy projects.

Osman M. Karatepe, PhD, is Professor of Marketing in the Faculty of Tourism at Eastern Mediterranean University (Gazimagusa, TRNC, via Mersin 10, 99628, TURKEY) and Fellow of the Hospitality and Tourism Management Academy: A Research Community. He obtained his PhD degree in business administration from Hacettepe University in TURKEY. His research interests are in the areas of services marketing and management, internal marketing, green management and marketing, and strategic management. Dr. Karatepe serves on the editorial review boards of top-tier journals such as Journal of Hospitality Marketing and Management, The Service Industries Journal, Cornell Hospitality Quarterly, and International Journal of Hospitality Management. He is the Senior Associate Editor of the Journal of Hospitality and Tourism Insights and the Regional Editor for the Middle East and Africa of the International Journal of Contemporary Hospitality Management. Dr. Karatepe who has an h-index of 68 according to Google Scholar citations database has authored or co-authored over 150 articles in various journals, including Journal of the Academy of Marketing Science, Journal of Business Research, Journal of Air Transport Management, Service Business, Journal of Services Marketing, Employee Relations, The Service Industries Journal, International Journal of Service Industry Management, Tourism Management, International Journal of Contemporary Hospitality Management, Cornell Hospitality Quarterly, Tourism Management Perspectives, and International Journal of Hospitality Management. Dr. Karatepe is frequently shown as one of the most prolific researchers in the field of hospitality management and has recently appeared within the 2% of the list of the world's most influential scientists.

Hasan Kiliç, Ph.D., is a professor of Tourism and Hospitality Management and the Dean of Tourism Faculty at Eastern Mediterranean.

Halyna Kushniruk, PhD in Economics, Associate Professor of the Department of Hotel and Restaurant Business and Food Technologies, Ivan Franko National University of Lviv (Lviv, Ukraine). Her research interests: theory and methodology of tourism; tourism economics; organization of tourism; tourism development in Ukraine; international tourism; gastronomic tourism; hotel and restaurant business. She is the co-editor of the GastroMedia Journal (Necmettin Erbakan University, Turkey). She is a member of the Hospitality Industry Association of Ukraine (HIAU).

Paulo Macedo is Professor at the European Business School (ISAG/EBS) and Polytechnic Institute of Cávado and Ave (IPCA). Research Center in Business Sciences and Tourism (CICET-FCVC).

Ramya Mahendran is the an managed innovation consultant. She is certified in design thinking, business modeling and jobs to be done. She has over 10 years of experience in the fields of managed innovation, startup incubation and acceleration, crowd sources idea management systems, design thinking and sustainability. She works with student entrepreneurs to build their business ideas into a successful business model. She works with some of India's leading Technology and Business Incubators, Institution Innovation Councils and Entrepreneurship Cells. She specializes in setting up innovation strategy, opportunity identification, large scale ideation campaigns and facilitating rapid prototyping events like design service jams and hackathons, organizing large-scale Innovation summits and global idea crowdsourcing events. Her current area of research explores how empathy can be taught with the help of design tools for product, service, and policy designers.

Mafalda Nogueira is currently a lecturer of Marketing and coordinator of the Marketing Management Master in IPAM - Portuguese Institute of Marketing Administration, Porto (Portugal). She holds a degree in Economics from Universidade Portucalense Infante D. Henrique, a Master in Marketing from the University of Porto, and a PhD in Marketing from Lancaster University. Having started her professional life in 2000, as a business consultant for a Swedish multinational, she embarked on the path of research and teaching in 2006, having spent the last 16 years engaging in visiting professorship roles in several higher education institutions namely at Lancaster University, London School of Design and Marketing, Polytechnic Institute of Management and Technology, Universidade Portucalense Infante D. Henrique. Her research interests are fundamentally focused on the relevance of marketing for good, having participated in projects that genuinely seek to create knowledge about social marketing, sustainability and social responsibility and socially responsible consumer behaviour. Her work has been published in different academic journals including Journal of Cleaner Production, Journal of Food Products Marketing, Journal of Nonprofit & Public-Sector Marketing, Journal of Purchasing and Supply Management.

Paulo Ramos is a PhD in Management Sciences Faculty of Economics of the University of Porto, MSc in Marketing and Product Management, Cranfield University, UK, BA in International Relations, University of Minho. Assistant Professor at the University Fernando Pessoa (member of the Scientific Board) Past Management Departmental Coordinator. Assistant Professor in University Lusíada Norte. Guest lecturer in: Catholic University, Faculty of Sciences of University of Porto School of Biotechnology, University of Lisbon and in the Polytechnic Institute of Viana do Castelo. He has lectured has a guest in: International School of Management, Dortmund, Germany. Lutheran University - Brazil ULBRA, Rio Grande do Sul. Member of the Editorial Board of the "Journal of Organisational Studies and Innovation" UK. Partner and board member of several companies in the areas of Agro food (Wine),

Education, Real Estate and Consultancy. External expert and adviser of the Coordinating Commission for Northern Regional Development of the Portuguese Government. He has several research papers in journals and proceedings of national and international conferences in the fields of Wine, Tourism, Brands and Services. Keynote speaker, guest speaker, Member of the Scientific Committee, Session Chair in international conferences. Awarded in 2012: "best short communication" in Economy and Law in OIV XXIV World Congress of Vine and Wine.

Helena Real is a Nutritionist; PhD in Food Consumption and Nutrition Sciences, in the Mediterranean Diet area. General-Secretary of the Portuguese Association of Nutrition. Guest Assistant Professor of the Degree in Nutrition Sciences at the University Institute of Health Sciences – Cespu. Element of the Toxicology Research Unit – TOXRUN – of the University Institute of Health Sciences – Cespu. Extensive experience in coordinating and managing programs and projects within the scope of Food Sustainability.

Hamed Rezapouraghdam is currently a member of the tourism faculty, Eastern Mediterranean University. He received his BA from the Faculty of Foreign Languages, Isfahan University, and his M.S. and Ph.D. from the Faculty of Tourism, Eastern Mediterranean University. His research interests are in the areas of corporate sustainability, behavioral psychology, and sustainable tourism.

Mykhailo Rutynskyi, PhD in Geography, Associate Professor of the Department of Hotel and Restaurant Business and Food Technologies, Ivan Franko National University of Lviv (Lviv, Ukraine). He is the author and co-author of 210 scientific works, including 7 monographs and 17 manuals and textbooks on geography, economics and history of tourism and hospitality industry. He is a member of the Hospitality Industry Association of Ukraine (HIAU).

Cigdem Sabbag received the master's degree in 2003 and the Ph.D in 2009 in Nutrition Sciences at Ankara University, Turkey. In 2010, she began working at Adiyaman University/Turkey as an assistant professor. She was appointed as an associate professor in 2015 and in 2021, she received professorship at Adıyaman University Tourism Faculty Tourism Management Department. Nutrition, healthy eating, gastronomy, and local food culture are among her research interests.

Heng Tang is Professor and Doctoral Supervisor. She is a Visiting Scholar of University of California, Los Angeles and University of Nevada. She is currently the Dean of the Intellectual Property School of Jiangsu University and the Executive Dean of the institute of New Structural Economics & Intellectual Property. She is also a member of the Expert Advisory Committee of China National Intellectual Property Administration (CNIPA), a member of China Technical Committee on Knowledge Management of Standardization Administration, and a Standing Director of Intellectual Property Research Association of China Universities. She has presided over more than 50 projects, including the National Natural Science Foundation of China, National Social Science Foundation of China, National Soft Science Research, national key R & D projects, and national intellectual property key research projects. And she has won one first prize of Jiangsu Provincial Philosophy and Social Science Outstanding Achievement Award, and six provincial and ministerial outstanding Achievement Awards such as the State Intellectual Property Office. She is mainly engaged in the research of intellectual property management, transformation of scientific and technological achievements and technology innovation management and takes the lead in the standardization research of intellectual property management in China. She led the preparation of

China's first intellectual property management standard DB32T 1204-2008, which was later upgraded to the national standard GB/T 29490-2013 by the national intellectual property administration, PRC.

Guido Valero is a Chef, Gastronomy and Hospitality professor, with over 20 years experience in the Hospitality and Food & Beverage areas.

Javier Vallejo is a hospitality professor and hospitality professional with over 20 years experience in hotels.

Dominyka Venciute is a Senior Assistant Professor and Postdoctoral Researcher at the ISM University of Management and Economics where she teaches Marketing Principles, Social Media Marketing and Integrated Marketing Communications. Besides her activities in academia, she runs a personal branding and marketing consultancy that aims to help professionals and organizations unleash the power of their personal and corporate brands.

Elvira Vieira is a PhD in Applied Economics, University of Santiago de Compostela. Professor, member of the Scientific Council at Institute of Marketing Management between 2007-2009 and Invited Professor at Minho University between 2009-2012. Professor at Polytechnic Institute of Viana do Castelo and Coordinating Professor at ISAG - European Business School since 2009. General Director at ISAG - European Business School since 2012, researcher in UNIAG, scientific coordinator of NIDISAG Research Center (between 2014 - 2021) and of Center in Business Sciences and Tourism (CICET - FCVC) since 2021. Author of several publications (papers and book chapters) indexed to the main bibliographic databases. Editor and one of the founding members of the European Journal of Applied Business and Management (EJABM). 2009 1st prize winner for research of the "Catedra da Euroregión Galicia – Norte of Portugal" and President of the European Grouping of Territorial Cooperation Galicia – North of Portugal, from 2010-2012.

Alejandra Ycaza is an Academic Coordinator, Hospitality professor with over 20 years experience in the Hospitality and Tourism sector.

Seydi Yıkmış was born in 1982 in Adıyaman. In 2005, he graduated from Inönü University, Faculty of Engineering, Department of Food Engineering. In 2013, he graduated from Süleyman Demirel University, Institute of Science, Department of Food Engineering. In 2018, he finished his PhD education at Istanbul University, Institute of Health Sciences, Department of Food Hygiene and Technology. He is still working as a Assoc. Prof. Dr. in Nutrition and Dietetics Department of Namık Kemal University.

Poshan (Sam) Yu is a Lecturer in Accounting and Finance in the International Cooperative Education Program of Soochow University (China). He is also an External Professor of FinTech and Finance at SKEMA Business School (China), a Visiting Professor at Krirk University (Thailand) and a Visiting Researcher at the Australian Studies Centre of Shanghai University (China). Sam leads FasterCapital (Dubai, UAE) as a Regional Partner (China) and serves as a Startup Mentor for AIC RAISE (Coimbatore, India). His research interests include financial technology, regulatory technology, public-private partnerships, mergers and acquisitions, private equity, venture capital, start-ups, intellectual property, art finance, and China's "One Belt One Road" policy.

Ding Zuo is an independent researcher. Her research interests include tourism economy, Chinese economy, and business analytics.

Index

H

I

K

L

M

N

O

P

R

S

T

U

W

www.ingramcontent.com/pod-product-compliance
Lightning Source LLC
LaVergne TN
LVHW062141300926
849161LV00010B/2597